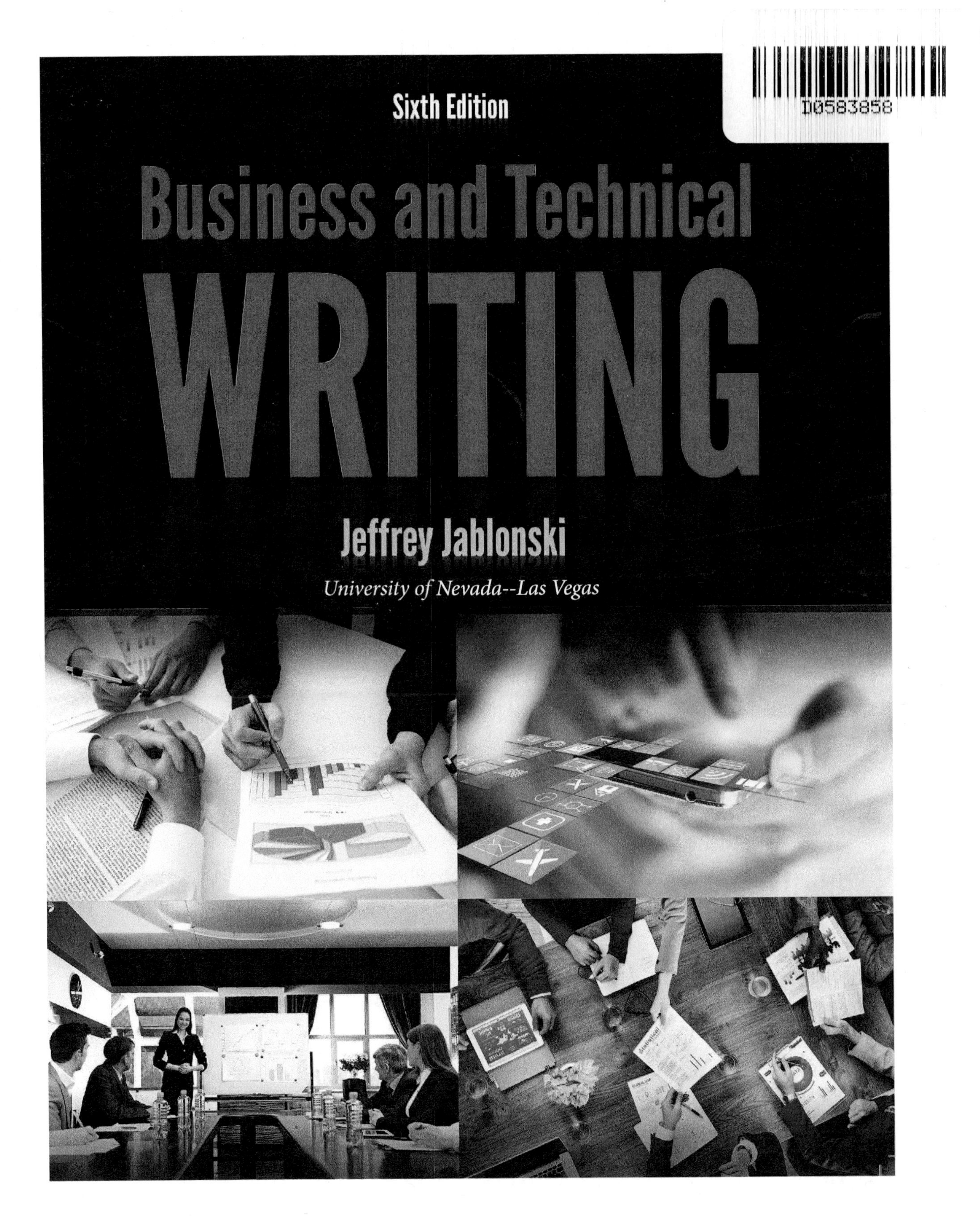
Sixth Edition
Business and Technical
WRITING
Jeffrey Jablonski
University of Nevada--Las Vegas

Kendall Hunt
publishing company

First and second edition was a Kendall/Hunt website component.
Textbook added to the third edition.

Kendall Hunt
publishing company
www.kendallhunt.com
Send all inquiries to:
4050 Westmark Drive
Dubuque, IA 52004-1840

ISBN 978-1-4652-8928-5

Printed in the United States of America

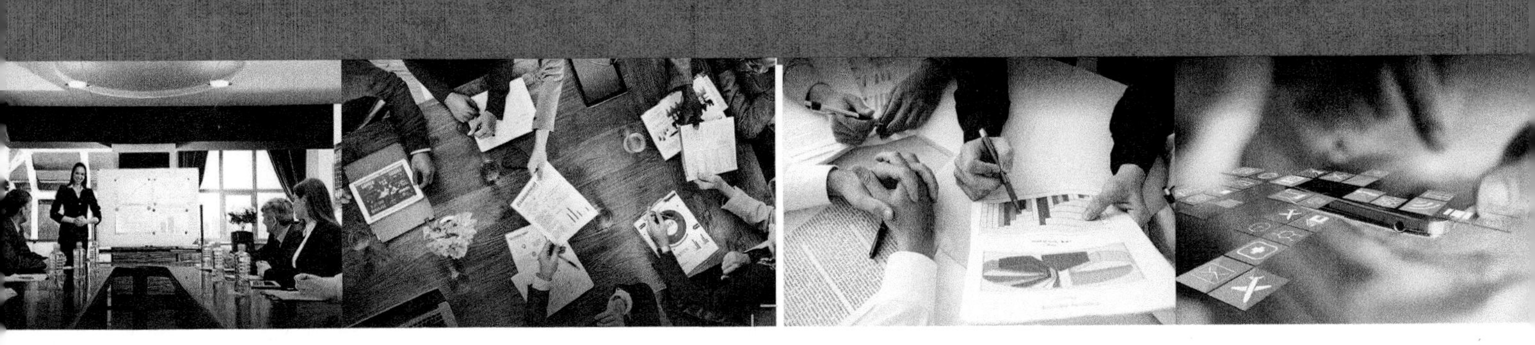

CONTENTS

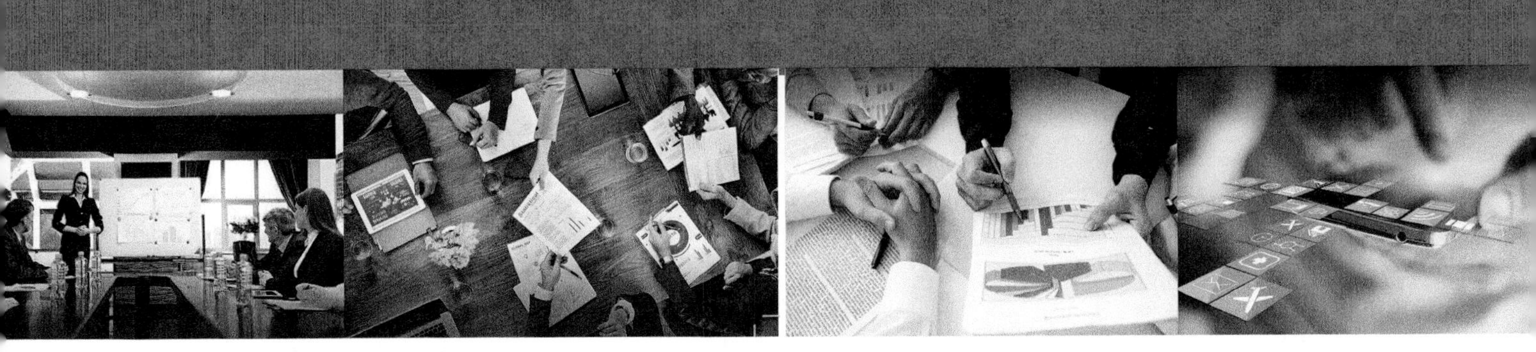

PREFACE

Experts now believe that multiple literacies are necessary for communicating effectively in today's business and technical settings. This text aims to develop the following multiple literacies:

- **Rhetorical Literacy:** You will develop an ability to analyze the demands of a given workplace writing situation and adapt your writing given your understanding of purpose and audience.
- **Visual Literacy:** You will learn how page layout, graphics, text (and hypertext) operate rhetorically and contribute to clear and effective communication.
- **Information Literacy:** You will improve your ability to find necessary information and integrate it effectively into your own writing without plagiarizing.
- **Computer Literacy:** You will learn how to use the computer as a tool for communication, including how to compose documents electronically and how to use the World Wide Web for finding and retrieving information.
- **Ethical Literacy:** You will learn how good writing is not only mechanically sound but also maintains positive relations and avoids harming others.

Part I of this textbook acquaints you with these literacies as well as the typical genres, or forms, of business and technical writing, including memos, letters, emails, resumes, reports, presentations, and social media. Part I also includes numerous strategies for planning, drafting, revising, and editing effective business and technical documents, both individually and in teams. This textbook also takes advantage of the Internet, providing access templates, worksheets, and links to online resources for each chapter.

Part II of this textbook offers a number of writing projects that require you to apply the relevant principles of effective written communication from Part I. This textbook takes a project-based approach to learning how to write. Traditionally, learning has all too often been passive; the teacher lectures about some subject matter, and the students passively take it all in. In traditional courses you're expected to memorize the material and are tested mostly on your ability to recall information rather than your ability to apply abstract concepts to concrete problems. Problem-based learning is much more active. Problem-based learning gives you a problem that requires knowledge and skill in certain principles to solve effectively. To solve the problems inherent in each project of Part II you will need to do more than just memorize principles of business and technical writing; you will need to apply them. You literally need to write your way out of the problems posed by the cases and projects in Part II. Research shows that a problem-based approach deepens your learning and increases the chance you'll internalize the strategies you used, so you will be more likely to apply principles learned to similar situations in the future as you write in the workplace.

Thus, Parts I and II of this book work in conjunction to build your business and technical writing ability. The first part gives you the necessary knowledge; the second part gives you opportunities to apply it. Assuming you put forth an honest effort as you work through the text, presumably under the guidance of an informed instructor, the combination of new information and practice should help you become a better writer.

ACKNOWLEDGMENTS

Many current and former UNLV business and technical writing faculty and staff have contributed to this textbook in various ways, including writing content, designing and testing assignments, and researching background information and links.

Elyse Arring
Jenny Bania
Suzanne Berfalk
Jeff Burbank
Joseph Cameron
Don Capp
Rebecca Colbert
Jason Coley
Anish Dave
Paris Drebes
Jaq Greenspun
Daniel Hernandez
Vanessa Huff
Chad Leitz
Carl Linder
Karen Lovat
Heather Lusty
Ed Nagelhout
Michael McCombs
Clancy McGilligan
Cory Ness
Jonathan Peters
Constance Pruss
Lara Ramsey
Kelle Schillaci
Leeanne Sorenson
Homer Simms
Julie Staggers
Paul Straus
Denise Tillery
Jenny Toups

PRINCIPLES

BUSINESS WRITING

CHAPTER 1

It is no secret that strong written communication skills are essential for business success. That's why a course in business writing is required for most of you reading this: your advisors do not want to leave it up to chance that you would *elect* to take a course that's focused on writing. To help you see just how important and valuable strong business writing skills are, this chapter discusses the following:

- Facts that demonstrate the importance of strong business writing skills
- Two examples of bad business writing
- The fundamental principles of business writing stressed in this course

Good Communication Skills Needed to Get Ahead in Business

Most employers complain that their employees can't write. Thirty percent of all business memos and emails are written solely to clarify earlier written communication that didn't make sense. This translates into an estimated $225 billion a year in lost productivity.[1] The rise of the Internet and the use of email has increased the need for strong writing skills, as many companies have switched to email and social media as the main means of communication within the company and to outside customers.

Survey after survey indicates that managers believe their employees lack the basic writing skills needed in today's information economy.[2,3,4,5] A widely publicized 2004 report titled

Writing: A Ticket to Work…Or a Ticket Out found that workplace writing is viewed by managers as a threshold skill used to make hiring and promotion decisions. This survey of 120 major American companies concluded the following:

- Employees "frequently" to "almost always" produce memos, letters, and reports, and communication through email and PowerPoint presentations is "almost universal"
- People with weak communication and writing skills will not be hired and are more likely to be fired than promoted
- The large majority of companies (80%) assess writing ability during the hiring process
- Many companies (50%) take writing into account when making promotion decisions
- American companies spend over 3 billion annually training employees with weak writing skills[6]

The above report was produced by the National Commission on Writing, an organization formed in 2002 to address the "growing concern within the education, business, and policy-making communities that the level of writing in the United States is not what it should be." In 2005 the Commission issued a follow-up report based on a survey of government sector managers titled *Writing: A Powerful Message from State Government*. The report amplified the findings of the previous survey, claiming that writing was an even more important consideration in government work, given the need for state employees to clearly communicate information about laws, regulations, and policies to citizens.[7]

Not convinced yet that it's important to have strong business writing skills? Consider these statistics:

- Almost 90% of business school graduates credited their writing skills as helping to accelerate their advancement
- Almost 80% of upper level managers identified business writing as the most useful college course they had taken[8]

To prepare you for writing in the workplace, this textbook emphasizes the ways that business writing differs from the way you've been taught to write for school. For instance, you'll learn that, unlike college essays, it is preferable to use first person "I" and second person "you" voice in most business writing. Unlike college essays, you should write memos and letters using single-spacing and shorter paragraphs. You should use headings and other visual elements like bulleted lists and graphic aids. To suit busy readers such as executives and customers, your business writing should generally be direct and to the point, unlike a lot of college writing. And, you may or may not be heartened to learn this, you will not be rewarded in business writing for using big words and fluff filler—the type of writing some professors may have encouraged in your past college experience. This habit of "academic-ese" that college graduates bring to their workplace writing is a big reason employers complain that employees can't write.

This textbook asks you do a lot of writing. Lack of practice is another reason people complain that no one can write anymore. Like any skill, writing is an ability that atrophies without use, and students are rarely asked to write in school. The more you write, the better your writing becomes.

The next two sections discuss examples of bad business writing to make the point that good business writing involves more than attention to surface features of grammar and mechanics.

No Microwave Food–Ever!

What do you think of the following memorandum (based on an actual memo, written at a real company)?

From: Standish, Royce (Operations, Acme Widgets Inc.)
Sent: Wednesday, February 18, 2004
Subject: Fire policy

Attention everyone: I have nothing against warm food... However, from here on out if you must reheat food you must do so in a microwave safe container. If you are found reheating food in the microwave using foil or other combustible materials you will be terminated, period! The reason for this is that most of all fire alarms in office buildings are because some moron set the microwave on fire with aluminum foil. There was another incident last night... There have been 4 alarms in the last 2 months. These microwave fires cost us between $5,000 and $8,000 each alarm. If you must have warm food; order out or use plastic, or you're fired. Thank you.

Royce Standish
Vice President,
Operations
1-800-xxx-xxxx Ext xxxxx

Laughable, right? But would you want to work at this company? What words would you use to describe Mr. Standish, the author of the memo? Would you want to work for Mr. Standish?

As with all business writing, the Fire Policy example serves many purposes, some explicit, others less obvious. The memo's main purpose is obviously to *inform* all employees of the ban on combustible materials in the microwave and the consequences for violating the new policy. Mr. Standish includes details about the costs associated with microwave fires to *persuade* employees to adhere to the policy. As another persuasive tactic Standish *invokes his authority* to terminate employees. This memo also has other, less explicit purposes. It *conveys an image* of the writer and *influences the culture* of the company. How did you describe Standish? Sarcastic, condescending, heavy-handed? Is that the kind of image a business executive should project? Do you want to work in a business environment where a vice president casually refers to his employees as morons in writing? It makes one wonder what else he thinks of his employees.

The Fire Policy memo also functions as a *record* of Mr. Standish's new policy—and his communication style—within his company. Perhaps Mr. Standish regrets writing this memo. But it's too late. The memo was emailed to hundreds of employees.

One thing is likely: Mr. Standish never took a business writing class. While he may have risen to the rank of vice president, he certainly hasn't done it in a way that would endear himself to his employees. Maybe he's not even the best person for the job. Is a ban on "combustible materials" the best solution to the problem at this company? There's more wrong with the Fire Policy memo than just formalistic concerns of tone and punctuation. Good business writing maintains positive relations and demonstrates careful, critical thinking.

Pancakes on Fleek

The biggest change in business communication over the last several years has been social media, or websites and applications that enable users to create and share content or to participate in social networking, such as Facebook, Twitter, LinkedIn, and Instagram. According to the Pew Research Center, nearly 65% of all American adults use social networking sites, 90% of young adults age 18–29 use social media, and 77% of adults age 30–49 use it.[9] Women and men use social media at comparable rates, and while those with higher education are more likely to use social media, there are no notable differences by racial or ethnic group, as 65% of whites, 65% of Hispanics, and 56% African-Americans use social media.[10]

On the surface, it doesn't seem too difficult to compose messages for sharing on the various social media sites. Twitter is limited to 120 characters, which in turn deemphasizes traditional grammar and encourages the use of acronyms and other text message–like shorthand. However, social media increases the frequency and level of communication between a business and its customers. Businesses are measured by their "engagement," the quality of interaction with consumers, and "responsiveness," how quickly the company responds to individual comments.[11] The interactive qualities of social media put additional pressure on effective communication. There are numerous examples of businesses misjudging their audience and posting messages that quickly backfire and draw criticism. The company then is forced to strategize how to minimize the damage and regain customer loyalty.

In one recent example, the restaurant chain IHOP took to Twitter to attract a younger customer base. In September 2014, IHOP tweeted "Pancakes on Fleek," which is slang for "these pancakes are on point." The post received over 25,000 retweets and 17,000 likes and subsequent media coverage over its cheeky campaign to attract a younger demographic.[12] IHOP followed up with more images of pancakes accompanied by such tweets as "Pancakes, you look good, won't you back that stack up" and "DEEZ are NUTs." However, IHOP apparently took things too far when it made a now-deleted tweet that featured a potentially suggestive-looking stack of pancakes with a caption derogatory to flat-chested women.

Find the tweet here: http://www.dailymail.co.uk/news/article-3281218/Flat-GREAT-personality-IHOP-lands-hot-water-tweeting-joke-comparing-pancakes-women-s-breasts.html

Many customers quickly expressed outrage over the sexist tweet. One Twitter user wrote, "Can't teach my Girl Scouts that casual misogyny is okay." Another user wrote, "Uh, @ IHOP I really must advise against," and yet another, "Someone on the @IHOP social media team don' gonna get fired, yo."[13]

IHOP quickly removed the tweet and posted the following apology: "Earlier today we tweeted something dumb and immature that does not reflect what IHOP stands for. We're sorry."

Find the tweet here: https://twitter.com/IHOP/status/655941856375779329?ref_src=twsrc%5Etfw

A few followers of IHOP on Twitter expressed support for the restaurant's attempt at humor, but many others did not, and the backlash received media coverage of its own.[14] Was IHOP's response and apology tweet appropriate? Should it have issued a more detailed apology or explanation? When using humor in corporate social media communication, how can one judge the difference between what is funny or edgy, and what may be offensive or hurtful?

The next section will outline key principles that should help you avoid mistakes like those in the Fire Policy memo and the IHOP tweet.

Rhetorical Principles for Effective Business Writing

Rhetoric is the ancient art of effective communication. Today, rhetoric all too often has negative connotations, as the word is sometimes used to describe deceptive or evasive language. But there is a more positive side to rhetoric. The ancients recognized that speaking and writing abilities could be improved through the study of effective communication and deliberate practice, which came to be known as the discipline of rhetoric.

Paying attention to rhetoric as you plan and compose your documents raises the following key questions:

- What is the *purpose* of my writing, or what am I trying to accomplish?
- Who is being addressed, or who is my *audience*?
- How does the *context*, or circumstances within which the communication occurs, affect what I write?
- What *image*, or impression of myself, do I want to create with my reader, what should the reader think of me?

Purpose, audience, context, image. As you prepare to write, the more you are aware of and account for what your purposes are, who your audience is, what context or situation you're writing in, and what image of yourself you want to convey, the more effective your written communication will be.

Know Your Purposes

The main goal of all business communication is to *develop and maintain positive relations between people.* Even if a sales letter doesn't make an immediate sale, that doesn't mean it failed—the letter helps develop a relationship with the customer. Clichés like "the customer is always right" are designed to help employees maintain positive business relations. Far too often, people forget this and communication dissolves into hostile, threatening, or condescending exchanges. There's a negative stereotype often associated with business. Portrayals of businessmen from Charles Dickens' Scrooge to Gordon Gekko, the greedy broker played by Michael Douglas in the movie *Wall Street*, reflect the belief that to be successful in business one must be ruthless and cutthroat. While some may have enough power or influence to act this way, few everyday business writers can afford to risk damaging relations with their readers, be they customers, clients, supervisors, or co-workers. And even if you could get away with heartlessness, is that really the way business *should* be done?

Other purposes important to business writing include:

- **To inform:** Most business writing communicates information to readers that will allow them to do something better, whether it be to act or think differently. The challenge is to understand what information is essential to helping readers act. It's also essential to include information about harms or risks that readers need to make informed decisions.[15]
- **To persuade:** Information alone rarely compels readers to act. They must be persuaded about what to do with the information. After asking "*what* is this information?" readers want to know "*why* should I act on this information?" Information frequently must include arguments, assertions, and opinions about its usefulness. Effective automobile advertisements make the connection between why good handling is desirable (to swerve out of the way of earthquake faults and falling boulders). Sales letters aren't the only occasion for persuasion. A worker's performance evaluation or progress report attempts to persuade a supervisor that they're being productive. When attempting to persuade, writers must be careful not to manipulate information for their own ends. Writers must be fair and honest and make sure communication is not limited to their own interests, but also considers the best interests of all parties involved.[16]
- **To establish a legal record:** Most writing in business *documents*, or creates a record, of the communication between two or more parties.[17] Writing in business is often legally binding and always has legal implications. Nearly every year there is a big news story that involves some "smoking memo" implicating a major corporation in wrongdoing of some sort. In 2000, Firestone Tires recalled 6.5 million tires because they were associated with numerous fatalities caused by Ford Explorer rollover accidents. Subsequent investigations revealed early memo reports written by engineers within Ford Motor Co. alerting management to design flaws contributing to the rollover problems. These memos suggested some negligence on the part of Ford. Does this mean that the engineers should not have written about the problems? Does it mean that Ford management should have destroyed the memos? No, the information is part of the ongoing work of the organization and serves as the organization's memory. Ford engineers and managers can retrieve memos like this and familiarize themselves with the history of the project. But so can lawyers and the public.

Pay attention to the record-keeping function of business writing and the reality that future, unforeseen readers have access to an organization's documents. Frequently, documents need additional background information to accommodate future readers. This is why you should pay attention to details such as dates and facts when you write. Always ask yourself if what you've written is something you want to put "in print." Email is frequently treated as an informal communication medium like speech, but high-profile lawsuits have demonstrated that email is not a secure, private way of communicating. It is typically monitored by an organization, and even if deleted, it is typically backed up and recoverable.

Analyzing your purposes for writing also emphasizes that business writing is an *action* done to cause further action. Most students associate (school) writing with meaningless busywork. But writing is central to the activity and goals of business organizations. People write in business for many reasons:

- Announce a new or updated product or service
- Establish a new policy
- Return credit to clients
- Request information or meetings
- Complain about services or products
- Inform clients of rate increases
- Praise or compliment services or products
- Express thanks
- Respond to a complaint (called an "adjustment letter")
- Refuse requests[18]

Can you think of other purposes not listed here? Notice how each of the reasons listed above begins with an *action verb*: to *sell*, to *thank*, to *request*, etc. Before you write, brainstorm a list of purposes and think about which will be most important. The more you can articulate the aims of your writing, based on your analysis of the situation, the more your writing will improve because it will give you goals and a means to measure how well your draft accomplishes your goals. Keep in mind that in any typical business situation **there are multiple purposes** and some may even conflict. And always check that your writing doesn't risk damaging relations, creates a suitable written record, and doesn't misrepresent or mislead.

Understand Your Audience

Audience is the key factor that shapes texts. Who your audience is will determine the level of formality of the writing, what terms should be defined, what information should be included, and how that information should be arranged. You should also consider the social and organizational status of your audience. Is your audience superior to you, such as your boss, or are they inferior to you, such as employees you manage? Are they insiders or outsiders? Insiders might be company or professional peers who are more familiar with your subject and the language you want to use. Outsiders might be the public or clients with less knowledge about your topic who may require simpler language or need technical terms defined so that they can understand. Writers always have to make a choice about how much they can assume an audience knows about their topic. Other factors such as age, gender, class, race, nationality, occupation, and political affiliation influence the receptivity of an audience to a text. As you plan your writing, you should always try to shift your perspective from "what do I want to say?" to "what information does my audience need to understand my message?"[19]

As the discussion of business writing as a legal record suggests, when you write in business you will never have only one reader, even if you are addressing only one reader.[20] You must always consider how your writing accommodates important secondary readers, anyone who is not the primary or immediate addressee of the document. A new policy written by an HR manager may be addressed to an immediate supervisor for approval but have HR office staff in mind as its primary audience. The policy, because it affects employees, may need approval by lawyers. The HR manager may have written the policy with her professional peers in mind because she wants to ensure that her organization manages human resources in line with professional standards, but the policy must also be written for superiors, lawyers, and employees and managers who aren't even in the company yet. It must also be intelligible to fellow HR staff in the event that the HR manager is promoted or moves to another company.

Analyze the Writing Context

Context as a rhetorical principle is a complex notion that includes an awareness of the immediate physical situation in which a person is writing. It also includes a more abstract situational awareness of other factors that influence the production and reception of text, including the organizational, cultural, historical, ethical, and legal contexts or perspectives that shape documents. Successful managers are often considered to possess a certain business savvy about what's feasible within an organization. This savvy reflects a high degree of contextual awareness and an ability to anticipate appropriate actions given the constraints of a particular situation. When writing, you should always try to view the outcomes of what you want to accomplish and how you want to accomplish it from a variety of perspectives.

writing context

Writing context involves situational factors that influence the writer's decisions, such as corporate culture, individuals including supervisors or clients, laws, social customs, cultural mores, and standards of ethics.

In business writing, the organization's *corporate culture*, its shared values, norms, roles, rituals, and beliefs, shapes permissible action. Many writers get into trouble because they don't think through the implications of what they write. Corporate culture influences writers' choices of content, persuasive approach, and word choice. For instance, one writer was told to delete the word *hope* from a draft. "We don't *hope* for anything around here," she was advised, "we *decide* what we want and then we make it happen."[21] Many organizations have clear chains of command and lines of communication. Ignoring either of these may create serious problems for someone wanting to accomplish something in writing, be it to complain about something or to change a way of doing something. The corporate culture determines when and why who speaks to whom, and who listens to whom.

Individuals within companies also affect what's written. Powerful executives' preferences evolve into maxims for behavior within the company. External factors also affect business communication, including governments, media, markets, unions, regulatory bodies, suppliers, trade organizations, competitors, and the public. Often, what may seem permissible within the bounded context of the organization is unworkable in light of government standards, acceptable market practice, union regulations, or public opinion.[22]

For more on using non-exist or gender neutral language see https://www.noslangues-our-languages.gc.ca/bien-well/fra-eng/style/nonsexist-guidelines-eng.html

More abstract factors such as the historical, cultural, and ethical contexts of communication are also important contextual considerations. Some action or argument may not be feasible because of recent events within an organization or larger society. After the dotcom bubble burst in the late 1990s, entrepreneurs had to change their tactics for raising money for business ventures. The Internet gold rush had ended and venture capitalists became much more skeptical about e-commerce business plans. Cultural considerations also affect communication. Since the civil rights movement of the 1960s more women and minorities have entered into the workplace, influencing all sorts of practices including acceptable communication. Now communication must be nonsexist and recognize diversity.

All communication must consider moral and ethical standards of conduct. What would you do if you were asked to contribute to a document that you knew could lead to harm of some sort for others? Would you falsify information if you knew you stood to gain a great deal of money? The impersonal bureaucracies of modern organizations give rise to ethical challenges that writers must confront. Everything you write is subject to the legal and ethical constraints of society. As James Porter notes, "Whenever you write, you have an obligation as a citizen and as a member of a community to write with the good of that community in mind. You are not just a worker in a business, you are also always a member of larger civic and social communities, whose interests you must keep in mind. 'I just did it because the boss told me' is not a valid defense if what you do is illegal or unethical."[23]

Establish an Appropriate Image

Everything you write within an organization contributes to readers' opinions about who you are—what kind of writer you are, what kind of thinker you are, and what kind of person you are. Frequently, people will *only* use what you write to form an opinion of you and your capabilities. Think about the job application process. Employers screen dozens to hundreds of resumes and cover letters submitted by hopeful job seekers, narrowing the pool down to a handful of applicants to interview in person. These employers make decisions about whether applicants can do the job based on what's written in the job application documents.

ethos

Ancient Greek rhetorical term referring to persuasive tactics based on the credibility or trustworthiness of the writer or speaker. Examples include expert testimony in courts and celebrity product endorsements.

Ancient rhetoricians were conscious of how a person who registered a strong moral character and a sense of goodwill established strong *credibility*, or trustworthiness and believability, which could have a powerful persuasive influence on an audience. The Greeks termed the character or moral nature of a person *ethos*. Ethos is related to the image that writers project in their writing. The following elements of writing affect a writer's ethos, either negatively or positively:

- **Errors in a text:** Have you ever sent an email to an employer or professor that your eighth grade English teacher would have marked up with a red pen? Readers tend to attach negative traits to writers of sloppy compositions, especially in business. Human resource managers will often stop reading resumes with just a few obvious errors.

- **Tone:** Customer service managers will respond more positively to an even-tempered complaint than a hostile one. In business you should always aspire to create a *professional ethos*. Do you want readers to see you as hostile, uncaring, and inattentive, or do you want readers to see you as even-tempered, considerate, and thorough?
- **Evidence used to support arguments:** A request for a raise would be more credible if the writer included specific details about the ways the writer improved productivity and sales within the past year. In academic writing, writers cite published sources to support their arguments and demonstrate to readers that they are familiar with what's already been written on a topic.
- **Status and reputation:** Right or wrong, a senior worker is more likely to be taken seriously than a new worker.

Another ethos-related factor to consider is that your business writing often represents not just your own views but those of the organization for which you write. The writing you do in business is often "ghostwritten," or not associated with your name. You may be asked to draft a letter for a supervisor's signature. You may write company promotional materials. You may be asked to write parts of documents that are primarily written by someone else or that represent the whole company. When you write as a representative of your workplace your identity is complicated: it overlaps with the identities of your organization and your profession. You can't just write what you want without regard to how it will reflect on the company to which you belong.

Your business writing simultaneously represents the viewpoints, values, concerns, and ethics of yourself and your organization and your profession. In this regard, your writing has to mediate those interests, which can conflict. Suppose you have to write a letter of apology to a client for the behavior of one of your staff members. You have obligations to yourself, your organization, and your client. You will want to apologize and restore the damaged relationship with your client without making your client feel wrong for complaining, without making yourself look overly apologetic, without making your employee look incompetent (for you hired him or her), and without making your company look desperate.[24]

Conclusion: "Good Writing" Is More Than "Good Grammar"

This chapter began by demonstrating the importance of written communication skills, particularly the value placed on such skills by employers. The second part presented a typical example of bad business writing to show that content is just as important as form in evaluating whether a document is well written. The last part outlined basic principles of rhetoric that allow writers to more carefully plan the content of their writing based on the rhetorical situation. This textbook aims to expand your notion of good writing beyond simply a command of the mechanics of writing. A facility with grammar, spelling, and punctuation is important, but it alone does not guarantee that a document is well written. In the complicated contexts of business writing, where the practical (read as "money"), legal, and ethical stakes are high, writers must be able to analyze and actively adjust their writing according to the purposes, audiences, contexts, and identities called for in each new writing situation.

There are other elements of good business writing not discussed in this introduction that this textbook helps develop. Effective writers also need *visual literacy*, the ability to recognize visual design as an element of written communication. Given the multimedia nature of the Internet and the advanced desktop publishing capabilities of word processing software, writers must be able to use page layout, typefaces, and graphics to make their documents more appealing and persuasive.

Writers also require *information literacy*, the understanding of how knowledge is constructed, organized, disseminated, and stored. Early visions of an Internet came about as theorists and scientists grappled with the idea of information overload, the realization that the rate of knowledge production was eclipsing our ability to process it. Effective business writers can find, analyze, and integrate information from a variety of sources (Internet, libraries, experts, surveys, users) into written products. Especially given the democratizing force of the Internet, another aspect of information literacy is the ability to evaluate the credibility, validity, and reliability of information. The temptation to copy and paste information from electronic sources is high; good business writers must know how to appropriately incorporate ideas, words, and images from a variety of sources into their writing.

Finally, writers must possess *ethical literacy*, the ability to understand and apply social and professional standards of appropriate action in everyday business situations. Steven Katz, a contemporary rhetorician, analyzed a memo written by a Nazi engineer during World War II according to traditional criteria for effective business writing. In terms of its *form* (its organization, style, tone, and argumentation) this memo is nearly perfect. However, the *content* of the memo is quite another story. It attempts to persuade a superior that technical changes need to be made to improve the efficiency of large trucks being used to kill prison camp detainees during the Final Solution, the Nazi's plan to exterminate European Jews and other so-called undesirables. The horrific subject matter of the memo was obfuscated by references to the "merchandise" and the "load."[25] This example underscores the importance of not reducing definitions of good writing to formalistic concerns of mechanics and style.

Whether they can put it into words or not, when employers complain that their workers can't write they are generally pointing to bigger problems than just spelling and punctuation; they are referring to writers' abilities to think critically about complex problems and respond appropriately in writing. These are the elements of writing constituted in the multiple literacies of business writing and emphasized in this course.

Exercises

1. Revise the Fire Policy memo. Consider not just the language and tone but the appropriateness of the solution to the Fire Policy problem as well.

2. Individually or in a group, find a current, recent, or past event in the news that relates to business writing/communication, similar to the case of Ford/Firestone or the IHOP Twitter campaign backlash. Use the Internet and your library's periodical indexes. Reflect on how the news story relates to issues discussed in this introduction.

3. Interview someone you know who writes in the workplace. Ideally, interview someone in the field you plan to enter. Brainstorm a list of interview questions in class, such as what types of documents are written, how much time is spent writing (including planning, drafting, and revising), and the type of training the writer received (in school and at work). Write up a profile of a business writer based on your interview and share it with the class.

End Notes

[1] "Email Exposes the Literacy Gap." *Workforce*, November 2002, 15.

[2] Plain Language Network, "Writing and Oral Communication Skills: Career Building Assets." Accessed January 1, 2003, http://www.plainlanguage.gov/Summit/writing .htm

[3] "More than a Third of Applicants Lack Basic Skills," *HRFocus*, July 2000, 9.

[4] "Wanted: Leaders Who Can Lead and Write," *Workforce*, December 1997, 21.

[5] Ann Fisher, "The High Cost of Living and Not Writing Well," *Fortune*, December 7, 1998, 244.

[6] *Writing: A Ticket to Work...Or a Ticket Out, A Survey of Business Leaders*, National Commission on Writing, September 2004. Accessed August 3, 2015, http://www .collegeboard.com/prod_downloads/writingcom/writing-ticket-to-work.pdf

[7] *Writing: A Powerful Message from State Government*, National Commission on Writing, July 2005. Accessed August 8, 2015, http://www.collegeboard.com/ prod_downloads/writingcom/powerful-message-from-state.pdf

[8] Plain Language Network.

[9] Andrew Perrin, "Social Media Usage: 2005-2015," Pew Research Center, October 8, 2015, http://www.pewinternet.org/2015/10/08/social-networking-usage-2005-2015/

[10] Ibid.

[11] Fiona Briggs, "Nordstrom Is Top Performer on Social Media, New Ranking of U.S. Retailers Reveals," *Forbes*, November 23, 2015, http://www.forbes.com/sites/fionabriggs/2015/11/23/nordstrom-is-top-performer-on-social-media-new-ranking-of-u-s-retailers-reveals/2/#34dfdabcb7f3

[12] Garet Slone, "Are Brands on Fleek with Slangly Tweets," *Adweek,* October 23, 2014, http://www.adweek.com/news/technology/whos-behind-these-crazy-ihop-tweets-160955

[13] Dominique Mosbergen, "IHOP Tweeted a Joke About Breasts. It Didn't Go Too Well," *Huffington Post,* October 19, 2015, http://www.huffingtonpost.com/entry/ihop-tweet-breast-joke_us_56249e61e4b08589ef47eacb

[14] Ibid.

[15] Lisa Toner, "Writing Business Letters," Purdue Business Writing Coursepack, 1995.

[16] Ibid.

[17] James Porter, "What is Business Writing? Why Do It?" Purdue Business Writing Coursepack, 1994.

[18] Toner.

[19] "Revision in Business Writing." Purdue Online Writing Lab. Accessed August 3, 2015, https://owl.english.purdue.edu/owl/resource/648/1/

[20] Toner.

[21] Linda Driskill, "Understanding the Writing Context in Organizations," in *Writing in the Business Professions*, ed. Myra Kogen (Urbana, IL: National Council of Teachers of English, 1989), 125–45.

[22] Driskill.

[23] Porter.

[24] Toner.

[25] Steven Katz, "The Ethic of Expediency: Classical Rhetoric, Technology, and the Holocaust," *College English* 54.3 (1992): 255.

TECHNICAL WRITING

CHAPTER 2

Technical writing is anything written that conveys highly specialized information. Specialized information is that which is related to a profession or academic discipline that has its own language for communicating the ideas and practices that are important to that field. To insiders—the engineers and architects, computer technicians, scientists, medical professionals, and lawyers that make up a field—this language is necessary for dealing with highly complex and abstract problems. To outsiders, this language, called *jargon*, is often difficult to understand.

Frequently, the cause of bad technical writing is that technical insiders fail to translate their specialized language for the benefit of outsiders with whom they are communicating. The most notorious examples of this are electronics manuals and software instructions, which are often written with such insider jargon as, "To initiate application initialization, populate the user ID field with the designated user identification authorization numeral." "What does that mean?" asks the befuddled couple trying to install the new accounting software they just purchased. What the technical writer should have written was, "To install the program, type the 15-digit number printed on page 2 of your manual into the box labeled 'Enter user ID here.'"

There are other reasons for bad technical writing, including lack of time, sloppiness, and deliberate obfuscation. Perhaps the writer of the example above didn't have time to check to see if that instruction made sense to an actual reader because it was part of a software upgrade that needed to be released in time for tax season. Or maybe the writer didn't check to see if the instruction made sense because he didn't care if anyone could actually use the product his or her company spent thousands of dollars developing. Or perhaps the writer deliberately wanted the statement to be so overly technical that users would have to hire a trained technician, likely from the company that produced the software, to install it for them.

Good technical writing effectively conveys information from a specialized field to specific readers, in specific situations, in such a way that the readers find the information useful. In this sense, *user friendliness* applies not only to computer software but to any document written for readers' instrumental use. Furthermore, good technical writing is that which is carefully produced, takes into consideration the needs and knowledge level of the reader, and doesn't deliberately misrepresent ideas for the writer's own benefit.

A common misconception is that science, engineering, and technology fields are neutral and that people who communicate in technical professions must be as neutral and objective as possible in their communications. Far from being neutral, technical professional activity occurs within social, cultural, historical, political, ethical, and other contexts. Communication never occurs in a vacuum. Whenever writers consciously (or unconsciously) adjust their message to the constraints of a given situation—and all writers must, to be effective—they are using rhetoric, the art of speaking or writing well. Techniques for making written information useful to actual readers include global, big picture issues such as understanding what readers will need to do with the information and adjusting the message accordingly, as well as micro, surface-level issues such as formatting the document with lots of spacing and easy-to-see section headings. This course will focus on these and many more techniques for effectively communicating technical information.

Why Study Technical Writing?

You may have heard we are in the information age. This means that ideas are considered tools—things that make our lives easier or help us get things done—that are just as important as the machines that replaced human labor in the industrial age. In other words, creating a database that can retrieve specific digital images captured during a continuous remote sensing experiment (e.g., a satellite that continuously takes pictures of a semi-dormant volcano's thermal emissions) is just as important as creating the digital equipment that generates the thousands of volcano images each day. While production jobs, like factory worker, will always be necessary, it is the symbolic-analytic jobs where people take symbols (i.e., words, numbers, and pictures) and create new ideas, like database solutions, that are most valued in an information economy. It is only through our language that information can be created, conveyed, and stored. You've heard of information overload. Well, we've gotten very efficient at creating and sharing new ideas, in large part due to computers and the Internet.

Those of you preparing for careers in technical occupations such as engineer or scientist will be writing a lot because your jobs will entail manipulating language (including mathematic symbols) to generate new ideas. Simply put, much of your daily activity will be mediated through written communication. One study of engineers and scientists who worked at a plastics company demonstrated that the technical employees at all levels of the organization—from staff, to supervisors, to upper management—spent 40 to 50 percent of their weekly labor engaged in writing-related activities, including drafting, revising, and editing.[1] The staff engineers and scientists spent a significant portion of their writing-related time *drafting* reports about their technical work. The upper mangers, on the other hand, spent most of their time *editing* the proposals and reports written by staff and supervisors. All levels of engineers and scientists at this company also spent a significant portion of their time in other writing-related activities including reading and responding to emails and letters, planning and recording meetings, and preparing oral presentations.

If you want to be successful, you had better be a good writer. If you desire to eventually be promoted to upper management, you not only need to be able to write but to critique and improve the writing of others.

 Effective Technical Writing Is Rhetorical

The best way to improve your writing is to carefully and thoroughly understand your writing situation, including who your readers are. Techniques for analyzing your writing situation come from the field of rhetoric. Achieving good writing, as you read above, takes more than paying attention to correct spelling and grammar. This course assumes, by the way, that you already possess a certain level of basic literacy that includes knowledge of spelling and grammar. (You've presumably taken both high school and freshman level English classes.) Certainly, grammar and mechanics are important for effective communication. Too many basic mistakes and your readers may have trouble understanding your message or, worse, they may stop trying to understand it. However, even the most error-free prose could fail to convey technical information if it doesn't take into consideration the following:

- What is the purpose for writing?
- Who is the reader?
- What is the context?
- Who is the writer?

What Is the Purpose?

Technical communication occurs in social contexts and therefore is a social action. People write technical documents because they want to do things in the world: the social action involved in technical writing connects the intangible sphere of thought with the concrete realm of human activity. A person authoring online help for a word processing program, for example, aims to help users understand a specific computer program and, in turn, write more effectively. Likewise, an engineer writing a 40-page recommendation report detailing a six-month-long study of groundwater patterns near a proposed factory informs crucial decisions about the factory's construction—decisions that will impact multiple individuals and groups such as the local community where the proposed factory will be built, the global company looking for U.S. operations, and the engineering firm conducting the study. If a technical document did not intend to initiate, effect, or affect action, then it probably wouldn't be written. Because technical communication is social, it is often understood differently by different people for different reasons. Technical writing often fails because writers have ignored the social nature of technical communication.

All technical writing has multiple purposes, and the more the writer is aware of the multiple purposes of a given writing task the better the writer can evaluate the document's effectiveness. These purposes include the following:

- To inform
- To describe
- To persuade
- To educate
- To summarize
- To warn
- To indemnify
- To document/create a record
- To maintain goodwill

The most common purpose of technical writing is to convey information. Scientific research articles present new knowledge based on original research. Engineering specifications describe a product in precise detail so it can be used, reproduced, repaired, manufactured, or tested. Instruction manuals inform users of the procedure for assembling a product and how to use it safely, including warning of any hazards associated with improper use of the product. There are other purposes of technical writing. Manuals also train or educate users about the best ways to accomplish a task using the technology documented in the manual. (Our culture, however, is very suspicious of manuals because of their history of being so badly written. Ever hear the expression, "When all else fails, consult the manual?") In this class you'll learn how to analyze your writing situation by first asking, "What is my purpose for writing?"

It is helpful to also be aware of secondary purposes. Think of product warnings. A warning's primary purpose is to inform the reader of a danger or hazard associated with misuse of the product. However, in addition to informing the user of the hazard, the warning must educate the user about how to avoid the hazard. Moreover, the warning serves to indemnify the company from product liability, protecting the company from litigation if a user gets injured using the product. Thus, it is in the best interest of the company to write a warning with sufficient information so that it persuades the reader to use the product properly. Oftentimes companies are afraid to include too much information in warnings, particularly about the potential for serious injury, for fear it will scare customers away. However, it is safer—for the company and the users—to be up front and honest about the risks associated with using a product. (Can you identify the multiple purposes of product warnings discussed in this paragraph?)

Another way to think about a document's purpose is to consider how the reader will use the information. Think of the bad software documentation example used previously. If the writer had imagined how the reader would be using the information, i.e., sitting in front of the computer and wondering how to get the new program working as soon as possible, he or she might have written the instruction more directly, in a way that the reader could understand.

You should also always consider as one of your purposes the following: to maintain goodwill, or to keep positive relations with the reader. Poorly written technical documents alienate readers from the technology and disincline them to the makers of the technology. Whose fault is it that people have trouble setting the clock on a VCR or using all the features of a cell phone, the instruction writer's or the user's? A lot of technical writing is about relations: who will do what job, who will fund which project, who will give you the time you need to finish a task, who will finish the test you need to confirm your product modifications? Engineers and scientists interact with all types of people day in and day out. Technical professionals must treat readers humanely, with respect and dignity, regardless of the reader's technical background and level of specialized knowledge.

Who Is the Audience?

The biggest factor to consider when relating technical information is the audience's level of knowledge or expertise about the information (see Figure 2.1). Are your readers experts who have the same background and familiarity with your subject as you, or are they a laypeople who are unfamiliar with the technology you are writing about and who do not have the same technical training as you? The less your audience knows about the subject, the less technical your document should be. The less your readers know about your subject, the more you will have to explain and the more you should avoid highly specialized language. It's impossible to avoid all technical language, so you may have to define certain terms.

Type of Audience	Lay audience		Expert/peer
Level of Technicality	Low technicality	⟷	High technicality
Type of Language	Avoid specialized language & explain unavoidable jargon		Use specialized language

FIGURE 2.1. Adjust the language of the message to the type of audience

It is also helpful to judge your audience's receptivity to your message. You might be writing a progress report about some product testing that is behind schedule and over budget. How will you convince your skeptical boss to allow you more time and money on the project? Some audiences are predisposed one way or the other to your message. For instance, writers of environmental impact statements have to contend with both supporters and opponents to proposed land-use projects. Anticipating the kinds of questions that readers will ask of your documents (e.g., How is your costly product testing going to save us money in the long run? How will the landfill affect the nearby housing development?) will help you adjust your message accordingly. If your audience is a group of people, or if you don't know your immediate audience, sometimes you have to make educated guesses about the audience's general needs and interests.

In today's global economy it is also important to take into consideration cultural factors that may affect an audience's ability to understand your message. Will your document be read by an international audience? Do you need to tailor your document to accommodate a worldwide audience?

Finally, there are always primary and secondary audiences to a document. Your boss may be the primary audience of a report you write, but your colleagues may also read the report. You might also be thinking of your professional peers as an audience if you wish to reuse some of the information in the report for a journal article you hope to publish. You must also always consider lawyers, the public (people outside the company you work for), and the media as audiences. Several scientists at the controversial Yucca Mountain Nuclear Waste Repository were found to have admitted falsifying important quality assurance records in emails related to research on the safety of storing nuclear waste 100 miles away from the city of Las Vegas, Nevada. Whether these scientists' actual research is valid is being debated. However, legislators, the public, and the media learned of these emails, calling into question the integrity of the scientists conducting these important safety studies.

What Is the Context?

Context means considering any factors related to your writing situation—the setting you are writing in—that might affect your document. For instance, you should consider when the document is due and what form the document should take. Should it be written as a lab report, specification, manual, tutorial, memo, or letter? Given the type of document called for, how should it be organized and formatted? Each research field, for instance, has different standards for citing previous research and they tend to differ from one another. Each company has slightly different ways of writing various documents. Many companies keep style guides that dictate document specifications such as margins, headings, and proper spelling and abbreviations. While this course will teach you some basic guidelines for good technical writing, it cannot possibly anticipate and teach all the variations among companies. You can often learn formats of company-specific documents by collecting samples of previously written documents.

When considering your writing context you should also consider more abstract cultural, ethical, and political factors related to your situation. There are more women than ever in technical fields. Technical communication must therefore be written using nonsexist language. The audiences of technical communication are more diverse and international than ever. Technical writing also often deals with subjects that have consequences for real human beings. Medical professionals must keep accurate patient histories. Airline technicians must keep accurate airplane maintenance records. Engineers must determine the safety of everyday products. Scientists must not falsify experimental procedures to achieve desired results. Writers of technical documents are often confronted with difficult ethical decisions that have financial, emotional, and physical health implications for the writer, the writer's organization, and the public. Lastly, consider if there are any political factors that could affect the document. Will you be stepping on anyone's toes? How will you report a costly error to your boss? How will your message impact others, positively or negatively? Are there any financial or legal factors that you should consider?

Who Is the Writer?

This sounds like an odd question to ask yourself before you write, but this question is designed to get you to think about your relationship as a writer (and a person) to your reader. The writer of the confusing accounting software manual forgot that he or she was a computer scientist writing to a layperson. He did not consider that the word *populate* means something different to non-computer scientists. Consider what relationship you have, or want to have, with your audience and then consider how you should adjust your message:

- Who are you writing to, your boss, your employees, the public, your profession, your teacher?
- Who are you writing as, a technical specialist, a manager, an employee, or a student?
- What tone (or mood/personality) should you use to convey your message?
- What language is acceptable for you to employ as a writer?

In general, your tone should always be balanced and professional, never hostile or arrogant. But should your writing be more formal, avoiding first-person "I" and second-person "you," or can it be more informal, addressing the reader directly? Most memos and letters are written using first and second person. It is also becoming more acceptable to use first person in technical reports (e.g., instead of writing, "The reaction of silicon dioxide films were studied," write, "We studied the reaction of silicon dioxide films.") When considering how you will communicate your message as a writer you can also think about any specific terms that will need to be translated or avoided.

Conclusion

In the previously mentioned study of technical writing at a plastics factory the researchers found that managers used the engineers' and scientists' writings to evaluate overall thinking ability, not just writing skills. "Their writing reflects the quality of my staff's minds," one supervisor commented. "Management wants original ideas...and weak or confused documents simply call attention to incompetence."[2] Since most of the work of a technical job is conveyed and documented in proposals, reports, and other memoranda, it should be no surprise that your writing will play a crucial role in how others judge you as a coworker and employee. In this sense, you should always aspire to produce good technical writing, writing that accomplishes its purpose, is sensitive to the needs and background of its readers, shows awareness of the context of the document, and is written in a manner that causes readers to perceive the writer as thorough, professional, and humane.

Exercises

1. In class, teams of 4 or 5 create a simple Lego object and write instructions on how to assemble it. Exchange the disassembled Legos and instructions to see which teams can produce the desired object. Discuss what made teams' instructions effective or ineffective.

2. Individually or in a group, find a current or past event in the news that relates to technical writing/communication. Use the Internet and your library's periodical indexes. Reflect on how the news story relates to issues discussed in this chapter.

3. Interview someone you know who writes in the workplace. Ideally, interview someone in the field you plan to enter. Brainstorm a list of interview questions in class such as what types of documents are written, how much time is spent writing (including planning, drafting, and revising), and the type of training the writer received (in school and at work). Write a profile of that workplace writer based on your interview and share it with the class.

End Notes

[1] James Paradis, David Dobrin, and Richard Miller, "Writing at Exxon ITD: Notes on the Writing Environment of an R&D Organization," in *Writing in Nonacademic Settings*, ed. Lee Odell and Dixie Goswami (New York: The Guilford Press, 1985), 285.

[2] Pradis, Dobrin, and Miller, "Writing at Exxon ITD," 295–96.

PROFESSIONAL WRITING STYLE

CHAPTER 3

Style in writing refers to **rhetorical form** rather than grammatical form. That is, style is a matter of choosing words and constructing sentences in a way that suits the writer's and, more importantly, the readers' tastes. These tastes have developed over time and are more about social custom and habit than any abstract set of grammatical principles.

style

Style is a matter of choosing words and sentences that suit the writer's and readers' tastes. Unlike grammar, which is bound by certain rules, style is more a matter of choice and convention. There are many types of style, such as academic, literary, and business writing style, and what counts as good writing in one context may not be effective in another. Business writing style generally aims to be clear and concise, uses readability devices such as shorter paragraphs, lists, and headings, and uses "I" and "you" to address the reader in everyday correspondence.

Effective business and technical writing style should be clear and concise. Because busy business readers—managers, workers, clients, and consumers—don't have the time and patience to deal with long, hard-to-read prose you want your writing style to be direct, readable, and easily understandable. You want readers to comprehend your points without having to waste time deciphering confusing sentence structure, word choice, or grammar.

Communicating effectively begins foremost with having a firm understanding of your *purpose* (what you are trying to achieve with your writing) and your *audience* (who you are communicating with). Once you clarify these for yourself, as a way to measure if you have communicated your purpose given a particular audience's knowledge and ideals, you can work on organizing your message, writing a rough draft, determining how well it accomplishes its purpose, then editing it for clarity and concision.

Revision versus Editing

Remember the difference between **revision** and **editing**. Revision happens when you make changes to the higher-order issues of purpose, organization, and development of details/content. Writing effectively means first getting your initial plans on paper, then revising them to improve their organization and development based on the reader's needs and anticipated questions.

Editing, on the other hand, represents lower-order changes to style, grammar, punctuation, and spelling. When first putting your thoughts to paper or computer screen don't worry about lower order concerns. Focus on getting your thoughts on paper in whatever form they come to you. Write out your ideas as if you were explaining them orally to a close, respected friend in a casual setting. Novice writers get hung up on crafting perfect prose, word for word, sentence by sentence. Rarely should you have to stop a thought in midsentence to search for the perfect word. Put down whatever word first comes to your mind and then, during the revision and editing stages, go back and change it if necessary. Worry first about the big picture, about getting your point out of your head and into the document and then go back and revise the text for clarity and concision.

Principles of Effective Professional Writing

After revising your document—be it a letter, memo, case analysis, or report—for purpose and audience, edit your prose carefully to conform to the following principles of effective professional writing style:

1. Use plain, conversational language (write clearly)
2. Eliminate unnecessary words (write concisely)
3. Guide readers through your documents
4. Use direct presentation, generally
5. Use active voice, generally
6. Use parallel grammatical structure
7. Keep your tone polite and professional
8. Write like a human being
9. Be diplomatic
10. Pay attention to usage, spelling, grammar, punctuation, names, and numbers

Use Plain, Conversational Language (Write Clearly)

Writing clearly means organizing your message and choosing your words in a way that optimizes the readers' chances of easily understanding your message. To write clearly: (1) keep it simple and (2) keep it conversational. You're not writing to show off or impress, but to inform and persuade. Lack of clarity, variously called gobbledygook, affectation, or double-speak, has crept into our way of communicating for many reasons, including the fear that plain messages don't sound important enough or won't impress subordinates or superiors. Gobbledygook, a term coined in 1944 by Congressional Representative Maury Maverick, refers to impenetrable corporate or government prose that is used to variously cover up a poorly conceived message, needlessly impress, deliver unpleasant information or, as in the case of jargon, to alienate or confuse laypeople (see Figure 3.1).

Instead of:

To ensure that the new system being developed, or the existing system being modified, will provide users with the timely, accurate, and complete information they require to properly perform their functions and responsibilities, it is necessary to assure that the new or modified system will cover all necessary aspects of the present automated or manual systems being replaced. To gain this assurance, it is essential that documentation be made of the entities of the present systems which will be modified or eliminated.

Write:

Make sure to document all changes to the current system so any mistakes can be corrected.

FIGURE 3.1. To fix gobbledygook, say what you mean and get to the point[1]

The easiest way to write clearly (unambiguously, in a way that is easy for the reader to understand) is to write as you talk. You want the *voice* of your writing to sound professional, yet not artificial. It is acceptable to use first person "I" and second person "you" in business memos and letters and some reports. You can also use contractions (two words joined by an apostrophe, e.g., *can't* and *don't*) in all but the most formal of correspondence (e.g., when applying for a job). Write as you talk and your writing will significantly improve, but avoid slang, which is unprofessional.

Cut these signs of gobbledygook from your writing:

- **Overly inflated or pompous word choices.** Instead of *effectuate*, write *do*. Instead of *fabricate*, write *make*. Instead of *initiate*, write *start*.
- **Clichés.** Clichés are those trite, wordy expressions used so often that we don't even know what they mean. Instead of *as per your request*, write *I have completed* or *Here is the information you requested*. Instead of *pursuant to*, write *about* or *regarding*.
- **Jargon with lay audiences.** Jargon is (1) confused, unintelligible language; strange, outlandish, or barbarous language; (2) technical terminology or characteristic idiom

of a special activity or group; (3) obscure and often pretentious language marked by circumlocutions and long words.

- **Buzzwords.** Buzzwords are trendy business terms that slip into an industry or profession, such as *robust*, *metrics*, and *synergy*.

Eliminate Unnecessary Words (Write Concisely)

Writing concisely is a way to achieve clarity by making your point in the fewest possible words. It means eliminating unnecessary and redundant words. This is often referred to as eliminating clutter, deadwood, or flabbiness. Novice writers tend to inflate their prose with extra adjectives, adverbs, and phrases. Efficient, uncluttered writing is more easily understood and appreciated because it doesn't waste the reader's time. Words worth cutting include:

- Passive constructions, verb phrases that include a form of *to be*, such as *am*, *is*, *was*, *were*, *are*, or *been*
  - Instead of *The report was distributed by the committee*, write *The committee issued the report*
- Most adjectives and adverbs
  - Instead of *Rogers convincingly explained his position*, write *Rogers explained his position*
  - Instead of *positively certain*, write *certain*
- Weak modifiers
  - Cut words like *very*, *rather*, or *little*
  - Instead of writing *you can do a little better*, write *you can do better*
- Meaningless words
  - Cut words like *kind of*, *actually*, *really*, and *basically*
- One part of a doubled phrase
  - *~~full and~~ complete*, *first ~~and foremost~~*, *~~any and~~ all*, *each ~~and every~~*
- Redundant phrases
  - *~~period of~~ time*, *~~basic~~ fundamentals*, *~~past~~ history*, *~~future~~ plans*, *~~free~~ gift*, *~~terrible~~ tragedy*
- Compound prepositions
  - Instead of *at this point in time*, write *now*.
  - Instead of *despite the fact that*, write *although*.
- Obvious words
  - *~~When it arrived~~, I cashed your check immediately*
  - *Our company president, ~~who ultimately makes all of the decisions~~, should be notified about this.*
- Wordy clichés
  - Use *because* instead of *due to the fact that*
  - Use *if* instead of *in the event that*
  - Use *must* instead of *it is incumbent upon*
  - Use *that you requested* instead of *as per your request*

For more help, particularly advice about when passive constructions are preferable, see the Purdue Online Writing Lab's "Active and Passive Voice" handout, which can be found at: https://owl.english.purdue.edu/owl/resource/539/01/.

A good habit to get into is this: after you have revised your draft to a point where you think you're done, **try to eliminate at least 20% of its total word count**. Does this sound excessive? It is the equivalent of editing a 10-word sentence down to 8 words. This percentage is just a rule of thumb. As you become more adept at eliminating useless words from your writing the percentage you cut may increase. The point is that good writers always pay attention to eliminating clutter. It is a service paid to readers.

Guide Readers through Your Documents

One of the best ways to improve comprehension is to add graphic highlighting and other subtle page design elements to make your documents more readable and scannable (faster to read through). Novice writers see these elements all the time but often fail to use them to improve their own documents. Add elements such as the following:

- **Summaries**, or overviews at the beginning of documents and sections within documents
- **Transitions**, words that help the reader see connections between ideas, such as *however*, *another*, *next*, and *thirdly*
- **Letters** such as (a), (b), and (c) within text
- **Numerals** such as 1, 2, and 3 listed vertically
- **Bullets**, raised periods, black squares, or other figures
- **Headings and subheadings** signaling the topic or main idea of particular sections
- **Emphasis** by creating contrast through **boldfacing**, *italicizing*, or <u>underscoring</u> can highlight main ideas or important points. You can also alter the size of text or use ALL CAPS. Be careful not to overuse emphasis.
- **Smaller paragraphs**, as readers don't like to struggle through dense, packed prose. Break up larger-sized paragraphs. It's okay to have one, two, or three sentence paragraphs.

Use Active Voice, Generally

Verbs are either *active* (The manager *assigned* the task) or *passive* (The task *was assigned* by the manager). Active verbs immediately follow the subject and make clear who or what the center of action in the sentence is: "*The manager* assigned…" Passive sentences use verb forms of *to be*, including *is*, *am*, *are*, *was*, and *were*. Sentences written in active voice are usually more concise and easier to understand than sentences written in passive voice.

Computerized grammar checkers can identify passive voice constructions and ask you to revise them to a more active construction. There is nothing inherently wrong with the passive voice, but if you can say the same thing in the active mode, do so.

Use Direct Presentation, Generally

Busy readers want to quickly determine whether they should read something now or later, if at all. Let readers know up front what they're reading. State and summarize your main points before developing them. Always include an opening or overview section in your documents that states your point in the first sentence or two. Don't wait until the end of the opening, as you were taught in writing introductions to academic essays. Headings and bulleted points can help present arguments directly, e.g., "Because of the recent corporate restructuring our firm must adopt the following new policies: 1) ... 2) ... 3) ...." Another example of directness in business writing is the type of organization called the *managerial style*, which presents recommendations at the beginning of a report. The name of this style derives from the decision-making manager's desire for clarity up front.

However, being direct isn't always the best tactic. In some cases, as when you have to deliver bad news or persuade a skeptical reader, a more indirect approach usually works better. In bad news letters this is called the buffer, a letter opening that begins with a positive but relevant statement before introducing the the bad news. Another instance where an indirect approach would be necessary would be in reports that deliver information to skeptical or hostile readers, such as a recommendation report calling for major increases in expenditures. In this case it would be preferable to first make your case and then present the recommendation toward the end of the report.

Use Parallel Grammatical Structure

Express similar ideas in balanced or parallel constructions. Writers often string together a number of items, activities, etc. into one sentence. For example, a writer might list goals for a project: "By quarter's end, we should increase sales, eliminate overhead, and cut losses." This sentence is parallel because each goal is stated using the **same verb form**, i.e., *increase, eliminate,* and *cut.* A sentence is unparallel when it fails to present similar items in the same grammatical fashion.

You can use this principle to present several points within a paragraph as a bulleted or numbered list:

> *The position is available to applicants that meet the following criteria:*
>
> - *Business-related degree, preferably a BS/MBA*
> - *5–6 years working in computer-related industry*
> - *High degree of experience interfacing directly (face-to-face) with customers and middle-level management*

Keep Your Tone Polite and Professional

The principle aim of all professional communication is to establish and maintain relationships. Without the goodwill of your audience you cannot motivate, persuade, sell, etc.—all that you're trying to accomplish in business. Avoid language choices that are angry, condescending, arrogant, or rude. Avoid words with negative connotations such as *cannot, forbid, fail, prohibit,* and *deny.* Instead of *poor service,* write *service,* as in, "I'm writing to ask your help in addressing the recent service I received from your agency." This often requires a shift in perspective. Instead of complaining, why not set out to inform your reader about the situation and politely ask for a remedy? This is often called taking a *you* approach. Avoid sounding self-centered; make the reader the main focus. Instead of, "This change addresses our labor costs associated with sloppy renters," write, "This change enables us to clean vacated properties faster, thereby insuring prompt service to you the customer."

Write Like a Human Being

As language expert William Zinsser writes, "Just because people work for an institution they don't have to write like one."[1] Your writing will be better if it treats readers with respect and attempts to connect with them as humans. The words you choose should project an image of yourself as respectful, thoughtful, warm, and personal. Your writing should focus on the needs and concerns of the reader. This is called the ***you* approach**. Show empathy for the readers' point of view, provide reasons for whatever it is you're asking them to do, and be polite and friendly. Instead of writing, "A sales receipt must accompany any refund," write, "To reduce the processing time of your refund, please provide a copy of the sales receipt." If you have to deliver bad news or compel the readers to act in ways to which they may react negatively, provide reasons why the bad news or unfavorable action is necessary.

you approach

The "you approach" refers to communicating benefits to readers prior to requesting actions and setting a positive tone in your business writing.

Be Diplomatic

Use positive words to express negative ideas. Instead of writing, "I want to complain about your poor customer service," write, "I wish to bring a customer service matter to your attention." Avoid words that suggest the reader is dishonest, careless, or mentally deficient. Be indirect in situations where you must deal with sensitive topics, persuade skeptical readers, or deliver bad news. Lastly, avoid putting anything in writing you might regret later.

Pay Attention to Usage, Spelling, Grammar, Punctuation, Names, and Numbers

Documents should always be proofread carefully. Remember, proofing a document should come at the very end, after you've sufficiently stated and developed your point and improved the style of its expression. The rule of thumb for checking surface-level issues like *usage* (using the right word—it's vs. its; affect vs. effect; principle vs. principal) or punctuation is this: if you're not 100% certain of the correct usage or rule, consult a reference guide or someone who knows the rule. Expert writers rely on reference aids as a basic tool when writing, whether to double-check a tricky rule, look up an unfamiliar word, or answer questions like, "Should I capitalize this word here?" With many reference tools online (e.g., Bartleby.com) this task has become faster.

To proofread more effectively:

- Allow adequate time to proofread. Build time for proofing into your schedule.
- Give yourself a break from the document. Come back to it after a period of rest, such as after lunch or the following day.
- Print a copy and read through it at least twice. Try reading backwards, bottom-to-top from the last page.
- Read the document out loud to yourself. You'll often catch constructions that read okay but don't sound right.
- For high-stakes documents read the draft to someone or have them read the draft back to you. It is commonplace for expert writers to give and receive feedback, especially in the workplace.
- Use, but *do not trust*, computerized spelling and grammar checkers. Have you read Jerrold Zar's "Ode to a Spell Checker"? See http://lingolero.com/2014/06/ode-to-a-spell-checker/

Draft, Revise, Take a Break...Then Revise Again

Experienced writers draft and revise rather than attempt to write a document all in one sitting. Few writers can craft perfect prose word for word. Initial drafts often contain what's known as **writer-based prose.** This is writing that satisfies the writer but not the reader. It contains:

- Long, rambling sentences that make sense only to the writer
- Lengthy paragraphs covering too many topics
- Sequential items in hard-to-read paragraph form rather than easier-to-read bulleted lists
- Ideas presented in a narrative, chronological structure rather than direct or indirect organization patterns

This kind of writing must be revised into **reader-based prose**, or writing that "foregrounds and makes explicit the information a reader needs or expects to find."[2] The more you can visualize an actual conversation with your reader, the more you can anticipate the kind of questions the reader might need answered, the kind of information the reader needs to know, and the best order to present that information to the reader.

Since you must revise to achieve effective reader-based business writing, here are some strategies for revising your initial draft:

1. **Test your draft against your initial plans:** Did all of your goals and good points from your notes make it into the draft? Did you forget any key ideas? Did any irrelevant ideas creep into the draft?

2. **Use clues that reveal your plan to your reader:** Did you include enough direct and explicit statements about your plans to your reader (e.g., opening purpose statement)? Did you include other cues like headings and transitions to help clarify your point to the reader?
3. **Keep the promises you made in your writing:** Did you follow through on any promises you made to readers. For instance, if your overview states that you will provide four recommendations each supported by an example, did you keep that promise?
4. **Shift your perspective to that of the reader:** According to one guide written by Purdue's Online Writing Lab, to revise effectively you must distance yourself from the draft and see it objectively from the readers' perspective. "Unless you divorce yourself from the paper," states the guide, "you will probably remain under its spell: that is, you will see only what you think is on the page instead of what is actually there." To see your writing objectively you need to build time into your writing process so that you can set your draft aside and return to it later, either after a break or the following day. The guide includes a detailed checklist for helping you assess your drafts from a reader's point of view. The following table summarizes that checklist and can be used to evaluate any business writing, from short memos or letters to longer reports:

Detail	Have I included all the information the reader needs to know, and nothing more?
Organization	Have I arranged topics in a way the reader expects the topics arranged, either directly or indirectly depending on the reader's favorable or negative disposition to my message?
	Have I compartmentalized the topics of my message?
	Have I included headings and transitions between topics to help guide my reader through the presentation of my message?
Language	Have I used the right words to get my message across, or are some words unnecessary, elevated, or in need of a definition?
Tone	Have I chosen words that present me to the reader as courteous, respectful, rational, professional, and ethical?
	Have I avoided negative language and used positive words to express negative ideas?
	Have I focused on the reader's needs and interests rather than my own (the you approach)?
Mechanics	Are there any glaring typos or mechanical mistakes that would diminish my credibility as a writer and businessperson in the eyes of the reader?
	Have I carefully proofread and corrected all spelling, grammar, and punctuation mistakes?

FIGURE 3.2. Reader-based revision checklist[3]

Exercises

1. Revise the following sentences for clarity and concision. Use the word counts in parentheses as a guide; see if you and a partner can revise the sentence to at or below the suggested word count *without altering the meaning of the original.*

   a. There are many words that are useless that can be eliminated through revision that is carefully done. (17 words to 5 words)

   b. After extensive exhaustive labors on the assigned project task, Roger needed to take a long respite. (16 to 10)

   c. In the event that customers who have questions about their automobiles call us, we have just created this really useful program that is intended to save time in our customer service responses. (33 to 7)

   d. The report was written by engineers out of the taskforce from Atlanta (12 to 6) *including parentheses as with other examples.

   e. Until such time as we are in possession of more information, we will be unable to offer a satisfactory reply to your inquiry. (12 to 6)

   f. We are unable to provide you with access to computerized laboratories in view of the fact that you are not in possession of personal identification. (26 to 8)

   g. Until such time as our company changes this policy, we are not in a position to supply information for the purpose of aiding your investigation of other companies. (29 to 11)

   h. In the event that employees fail to utilize the requisite safety precautions, our optimum course is to transmit this information to OSHA and request directives before deleterious effects ensue. (30 to 13)

   i. People renting our apartment units are concerned about whether we would provide for replacements of damaged personal property in the event of accidental discharge of the sprinkling system in the apartment complex. (33 to 15)

   j. Subsequent to my conversation with Mr. Jones via telephone, the physician at the hospital where Mr. Jones is hospitalized indicated that Mr. Jones has an enlarged kidney and must undergo immediate surgery if a cure is to be effectuated. (40 to 18)

End Notes

[1] William Zinnser, *On Writing Well*, 4th ed. (New York: Harper Collins), 147.

[2] Linda Flower and John Ackerman,"Evaluating and Testing as You Revise," in *Strategies for Business and Technical Writing*, 4th edition, ed. Kevin J. Harty (New York: Allyn and Bacon, 1999).

[3] "Revision in Business Writing," Purdue Online Writing Lab, accessed August 3, 2015, https://owl.english.purdue.edu/owl/resource/648/1/.

CHAPTER

4

EDITING

Revision and editing are often confused as being one and the same. Although it does include copyediting (meaning changes to spelling and grammar), revision is generally considered to include higher-order changes such as purpose, organization, and development of details and/or content.

Editing, or copyediting, includes lower-order changes such as spelling, grammar, punctuation, and changes to style or formatting. Both the higher-order process and the lower-order process are critical to the editing process and will hereby be referred to simply as editing.

Why Edit?

Editing ensures that documents are error-free, with the message presented as clearly, concisely, and effectively as possible in order to deliver the message to the intended audience. Editing removes all distractions and helps to satisfy the audience's need for information. Editing is a part of the reader-centered approach to writing that strives to tailor the message for the intended audience while making the readers' needs a priority over what the writer wants to say.

Chances are, if you are reading this chapter, you are in college aspiring to attain a professional position and eventually move into higher-paying management positions or run your own business. In either case, you will likely be managing employees. A landmark study of writing within one organization showed that managers and supervisors spent more time engaged in reviewing and editing their employees' documents than they did writing their own documents. Upper-level managers spent 72 percent of their writing-related activity editing versus just 14 percent writing.[1] Thus, effective managers must possess strong editing skills including how to

- Identify weaknesses in a written document
- Encourage employees to write more effectively
- Motivate employees to write faster
- Revise text to have a clear message and reader-centered content
- Help employees eliminate errors in grammar, mechanics, and usage

Uses of Editing

Editing is a valuable tool that should be used each time a writer produces text in order to deliver an error-free and understandable message.

In the business world, oftentimes projects are completed collaboratively by many writers. Several writers may contribute information, pieces, or chapters to the same document. Editing removes not only lower-order problems such as spelling, grammar, and punctuation errors, but ensures that all writers present the information as if written with one voice, that the writing is uniform in purpose, organization, details, and content.

In both business and academia, oftentimes writers will need to revise an article or story for publication. This revision process can include both revision and editing, and is oftentimes referred to simply as revising. Before an article is published, the document must go through several, if not many, revisions. Sometimes, revision includes only small changes such as copyediting, but oftentimes the same article or story will be revised with higher-order changes to create an almost different article altogether. Changing the audience or purpose of an article, yet delivering the same information, can make all the difference in the message delivered to an intended audience.

In academia, oftentimes peer editing is performed on documents prior to submittal. Peer editing is the practice of reading a peer writer's document and making editing suggestions. The peer-reviewed document is then corrected by the writer. Peer editing is a win–win situation as the reader will identify writing issues of their own when peer reviewing and be able to improve their writing skills, and the writer will benefit from the peer review and the incorporation of changes.

Any peer-editing process can involve hard copies (printed documents) using traditional editorial marks, or digital copies using MS Word's Review tools as well as Spelling and Grammar Checkers. Both of these editing processes are covered within this chapter.

Peer editing involves the processes of revision and editing performed simultaneously. Although it is a good idea to make several passes or "reads" of a document, with practice, most readers can find many editing problems with one pass that incorporates all levels of a top-down editing process. This top-down editing process is described next.

Levels of Edits in a Top-Down Editing Process

Top-down editing is a structured approach to editing a document that compartmentalizes review and prioritizes the big-picture, global or higher-order issues over sentence-level issues. All levels are important to the success of a document, but novice editors frequently place too much importance on sentence and word-level issues over bigger-picture rhetorical concerns.

Level 1: Global Editing—Document Level

This first step in the top-down editing process looks at the entire document, including title, introduction, format, layout, headings, message, subject, purpose, readers, and context of use. Some questions to consider when performing global editing on the document level include the following:

- Is the title reflective of the document?
- Does the opening introduce the information clearly?
- Does the information across sections follow a logical order or does any information appear out of place or missing?
- Are the layout, headings, and formatting clear, concise, and effective, as well as follow standard conventions/expectations of the document? In other words, does the book look like a book?
- Is the writer aware of the reader and using reader-centered writing techniques such as talking headings and section overviews?

Level 2: Substantive Editing—Section Level

The second step looks at each section individually, including topic sentences, bullet points, information clarity, and paragraph transitions. Some questions to consider include the following:

- Does the topic sentence introduce the information?
- Do the bullet points highlight the intended information?
- Does the information convey the necessary information based on the readers' needs?
- Are there any quotations, facts, or figures that need to be checked or cited?
- Are visuals relevant to the section or could a visual be added to clarify any details?

- Are there clear paragraph transitions that lead the writer to the next paragraph/ section?

Level 3: Copy Editing—Paragraph/Sentence Level

The third step in the top-down editing process concentrates on sentence structure in each paragraph/section for clarity, persuasion, and consistency.

- Are there any sentences that are unclear, hard to read, or that do not "sound right" when read aloud?
- Are lists in parallel form (the same grammatical structure, e.g., beginning with verb in same tense)?
- Does the writer make any inappropriate word choices related to tone or meaning?
- Are there any sentences that contain excessive repetition, wordiness, or redundancy?
- Are keywords and technical jargon being used properly?

Level 4: Proofreading

The final step includes copyediting and concentrates on grammatical errors including misspellings and grammar usage issues. This step typically occurs last because the higher-order issues will distract more from the intended message than a misspelled word or two. However, some mechanical errors such as run-on sentences can interfere with reader comprehension and obvious typos can diminish the credibility of the writer (or business).

The following is a checklist for common errors in business English:

- Are there any misspelled words or usage errors (i.e., choosing the wrong version of a word spelled different ways, e.g., effect vs. affect)?
- Are there any sentence fragments or run-on/fused sentences?
- Are there any missing or misused punctuation marks, particularly commas, semicolons, and colons?
- Are rules for capitalization properly followed?
- Are numbers being spelled or written correctly and consistently?
- Are people, place, and company names and other proper nouns spelled correctly?

Utilizing Word's Spelling- and Grammar-Checking Capabilities

To make documents appear professional, as well as to allow the message to reach the audience in a clear, concise, and effective manner without distractions, spelling and grammar errors must be eliminated. Microsoft Word provides several features for checking spelling and grammar.

AutoCorrect Spelling and Grammar Checking

MS Word automatically checks documents for spelling and grammar errors. These errors are indicated by colored, wavy lines and should be corrected when noted:

- Red line indicates a misspelled word.
- Green line indicates a grammar error.
- Blue line indicates a contextual spelling error (This feature is turned off by default and needs to be turned on to use. See File, Info, Options, Proofing, AutoCorrect Options.).
- The colored wavy lines will disappear from a document after a correction is made.
- Always proofread to be sure most lines have been addressed, although some remaining lines can be common. (See AutoCorrect Options following the above route to remove all lines.)

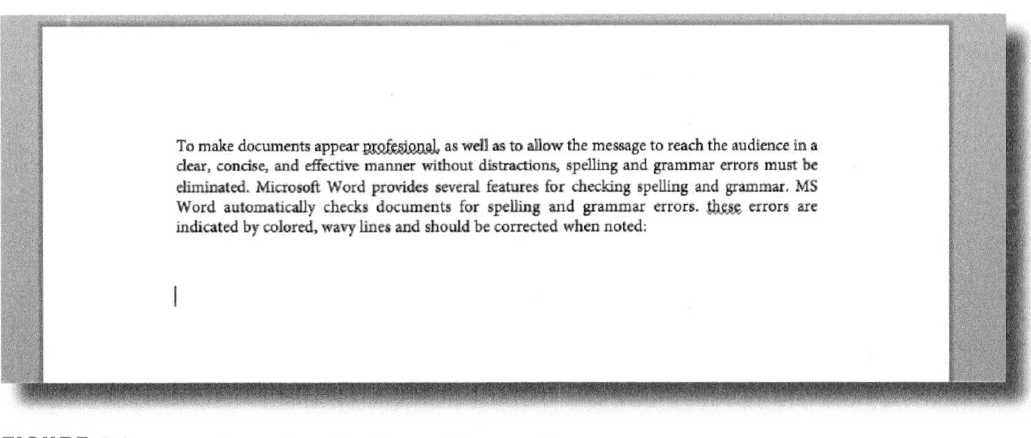

To make documents appear profesional, as well as to allow the message to reach the audience in a clear, concise, and effective manner without distractions, spelling and grammar errors must be eliminated. Microsoft Word provides several features for checking spelling and grammar. MS Word automatically checks documents for spelling and grammar errors. these errors are indicated by colored, wavy lines and should be corrected when noted:

FIGURE 4.1. Screenshot of Spelling and Grammar Features

Using the AutoCorrect Spelling Check Feature

- Right-click the underlined word. A pop-up menu will appear.
- Click on the correct spelling of the word from the listed suggestions.
- The word will be corrected and appear in the document.

Using the AutoCorrect Grammar Check Feature

- Right-click the underlined word or phrase. A pop-up menu will appear.
- Click on the correct phrase from the listed suggestions.
- The phrase will be corrected and appear in the document.
- This feature will also display the nature of the grammatical problem under the About This Sentence tab so that future errors can be avoided.

Using the Spelling & Grammar Function

Although not an automatic function, Word's Spelling & Grammar feature should always be used before any paper is submitted.

- On the Word toolbar, select Review.
- In the Proofing pane, click on Spelling & Grammar. Box will turn gold when Spelling & Grammar function is active.
- The Spelling & Grammar dialog box will open.
- Select a suggestion and click change to correct the error.

Suggestion: Investigate the Research, Thesaurus, and Word Count functions in the Proofing pane as well as the Translate and Language features in the Language pane.

Proofreading Strategies

Good writing involves revision, editing, and proofreading. Good writers always proofread, run the Spelling & Grammar Checker, and then proofread again to catch errors the Spelling & Grammar Checker didn't catch. Another habit of good writers is to put the document away for several hours or days, then come back and revise. A fresh set of eyes will reveal many writing problems.

Another good idea is to have someone else read your document. It is an unwritten rule in business to never proofread your own documents—always get a second reader to edit. Of course, this doesn't mean that you don't proofread at all. You are expected to produce a high-quality draft, complete your own proofreading, and then have a colleague proofread and offer editing suggestions before submittal. Remember, once a document is printed or sent electronically, the error is there forever! You don't necessarily require a coworker to edit every informal message such as setting up a meeting with colleagues, but important messages to internal audiences such as supervisors and external audiences such as clients should always be reviewed. The higher the stakes of the document, the more attention should be paid to editing. Even a few obvious typos in an informal message to internal audiences can potentially make a negative impression and be embarrassing.

When editing on a hard copy (printed) document, many experts suggest reading backward, starting with the bottom of the last page and reading toward the beginning of the document. You can also proofread paragraphs in random order, making sure to carefully examine one paragraph at a time while not getting distracted by reading in linear order. Many experts also suggest reading aloud; you'll pick up errors that you'll miss when reading to yourself.

Good writing involves writing, revising, editing, and more writing and more revising. Quality writing isn't affected by using either hard copy editing or digital editing as one process isn't any better than the other. However, in today's digital world, more and more writers and editors are utilizing electronic editing tools such as Word's Track Changes.

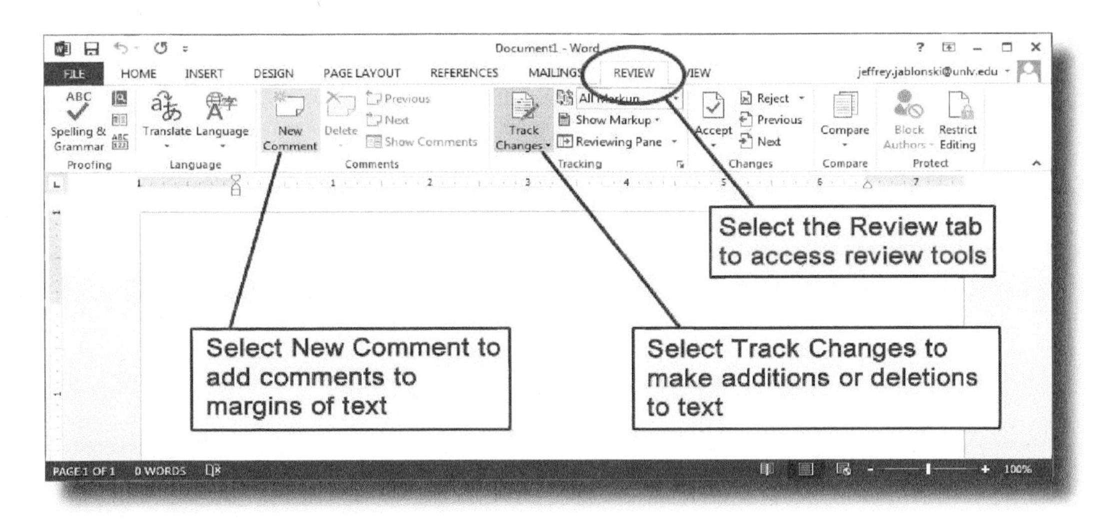

FIGURE 4.2. Microsoft Word's review tools

Digital Copyediting Using MS Word's Track Changes

Microsoft Word's Track Changes feature allows Word to keep track of changes made in a document. From proofreading to peer reviews to collaborating on the content of a document, Track Changes can make it easier to manage editing and collaborate with others.

Getting Started

Turn on Track Changes and Reviewing Pane:

- On the Word toolbar, select Review.
- In the Tracking pane, click on Track Changes. Box will turn gold when Track Changes is active.
- Click on the drop-down arrow, then click on Change User Name from the pop-up.
- Change the user name and initials to your name and initials in the Personalize Your Copy of Microsoft Office.
- Click OK.
- Click on the Reviewing Pane. Box will turn gold when this feature is active.
- Click on the drop-down arrow to select either Reviewing Pane Vertical or Reviewing Pane Horizontal. This is merely personal preference. This pane displays all revisions in summary form including insertions, deletions, moves, formatting, and comments, along with the user name and date.

To Make Changes

Delete, insert, move, format, or comment on information:

- To delete information, highlight word/sentence/paragraph and press Delete on keyboard. (A colored line will "strike out" highlighted information.)
- To insert a comment, place cursor before or after sentence, or highlight information. Click New Comment on toolbar. Type comment in pop-up box.
- To insert information, place cursor in area desired to add information and start typing.
- To move information, highlight information, copy and paste to desired area. (Shortcut: Ctrl + C to copy, Ctrl + V to paste.)
- To format information, highlight information, proceed with new formatting.
- Save document regularly.

Which Review Tool to Use

Track Changes - use for making any lower-order edits to spelling, word choice, or punctuation directly to the text.

Comments - use for making substantive questions or comments on higher-order issues of audience, organization, or content in the margins of the text.

Suggestion: While any changes can be made within the document itself or included in a new comment, it is advisable to make any lower-order changes to spelling, grammar, or mechanics directly to the text itself and reserve the comment feature for substantive comments or questions on higher-order issues of audience, organization, or development. This is just a guideline, because you might use both, for example, moving a chunk of text to make an organization-related suggestion and then adding a comment explaining why you rearranged the text.

To View Changes

Clicking Final does not remove Track Changes. This feature simply allows one to see the document with the changes.

- Click on the Display for Review box which states Final: Show Markup default displayed. Box will turn gold when this feature is active.
- Click on the drop-down arrow to select either Final: Show Markup, Final, Original: Show Markup, or Original. The differences between versions will become obvious as each selection is highlighted.
- Final: Show Markup is the version to be saved (File, Save on toolbar) and sent for peer reviews.

To Accept Changes

To create a finished document (one to be sent to the professor for reviewing or final grading), one must make and Accept all changes in the document as well as remove all New Comments and spacing lines.

- Click on the Accept box in the Changes pane. Box will turn gold when this feature is active.
- Click on the drop-down arrow to any of the four options. Proceed accordingly until all changes have been made.
- Also use this feature to remove lines along the left-hand side of the page displayed from respacing.
- Select the Accept all Changes in Document feature when finished with all corrections.
- Save document regularly.

To Remove Comments

After making and accepting all changes, all comments must also be removed from the corrected document.

- Click on the comment to be deleted. Boxes in the Comments pane at the top of the Review toolbar will turn gold when this feature is active.
- Click on the Delete icon, which will turn gold when active.
- Click on the drop-down arrow to any of the three options.
- Proceed accordingly.

For more help with Microsoft Word's Track Changes feature, see Word's Help function.

On the main toolbar under File, click on Help located on the left vertical toolbar. Click on Microsoft Office Help. Type in Track Changes in the search bar displayed at the top of the screen.

Copyediting Symbols and Their Uses

Although most editing is done electronically today, there are times when it will be required to know proofreaders' marks for the editing of documents in a hard copy format (written or printed rather than on-screen).

These proofreaders' marks are universally known and accepted (See Figure 4.3). Memorization of all of these marks is unnecessary, but one should be familiar with the most common marks such as Delete, Insert Space, Begin New Paragraph, etc., and keep the list on the following page as a reference.

Proofreaders' marks include special signs for the following:

- Operational errors such as those mentioned above as well as Center, Flush Left, and Spell Out
- Typographical errors such as Insert Here, Set in Italic Type, and Set in Lowercase
- Punctuation errors such as Insert Comma, Insert Quotation Marks, and Insert Period

Online Resources

For more help with Microsoft Word's AutoCorrect Spelling and Grammar features, see Word's Help function:

- "Word 2016: Checking Spelling and Grammar" by GCFLearnFree.org at http://www.gcflearnfree.org/word2016/checking-spelling-and-grammar/1/
- "Word 2016: Spelling and Grammar" video tutorial by GCFLearnFree.org at https://www.youtube.com/watch?v=h20EBvn1UCY
- "Checking Spelling and Grammar Microsoft Office Word 2016" by Install Help at https://www.youtube.com/watch?v=vW--_o5ek_U

For additional information concerning Track Changes in Microsoft's Help function:

- "Word 2016: Track Changes and Comments" by GCFLearnFree.org at http://www.gcflearnfree.org/word2016/track-changes-and-comments/1/
- "Word 2016: Track Changes and Comments" video tutorial by GCFLearnFree.org at https://www.youtube.com/watch?v=m7tmsWN6uH0
- Microsoft Word 2013: Track Changes and Copyediting in Word 2013 - Office 2013 Webinar" video tutorial by EPC Group.net at https://www.youtube.com/watch?v=O5OrMmCc3_4

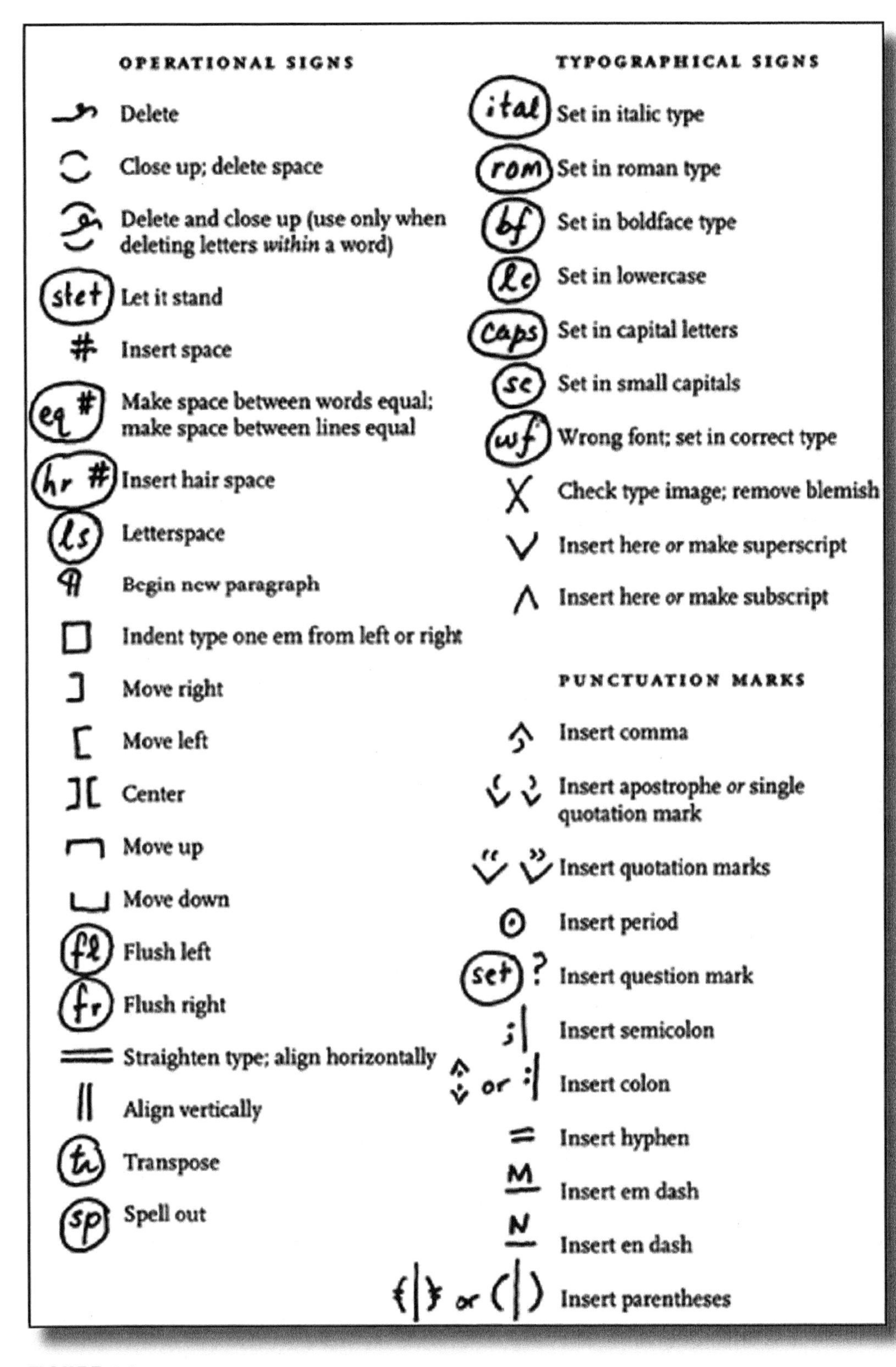

FIGURE 4.3. Proofreaders' marks

Source: The Chicago Manual of Style Online, http://www.chicagomanualofstyle.org/tools_proof.html (accessed August 3, 2015).

Exercises

1. Access your Introduction or Introductory Memo Project materials. Go through each document using the AutoCorrect Spelling function.

2. Repeat the AutoCorrect process on each Introduction Project deliverable using the AutoCorrect Grammar function, making note of repetitive grammar issues—name, problem, and solution.

3. Practice Track Changes on the Introduction Project deliverables. Be sure you are comfortable commenting, inserting, deleting, and formatting. Save edited version and put aside for a little while.

4. Open edited deliverables with Track Changes. Correct all editing. Remove all comments. Save and email to yourself. Open and proofread to check for accuracy.

End Notes

[1] Paradis, James, David Dobrin, and Richard Miller. "Writing at Exxon ITD: Notes on the Writing Environment of an R&D Organization." *Writing in Nonacademic Settings*. Ed. Lee Odell and Dixie Goswami. New York, NY: Guilford Press, 1985. 281–307.

BUSINESS CORRESPONDENCE

CHAPTER 5

The principles for writing letters and memos are very similar. The audience and occasion determines whether a letter, memo, or email is necessary. The advent of email has blurred the traditional separation of memos, which have been for internal communication, and letters, which have been for formal, external correspondence.

Memos

Memos have traditionally been used for *internal correspondence* within a company (see Figure 5.1). When you need to communicate in writing to someone within your own company, you'll write a memo. But memos are often used for external correspondence with customers and clients, especially as email blurs the boundaries between memos and letters.

	Memo	Letter	Email
Audience	Internal	External	Both
Formality	Informal	Formal	Informal
Social Cordialities (salutation, complimentary close, signature)	No	Yes	Yes

FIGURE 5.1. Memo, Letter, Email Comparison: This table lists the generic tendencies for each form of correspondence, but all three forms are more or less interchangeable depending on the situation and organization.

In heading, use complete names and position titles.

Use a short, specific subject line.

Add an opening that states purpose clearly.

In body, break up big "chunks" of text into smaller, more readable paragraphs.

Use headings, lists, and emphasis (bold) to increase readability.

Notations can appear in either the heading (e.g., the "CC" line) or at the end of a memo.

To sign a memo, either type or handwrite your name or initials after the closing. You can also initial by your name in the heading (as this sample does).

Add a clear closing.

To: All employees, SeaCorp Inc.
CC: Roger May, VP Operations
FR: Elaine Darling, Chief Executive Officer ED
DT: January 4, 2004
RE: **Email Use Policy**

The following policy covers appropriate use of any email sent from a SeaCorp email address and applies to all employees, vendors, and agents operating on behalf of SeaCorp.

Rationale

Our company needs to implement email use guidelines for the following reasons:

1. **Professionalism:** by using proper email language, our company will convey a more professional image and avoid tarnishing the company's public image.
2. **Efficiency:** emails that get to the point are much more effective than poorly worded emails.
3. **Protection from liability:** employee awareness of email risks will protect our company from costly lawsuits.

Email Etiquette

It is far too easy to treat email communications as an informal manner of communicating; however, it is a written record that is maintained in the ordinary course of business. You are reminded to maintain a business-like and professional decorum to your email correspondence:

- Format emails as you would print memos
- Use proper spelling, grammar, and punctuation.
- Never send messages containing derogatory, defamatory, obscene, or inappropriate content or attachments

A simple test to bear in mind is: if you would not write the email content in a formal business letter, then refrain from using such content in your email messages.

Company Monitoring

All electronic mail messages are the property of SeaCorp, and employees should have no expectation of privacy whenever they store, send, or receive electronic mail using the company's electronic mail system. Email is a business record and may be subject to review by your manager, other employees, the courts, government agencies, litigants and other persons who are not the intended recipients of the email. Deletion of a file on an employee's computer does not delete a record of the email from the company's system.

Personal Use

Using a reasonable amount of SeaCorp resources for personal email is acceptable, but non-work related email must be saved in a separate folder from work related email. Sending chain letters or joke emails from a SeaCorp email account is prohibited. Virus or other malware warnings and mass mailings from SeaCorp shall be approved by SeaCorp's VP of Operations before distribution. These restrictions also apply to the forwarding of mail received by a SeaCorp employee.

Enforcement

Any employee found to have violated this policy may be subject to disciplinary action, up to and including termination of employment.

Questions about this policy should be directed to the VP of Operations.

FIGURE 5.2. Sample Memo

Because they are primarily intended for in-house audiences, memos typically dispense with the social cordialities of business letters such as using a formal salutation ("Dear Mrs. Doe"). But that does not mean memos are any less important. Memos inform people of decisions, policies, procedures, problems, solutions, responsibilities, and deadlines—all while creating a record of a company's activity.

Print and electronic memos often replace oral meetings in instances where face-to-face communication is impractical or undesirable. While most memos range from one to three pages long, reports written in memo format can be dozens of pages long. For a sample memo, see Figure 5.2.

Letters

Letters are generally for *external correspondence* between a company and its clients, customers, or other interested parties. Letters are generally used when a more formal document is needed, such as legal proceedings, promotions, terminations, and thank yous. Thus, letters can also be written to members inside an organization when the occasion calls for it.

As noted in Figure 5.1, letters include the traditional social conventions of the salutation, complimentary close, and signature. These elements are what lend a formal air to the letter. (See Figure 5.3 for a sample letter.) These elements can also be used to make a memo or email seem more formal. Electronic mail is being used more and more for external correspondence, but it is still considered bad form to use email at times when very formal correspondence is needed. Don't fire anyone in an email!

Inside return address.

States Bank
Post Office Box 2130
New York, NY 11042

July 25, 2007

Recipient's address.

Sue M. Owens, Accounts Payable
Trico Cleaners
3211 N. Tower Ave.
Las Vegas, NV 89121

Subject line.

RE: Account number 198 211 14 2259 04

Salutation.

Dear Ms. Owens:

Opening that states purpose, in this case including a dated reference to the original correspondence.

Your July 1,2007 inquiry regarding your missing payment has been referred to my attention for review and reply.

A review of your account indicates that your May payment of $170.00 was misapplied.

I have corrected this error, and your payment has been credited to your account and back dated to reflect the correct receipt date. Any late charges incorrectly applied or adverse comments have been permanently removed from our records and those of credit reporting agencies used by National Bank.

Closing paragraph.

Please accept my apologies for this error. If I can be of further
assistance, please call the Customer Service Department at 1-800-226-5721, Monday through Friday, between 8 a.m. and 7 p.m., Eastern time.

Complimentary close.

Sincerely,

Signature block.

Michelle Nicolette
Research Representative
CS002

FIGURE 5.3. Sample business letter

Email

Perhaps one of the biggest changes the computer has brought to business is email. Electronic mail was invented in 1971 but didn't take hold as a ubiquitous communication tool until the late 1980s, when its use exploded. One source estimates that the average employee sends and receives 122 emails per day including clutter such as spam, unwanted messages, and graymail, or promotional messages that the user opted in for at one time.[1] Another study estimated that business managers send and receive twice as many emails as front-line employees.[2]

The problem with email is that it is rapidly replacing all other forms of communication, including phone and face-to-face conversations. Add in personal emails and spam, and employers are spending too much time sifting through new emails and hunting for important information in old emails. Many experts are writing about a general backlash against email.[3, 4, 5]

Email is best for simple requests, statements, questions, or acknowledgements. (See Figure 5.4). It is also good for sending large documents as attachments.

Email is not good for conveying mass amounts of information, subtle emotion, humor, or sarcasm. Discussing long and complex issues via email can actually delay decision

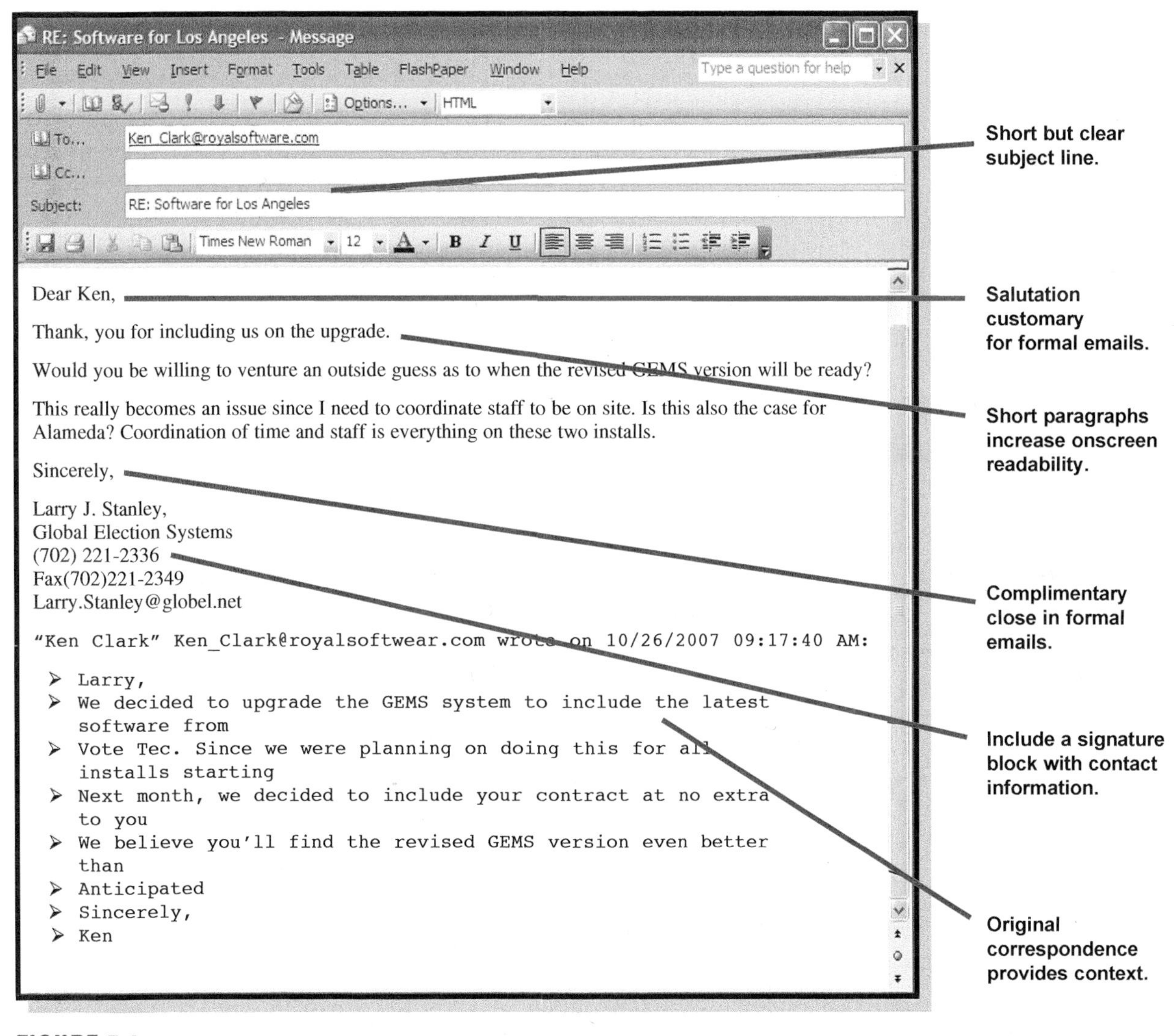

FIGURE 5.4. Sample email

making. It is better to have a phone conversation or meeting and use email to confirm points discussed in person.

Email is particularly poor at conveying subtle emotions. Hence, email should not be used for sensitive situations such as major announcements, firings, job evaluations, or negative criticism. Too many people treat email as an informal communication medium similar to speech, and use business email to send all kinds of unprofessional messages including tirades, whiny complaints, and personal attacks on co-workers.

Generic Structure of Memos, Letters, and Email

The basic structure, or organization, of letters and memos is based on how business readers approach such documents. When planning your letter or memo, always imagine a busy reader—busy boss, busy coworker, busy customer, or busy client. Readers don't have time to pore over your whole document to figure out why you're writing and what you want from them. The basic structure of a memo or letter includes the heading, opening, body, and closing. Any memo or letter you write should use this structure to as quickly and clearly as possible address basic questions readers have when reading any correspondence (see Figure 5.5).

Memo or Letter Parts	Answer Reader Questions
Heading and Opening:	Who is this from?
	What is this about?
Body:	What do I need to know about this?
	Why is this important to me?
Closing:	What do you want me to do, and when?

FIGURE 5.5. Basic structure of the reader-centered memo, letter, or email

The Heading

The first parts of the correspondence answer the reader's question, "What's this about?" The heading contains the particulars about the sender and receiver of the document. It also includes information about when it was written and what it is about. Heading formats vary from memo to letter, and from organization to organization, but most generally contain some or all of the following information:

- Complete and correctly spelled names
- Proper job titles
- Correct date
- Business addresses
- Short but informative subject line

You can compare Figures 5.2 and 5.4 to see the difference between the letter and memo headings. The letter heading includes the sender's and recipient's complete mailing addresses. The memo includes the "to" and "from" lines. Both letter and memo headings include a date and should include a subject line. Note that business mailing addresses include job titles and company names, in addition to the person and street address.

The reader uses the information in the heading to determine when, if ever, the memo or letter should be read. Business readers prioritize what to read based on the sender and subject lines. Blank, vague, or clichéd subject lines might cause confusion or be ignored. While you can add details such as "IMPORTANT" and "URGENT" to a subject line to catch a reader's attention, you should avoid doing this to every letter, memo, or email you send.

The Opening

The opening or overview section of a memo or letter is like the introduction to an essay. After scanning the heading, the reader looks to the first few paragraphs to further determine the document's purpose and whether it deserves the reader's full attention immediately, later, or ever. When writing the opening, be sure to:

- State the purpose for writing in the first sentence or two (Note: You don't have to write "The purpose is...", just state the purpose, e.g., "This memo presents plans for restructuring . . .")
- Summarize the main points of the memo
- If you're responding to a request or inquiry, reference the date and subject (e.g., Here is the report you asked for last Thursday, April 28...")
- If you're addressing a problem, include details about the problem
- If you're making recommendations or requesting readers take action, include details of the recommendations or request

The Body

The body or discussion section of the memo or letter develops the information noted in the opening. The body answers the reader's questions, "What do I need to know?" and "Why is this important?"

There are two basic ways to organize information in the body of a memo or letter:

1. **Direct organization:** This arrangement puts the most important points first, followed by supporting information. For instance, if you're writing a recommendation report to address a problem, first provide the recommendations and then discuss any research that you conducted to address the problem. (This plan assumes you briefly summarized the problem in the opening of the memo or letter. You might need to include a more detailed discussion of the problem in the body, after the recommendations section.)
2. **Indirect organization:** This arrangement puts supporting details first, leading to your main point. If you expect readers to react skeptically or negatively to your main points, put them at the end of the memo. This way, you can use supporting details like the statement of the problem and discussion of alternatives to prepare readers for your conclusion.

Make sure, as Figures 5.1 and 5.2 show, to use the following devices to help guide your readers through your discussion and make your memo easier to read:

- **Use shorter, single-spaced paragraphs.** Break up big chunks of text into smaller, more readable paragraphs. It is okay to have paragraphs that are one, two, or three sentences long. Memo paragraphs are usually single-spaced with a double-space between paragraphs. Paragraphs may or may not be indented, depending on a given company's format.
- **Compartmentalize** each different point or topic in a separate paragraph. As you revise your memo, if you find a paragraph that has more than one main point or is too long, consider splitting it into two or more paragraphs.
- **Use section headings** that indicate the main point of each section, helping readers scan the contents and organization of the document. You can use headings in documents of any length, and you should definitely use them in memos and letters longer than half a page.
- **Use bulleted or numbered lists** to present itemized or sequential topics, such as multiple recommendations, the steps in a procedure, or a series of questions. Avoid overusing bullets, however. A separate bullet for every paragraph of your memo can make it difficult to distinguish important points.
- **Use text emphasis** such as bold or italics to highlight key words or important information. Use text emphasis sparingly, however. Highlighting too much text in a document defeats the purpose of using emphasis because nothing stands out anymore.

The Closing

The closing reiterates for the reader key points discussed in the correspondence answering the questions, "What do you want me to do?" and "When do you want it done?" Since most business writing seeks some kind of action from the reader, make sure you state this action clearly in the closing. Include specific information about when you want this action accomplished. Avoid vague, clichéd endings such as "If you have any questions, please feel free to contact me." If you want a response, then state so unequivocally: "Please respond to my request by noon on Friday, April 28." Be polite and courteous in your closing, especially when asking people to do things for you or on your behalf.

Letters include a signature block that includes a complimentary close, ink signature, and signature line (see Figure 5.3). Memos do not have to have a signature block at the end because the writer's name appears in the heading. However, many memos include the writer's initials, first name, or full name either typed or handwritten at the end of the memo or next to the "From" line in the heading (see Figure 5.2). Always include a signature block in electronic correspondence (see Figure 5.4).

Notations

Memos and letters often contain additional information in either the heading or after the closing:

- **Copies:** "CC" (*carbon copy*) indicates other recipients of the memo. Include full names and position titles for all you copy. This way, others can see if, for instance, an attorney or senior executive received a copy. "BCC" (*blind carbon copy*) means those who received CC copies won't see who received the BCC.
- **Enclosures:** Notations such as *Enclosure*, *Encl.*, or *Encl.: Report (15 p.)* indicate and create a record of any additional documents included with the memo. It's best to include specific information about the enclosures such as title and number of pages. It's bad form to write "Attachment."
- **Initials:** Initials at the end of a document such as "GK/lm" indicate the writer (in all capital letters) and the typist (lowercase letters).
- **Multiple pages:** If your memo is longer than one page, include page numbers on every page after the first page. Other information such as the date of the correspondence, the company name, the subject of the memo, or the name of the writer is often added, which helps create a professional appearing document.

Correspondence Style

As discussed in Chapter 3, Professional Writing Style, your business writing should be conversational, clear and concise, human, and diplomatic.

Conversational

It's acceptable to use first person "I" and second person "you" in business memos and letters. You can also use contractions (two words joined by an apostrophe, e.g., *can't* or *don't*) in all but the most formal of correspondence. Write as you talk and your writing will significantly improve, but avoid slang, which is unprofessional.

Clear and Concise

To write clearly (unambiguously, in a way that is easy for the reader to understand), avoid letting gobbledygook slip into your writing. *Gobbledygook* is the term of art for overly pompous, affected language that is long-winded and difficult to understand. Instead use plain, everyday language. Rather than writing *utilization* write *use*. Depending on your audience's familiarity with technical terms, also limit your use of jargon and buzzwords.

A good habit to get into is this: after you have revised your draft to a point where you think you're done, try to **eliminate at least 20% of the total word count**. This percentage is just a rule of thumb. As you become more adept at eliminating useless words from your writing, the percentage you cut may increase. The point is that good writers always pay attention to eliminating clutter. It's a service paid to readers.

Human

As language expert William Zinsser writes, "Just because people work for an institution they don't have to write like one."[6] Your writing will be better if it treats readers with respect and attempts to connect with them as humans. The words you choose should project an image of yourself as respectful, thoughtful, warm, and personal. Your writing should focus on the needs and concerns of the reader. This is called **the *you* approach**. Show empathy for the reader's point of view, provide reasons for whatever it is you're asking them to do, be polite and friendly. Instead of writing, "A sales receipt must accompany any refund," write, "To reduce the processing time of your refund, please provide a copy of the sales receipt." Also, avoid using the "royal 'we'" voice. Use first person "I" instead. It is understood that you are writing on the behalf of an organization.

Diplomatic

Being direct isn't always the best tactic. Be especially tactful and polite when conveying bad news or attempting to persuade skeptical readers. Use positive words to express negative ideas. Instead of writing, "I want to complain about your poor customer service," write, "I wish to bring a customer service matter to your attention." Avoid words that suggest the reader is dishonest, careless, or mentally deficient. Lastly, avoid putting anything in writing you'd regret later.

Common Genres of Correspondence

Many memos and letters are written for occasions common enough that they can be classified into the following categories:

- Inquiries
- Claims and adjustments
- Rejections and acceptances
- Transmittals and cover letters

Inquiries

Common reasons for writing inquiries are to find more information about something from the recipient. You may want to learn more about a product or service, so you write the company requesting more information. You may want to know if a company can offer some service beyond what it ordinarily provides, such as helping a school or charitable organization with a donation of money, goods, or services. You may even want to know if a company has any unadvertised job openings. All of these are occasions that prompt an inquiry memo, letter, or email.

When writing an inquiry, follow these guidelines:

1. State your reason for writing in the opening of the letter.
2. Identify who you are and why you need the requested information.
3. Use a numbered or bulleted list to outline specific information you desire or questions you have.
4. If possible, offer to compensate the recipient for replying, e.g., by paying phone or mailing costs, to acknowledge the recipient in your work, or to provide the recipient with a copy of whatever you are working on, such as a report, charity program, etc.
5. If persuasion is necessary, provide a tactfully worded reason how the recipient will benefit by providing the information you desire.
6. In the closing, thank the recipient for their cooperation.

Claims and Adjustments

Common reasons for writing claims are to request some compensation that you are due or inform a recipient of some problem and attempt to persuade the recipient to provide some remedy. If you were not happy with a stay at a hotel, and weren't satisfied with how the front desk manager responded, you might write to the corporate headquarters requesting compensation. The "adjustment" is the response by the recipient that informs the claim-writer of how the claim is being resolved. Claims are sometimes also referred to as complaints, but not all situations that require a claim are necessarily occasions for complaint. Furthermore, as the Business Writing Style chapter indicates, the act of complaining has negative connotations and often puts recipients on the defensive. Instead of angrily "complaining about poor service," the goal should be to calmly "bring a customer service matter to the company's attention."

Claims almost always involve argumentation and subtle persuasion, since you are usually asking for a remedy that goes beyond what the recipient wants to provide. It is important you build a case for your claim by providing an accurate summary of the problem, reasons why you believe compensation should be provided, and a clear statement of the compensation you desire. Avoid overly negative, hostile, or insulting language. Most individuals want to be fair and provide quality service; however, people are less inclined to help irate customers and are even willing to lose the irrational customer. As the saying goes, you can catch more flies with honey than with vinegar.

When writing a claim, follow these guidelines:

1. State the reason you are writing, in generally positive language, e.g., to request compensation or bring a customer service matter to their attention.
2. Provide a detailed description of the situation or problem. Be specific about dates, incidents, model or part numbers, etc. (you may also wish to provide documentation as evidence to further support your claim, e.g., copies of receipts, contracts, correspondence, photos).
3. State what specific compensation you desire, making sure the compensation is in proportion to the claim.
4. Provide reasons why your request should be granted. You can appeal to the recipient's sense of fairness, commitment to quality, or desire for customer satisfaction.
5. Close politely.

When writing the adjustment, or reply to the claim, follow these guidelines:

1. Reference the date of the original claim letter and state whether you are granting or denying the claim in the opening.
2. Acknowledge the claim-writer's concerns and efforts for writing.
3. If you deny the request, explain the reasons why the request can not be granted.
4. If you deny the request, consider offering some partial or alternative compensation.
5. Conclude politely, reiterating you or your company's values relevant to the matter, such as customer service or product quality.
6. Maintain a professional and tactful tone throughout the letter. Avoid sounding defensive, defiant, arrogant, or resentful.

Rejections and Acceptances

Common occasions for rejection letters are having to inform someone of bad news such as not being hired in a job search, being denied a promotion or raise, or being laid off from a company. An adjustment letter denying a claim is another form of rejection. The same rules of tactfulness, professionalism, and sincerity apply. Most people can handle rejection or bad news if they are treated like a human being. Providing sincere reasons often lessons the sting of rejection.

When writing a rejection letter, follow these guidelines:

- Be thankful and appreciative in the opening.
- Provide a tactfully worded rejection in the opening or second paragraph.
- Provide specific reasons for the rejection, if possible.
- End politely, trying to leave the door open for a future relationship.
- Avoid using generic or canned language. Part of treating people human is not resorting to stereotypical or clichéd language. Be as concrete and sincere as possible when providing reasons for rejection or bad news. Be tactful and indirect, but don't "sugarcoat," as most readers can see through false sentiment.

Acceptance or good news letters are easier to write and often only require a few short words of congratulations. Recipients of good news letters usually shift gears to quickly wondering about the next step, such as when the job starts, if the company will pay for relocation, how the compensation will be delivered, etc. Try to anticipate such questions and provide suitable answers. Good news recipients also appreciate being given reasons for their good fortune. Giving details about the number of applicants reviewed or number or entries in a competition heighten the sense of accomplishment and make it possible to use the letter as documentation of achievements in future settings, such as a career portfolio.

Transmittals and Cover Letters

Transmittal correspondence informs the recipient of delivery of something else, usually a document such as report or a resume. More information about writing transmittals can be found in Chapter 31, The Client Project.

A cover letter is a specific form of transmittal letter used in the job search to accompany the resume. More information about writing cover letters can be found in Chapter 10, Cover Letters and Personal Statements.

Writing Electronic Correspondence

Remember that email is a permanent written record that is generally regarded as company property. Companies are becoming increasingly wary of how much time and money is wasted on poorly written or inappropriately used emails. To write more effective emails, follow these guidelines:

- **Use a clear subject line** that makes it easier for the reader to understand why they received the email and that make it easier to search in the future. It is particularly important to avoid vague subject lines in emails. Take a keywords approach that strings the key words of the message into a meaningful phrase.
- **Use all of the devices for readability** previously discussed for memos and letters. Use shorter paragraphs, lists, and text emphasis to make the email easier to read.
- **Format your emails similar to a letter.** Formal emails should include a salutation line such as, "Dear Ms. Smith:" and a signature block at the end that includes a complimentary closing, your name, title, and contact information.
- **Consider when a different communication medium would be preferable.** Many companies are encouraging employees to use the phone more to conduct business that is too complex to be dealt with efficiently in back-and-forth email exchanges. Use the phone when a personal touch is needed, the message is too nuanced to be conveyed in emotionless email, or the information is too sensitive to be recorded in an email. Many companies also encourage the use of Instant Messaging (IM) chat for activities such as meeting planning.
- **Keep the length of emails short.** Knowing that employees and managers receive dozens of emails daily should remind you to keep your emails short. Most emails should not be longer than one screen. Busy readers do not have the time or inclination to read dense email messages. Some companies use email to circulate denser reports, but longer documents should be attached to the email.
- **State the reason you are writing in the beginning,** just as you should with memos and letters. Don't bury key details in the middle of paragraphs and at the end of the message.
- **Remember the difference between personal email and business email.** Business emails should be treated as formal as print memos, in terms of style, language, and mechanics. Though it depends on how well you know your audience, you should generally avoid slang and IM speak (the abbreviations, acronyms, and phonetic spellings used in chat and text-messaging) and instead use proper grammar and punctuation.
- **Be professional.** Partly because email is used for personal reasons too, people confusingly think of email as a more conversational and less permanent medium than print memos and letters; however, it is just the opposite. Emails are easier to retrieve than most print documents, and they are potentially saved longer! Only write what you would feel comfortable saying to the person's face. Don't send confidential or sensitive emails that you don't want people other than the recipient to read. Don't write anything nasty about coworkers or bosses, and don't send venting emails that complain, whine, or reprimand.
- **Check for misspellings, poor grammar, and typos.** Be sure to edit your business emails, particularly when the message needs to be more formal and is being sent to someone who doesn't know you.
- **Double-check before hitting Send.** For very important emails you may even want to print the draft for closer editing. Most email applications allow you to save

drafts. You could save the draft, print it out for closer reading, and return to it after a break so you can re-see it from a fresh perspective.

- **Be familiar with your organization's email use policy.** The courts have been very clear that your company emails are not your private property. Most companies have email use policies and archive all emails (even when they are deleted from your work PC). Many companies monitor company email for inappropriate content.
- **Don't let email run your (work) life!** Turn off the auto-alert function so you don't find yourself checking emails constantly. Block out specific times of the day to check email, such as in the morning, after lunch, and before leaving the office. Don't get in the habit of replying to emails immediately. Decide if email is the best way to communicate and don't copy the world on all your messages.

Exercises

1. Revise the memo below:

   Acme Inc. Memorandum

   To: All Division Managers

   From: Stanley Rogers, Vice President, Admin.

   RE: Policies

   CC: Thomas Rickley, Manager, Building Services

   The garage on the south side is being renovated to include a new third level and 150 new parking spaces. Because of these renovations, a new parking policy will be in effect between the months of February and December of this year. Please circulate this memo to all of your division employees and make an announcement at your next Division meeting. Between February to July, Administration and Finance employees can continue to use the first-level of the south side garage. Textile and Packaging employees can continue to use the second-level of the south side garage between Feb. and July. Service and Manufacturing employees can use either the second level of the south side garage or park in the south side lot at this time. (It is recommended that first- and second-shift Manufacturing employees use the south side lot.) The south side garage will be closed to all employees between August and December. At that time, Administration must use the front lot. Finance employees must use the east side lot. Service and Manufacturing employees must use the east side lot, and Textile and Packaging employees must use the west side lot. Employees with special circumstances or handicaps should contact Thomas Rickley, Building Services Manager (ext. #7765), for a front lot permit during the renovations. All other questions should be directed to the Building Services office.

2. Find a sample business memo or letter from your work or personal life. (Make sure it does not contain any confidential or proprietary information.) Write an analysis of the document's rhetorical situation (context, writer, purpose, audience) and how well written the document seems to be given the rhetorical situation. Identify if the memo or letter is written in a common genre and include a critique of the document's format and style in your analysis. Consider whether the format and style contribute to the overall effectiveness of the document.

3. Individually or in small groups, copy the contents of the sample documents from exercise 2 into Word and run a word count. Then revise the memo to make it more concise. Aim to eliminate at least 20% of the word count without affecting the overall intended meaning of the memo. This may require eliminating some unnecessary words, re-writing excessively long sentences, and using other devices, such as shorter paragraphs, lists, and headings. OPTION: Choose one memo in particular and have the entire class work on and compare the same example.

End Notes

[1] The Radicati Group, Inc., "Email Statistics Report: 2015-2019," March 2015. http://www.radicati.com/wp/wp-content/uploads/2015/02/Email-Statistics-Report-2015-2019-Executive-Summary.pdf

[2] Ibid.

[3] Darren Dahl, "The End of Email," *Inc. Magazine*, February 2006, 41–42.

[4] Doug Beizer, "Email is Dead," *Fast Company*, July/August 2007, 46.

[5] Diane Brady, "*!#?@ the Email. Can We Talk?" Business *Week*, December 4, 2006, 109.

[6] William Zinsser. *On Writing Well: The Classic Guide to Writing Nonfiction*. New York: Harper- Collins, 167.

WRITING ELECTRONICALLY

CHAPTER 6

Learning Outcomes

After reading this chapter, you should be able to:

jannoon028/Shutterstock.com

1. Write effective email messages, instant messages, and web content for businesses and organizations.
2. Use the business writing rules in email messages.
3. Discuss when to send messages by email and when not to.
4. Manage your overloaded email inbox.
5. Describe why employers are implementing Internet use policies, describe what these policies generally cover, and discuss the ethics of the issues involved in implementing these policies in the workplace.
6. Discuss ways to recognize the email culture of the business organization for which you work. Avoid an email mistake or an email career killer (ECK).
7. Avoid creating confusion, misunderstandings, and hurt feelings with your emails. Ensure you get the busy reader's attention by thinking and planning before you draft and revise your message.
8. Follow the rules of email etiquette (netiquette) when writing email.

9. Describe the two defining characteristics of instant messaging that make it a compelling alternative to email and voicemail, both of which have a tendency to pile up.
10. Practice good instant messaging etiquette.
11. Discuss the role of instant messaging in organizations.
12. Develop readable, coherent, grammatical, and accurate online text.
13. Discuss how text messaging is used in the business place.
14. Define blogging and discuss ways blogs are used in businesses.
15. Implement good skills for planning, drafting, and proofreading user-friendly web content.
16. Describe ways to write effectively for websites.
17. Describe social networking sites and discuss ways they are used in businesses.
18. Discuss how Twitter is used in the business place.

Benefits of Learning about Electronic Writing

1. You will understand the impact of electronic writing on the way people communicate in the workplace.
2. You will learn how to communicate effectively using email, instant messaging, text messaging, and blogs.
3. You will understand the importance of netiquette when composing email and instant messages.
4. You will learn how to get the web browser's attention with clear, scannable, and credible content.
5. You will understand the growing roles of instant messaging, text messaging, blogging, websites, and social network sites in the business place.

electronic writing
Messages developed and transmitted via email, instant messaging, text messaging, blogging, websites, social network sites (e.g., Facebook, LinkedIn), and tweeting.

artellia/Shutterstock.com

The intent of this chapter is to provide you with information on writing electronically. This goal is realized through discussions of the following topics: electronic written communication in organizations, writing effective email messages, writing effective instant messages and text messages, instant messaging and text messaging etiquette, writing effective business blogs, writing for websites, writing effectively at social network sites, and choosing the right medium.

Electronic writing refers to messages developed and transmitted via email, instant messaging, text messaging, blogging, websites, social network sites (e.g., Facebook, LinkedIn), and tweeting. Speed and convenience are the two main reasons electronic communication has become so popular in U.S. organizations and elsewhere. However, writing for these media presents challenges. We describe three of these challenges here.

Allies Interactive/Shutterstock.com

1. **Honing Traditional Writing Skills and Acquiring New Skills.** Readers read differently online than offline, thus, you must adapt your online copy to meet readers' needs. Specific writing suggestions for to **email**, **instant messages**, **text messages**, **blogs**, **websites, tweets**, and social network sites appear throughout this chapter.
2. **Merging the Writing Process with the Speed, Convenience, and Protocols of Electronic Communication to Achieve Effective Documents.** Popular electronic media such as email, text messaging, instant messaging, blogging, tweeting, websites, and social network sites encourage the development of short messages and invite some to ignore grammar, punctuation, and spelling rules. In addition, many writers do not edit and revise their electronic messages before sending them. Too often the result is poorly written, confusing memos and letters that contain too little detail to achieve clarity, leaving readers with a poor perception of their communication partner's writing abilities. Be careful. Poorly written communications can be career killers.
3. **Being Realistic about the Importance of Building Relationships.** Before expanding on this challenge, understand that in the U.S. business place strong relationships with stakeholders (e.g., colleagues, clients, suppliers) are crucial to job stability and career growth. Building business relationships is best accomplished through face-to-face interaction. Herein lies the challenge. Avoid doing so much of your communicating (written or otherwise) electronically that you fail to build strong relationships with your stakeholders. Communicating electronically is appealing and efficient on several levels. However, communicating too frequently via email, instant messages, text messages, blogs, and tweets can compromise your ability to achieve your career goals. Despite our good intentions, we cannot always meet with others face-to-face.

email
Electronically generated messages sent over the Internet, intranets, and extranets.

instant messaging
(IM) A form of online chat.

text messages
Short, written messages sent to mobile phones.

blog
Short for "web log"; a user-generated website acting as a communication and networking medium.

tweets
Messages sent on Twitter.

When you cannot, but should, place a phone call or hold a videoconference instead of emailing or texting. Phone calls and videoconferences are more personal and better support relationship building than do emailing and texting.

You have probably noticed in your classes the occasional student who spends more time texting than participating in class discussions. (This certainly does not describe you!) Those students who text or search the Internet during class either do not know how inappropriate, rude, and distracting their actions are or they simply do not care. Unfortunately, they are forming bad habits that will be hard to break. Thus, they are likely to carry them forward into the workplace, where their job stability and career growth can be easily derailed by their unprofessional actions.

Forming perceptions and making judgments about others' actions and behaviors are commonplace in most settings, be they professional or otherwise. Inappropriate actions and behaviors are typically noted and often taken into account when making decisions regarding bonus distributions, pay raises, promotions, and dismissals. Remember that the majority of U.S. managers prefer that their employees do not text, tweet, and browse the Internet at inappropriate times and for nonwork-related purposes.

FIGURE 6.1. Job/Career Threats

In this chapter you will learn to write effective, reader-friendly email, instant and text messages, blogs, website content, and social networking site content. Competency in these areas will enable you to take advantage of the possibilities of each medium, while avoiding inherent pitfalls. Since electronic communication will certainly comprise a large percentage of your business communications, mastering the basics now will give you a competitive edge.

Currently email is the most widely-used communication medium in U.S. organizations, although it will likely not always be the case. Humans have a long history of eventually replacing one technology with a newer one. For example, PCs quickly replaced typewriters. Cell phones and smartphones are steadily replacing landline phones. This is not to suggest that texting and tweeting will replace email anytime soon. However, the rate at which electronic communication media ranging from texting and blogging to social network sites are growing in popularity is astounding. While email remains popular, will this always be the case, or will email eventually fall from favor? When? Time will tell.

Electronic Written Communication in Organizations

We write and transmit business documents in numerous ways. Common forms of written communication include memos, letters, reports and proposals, email, instant messages (IMs), text messages, website content, blogs, and tweets. In addition, organizations routinely post written information on company social networking sites such as LinkedIn and Facebook.

Minerva Studio/Shutterstock.com

In this chapter we focus on email, instant messaging, text messaging, blogging, websites, social networking sites, and tweeting. The other forms of written communication mentioned above are addressed elsewhere in the book; memos and letters are discussed in chapter 5 and reports are discussed in chapter 8.

Formality plays an important role in selecting the best medium for each writing situation. Written documents and messages are frequently viewed as being formal, informal, or semiformal. For example, letters are considered formal documents. Documents and messages developed and/or transmitted electronically (e.g., blogs, emails, e-memos, IMs, text messages, tweets) are considered informal. Awareness of such differences in perception is important because readers' formality expectations vary and must to be taken into consideration. For example, for an important message to an external client, you would be expected to send a formal document, in this case, a hardcopy letter. In contrast, if you need to send a brief message with routine, straightforward information to a subordinate within the company, an informal written media such as email is a good choice. Or, if you and a fellow worker, who are both on the same job level, need to discuss some points pertaining to a routine, noncontroversial matter, IM would be the informal media of choice. Before moving on, let's look at one more example that lands you midstream on the formality spectrum. If you need to send a message to subordinates about changes to internal procedures, a semiformal document is necessary. In this case, it would be a memo.

Writing Effective Email Messages

Email is the predominate communication medium in U.S. organizations, and elsewhere for that matter, and no wonder why. It is a fast, convenient, and inexpensive way to reach one or multiple recipients with a few clicks of the mouse. Owing to its popularity, email has replaced messages that were once transmitted exclusively through hardcopy letters and memos. Email has also replaced the need to send faxes as often. Some even send sensitive correspondence via email, although you must be careful to remember that email is not a private medium.

The prevalence of email means that more workers than ever are writing. Even entry-level employees who used to be exempt from most writing activities now write emails to accomplish a host of activities from explaining procedures to coworkers to apologizing to a customer for a billing error.[1] Where once we picked up the phone to communicate business and personal information, today we use email. In fact, more people used the Internet in its first 3 years than used the telephone in its first 30 years.[2]

Employees use email for personal as well as workplace communications. When asked by Vault.com how many nonwork-related emails they send in a day, 18 percent of office workers said none; 56 percent said 1–5; 13 percent said 6–10; 6 percent replied 11–20; and 7 percent said they sent more than 21 nonwork-related emails a day.[3] According to a survey by AmericanGreetings.com, 76 percent use email to send jokes, 75 percent to make social plans, 53 percent to send online greeting cards, and 42 percent to gossip.[4] By replacing phone calls with emails, we have created a process by which informal conversations are recorded and formal business decisions are documented electronically.[5] The result is an unprecedented amount of documented business communication that can be used against an employer or employee should a workplace lawsuit be filed.

The convenience of email comes at a price for employers: increased liability concerns. Many companies are reducing their exposure to legal "e-disasters" by establishing and enforcing policies that govern employees' electronic writing. These policies are discussed next.[6] On the employee side, the convenience of email means that much of your relationship with coworkers, customers, and supervisors may rest on your email exchanges.[7] In this writing-centered environment, email is no longer the equivalent of a casual lunch conversation, and your ability to quickly organize and clearly express ideas in email becomes important to your overall success in any organization. In today's electronic workplace, written communication makes a competitive difference.[8]

To write effective email messages, you *must* think before you send. Therefore, the standard rules that apply to all business writing (e.g., appropriate tone, logical organization, grammatical correctness) are required when you write emails. When written with consideration for your audience, email is a productive communication tool that allows you to make a direct, positive, and polite connection with your reader. This, in turn, will win you the attention and goodwill of coworkers, customers, and other stakeholders.[9] The trick is to develop an effective email style.

Rawpixel.com/Shutterstock.com

Selecting the Right Medium: To Email or Not to Email

Not all messages should be sent by email. While corporate email cultures differ, some rules on whether to email the message apply across the board.

Consider the formality or informality of the message context. For instance, when introducing yourself or your product to an organization, the proper route to take is still a formal letter, report, or proposal. In addition, even though emailing a thank-you note following a job interview is usually acceptable, sending a promptly-mailed letter would definitely garner more attention and probably would be more appropriate.[10]

Emailing a thank you for a job well done is always appreciated, but a face-to-face thank you is better. Likewise, important job requests, such as requests for a raise or promotion or a resignation announcement, should be done face-to-face.[11] An email that requests a raise would most likely be considered spineless or not be taken seriously. A request for a sick day should be made over the phone in most offices.[12]

On the other hand, email is widely used to communicate routine, daily business messages since it is cost efficient and fast. Direct requests or informative messages to coworkers; companywide announcements, like the time and place of the holiday party or the next staff meeting; or a change in an organization's travel policy can all be sent safely by email as long as everyone in the office has access to it.[13] Equally safe to send by email are major personal announcements, like the happy news of a new mom or dad or the sad news of a death in a coworker's family. Other announcements, like graduations or birthdays, probably do not warrant a companywide email.[14]

Do not use email, however, if your message meets one of the criteria listed here. Instead, make a phone call or immediately hold a meeting. Place a phone call or schedule a meeting if:

- Your message requires an immediate response. You cannot assume that your recipient will answer his or her email right away. It is inappropriate to send a follow-up message demanding to know why a recipient has not responded to your message.[15]
- You want to hear the tone of your communication partner's voice so you can read between the lines.
- Your message has many discussion points and needs an extended response or extended negotiations to resolve.
- You want your comments to be private. No message that is private, confidential, or sensitive should be sent by email.[16]

When deciding whether to email your message, ask yourself if your need is for speed and/or mass distribution. Speed and mass distribution are email's strengths. If you need your reader to receive your communication in a few seconds rather than a few days—especially if you have a tight deadline and want to work up to the last minute—email is the way to go. Remember, though, that a quick message does not necessarily mean a quick reply. The recipient is under no obligation to answer your message quickly even though good email etiquette recommends responding to email promptly.

If you have multiple recipients for that 20-page report, consider emailing it as an attachment rather than making 50 copies for distribution. Be sure to check with your recipient first to make sure he or she can open an email attachment and if there is a size limit to an attachment. With email, you can send your message to six recipients or to 6,000.[17]

Third, email is a good tool for communicating across time zones and continents. Rather than call your business contact in Germany in the middle of the night, email lets you and your international recipient conduct business during your respective normal business hours.[18]

Managing High-Volume Email

According to the findings of the 2005 Microsoft Office Personal Productivity Challenge survey of 38,000 people in 200 countries, U.S. office workers received an average of 56 emails daily while their counterparts abroad received 42. This study did not report how many email messages they sent daily. It is likely that on average they sent at least as many as they received, if not more. In contrast, a survey conducted by Pitney Bowes in 2000 reported that U.S. employees sent or received an average of 50 emails a day.[19] Both surveys go back a ways, so it is probably safe to assume that the average number of emails sent daily has increased. However, some of email's original workload is already being replaced by text messages, tweets, and the like.

Given the high volume of email activity, you are among the lucky few if you have an assistant who sorts through your incoming email, deletes the detritus, and forwards summaries of the important messages to you. If not, you alone must decide how to deal with all those emails that arrive in your inbox daily. Good luck with that! However, help has arrived. The following tips can help you manage that overloaded inbox:

- Follow the Last In, First Out rule. Read the last email in a series email in case the issues from the previous emails have been resolved. If you need to reply to the series, read them all first before you reply.[20]
- If you are away from the office for a few days and return to an overflowing inbox, respond to the last day's email first; these responses will not be late. As for the other emails, since you are already late in replying, adopt a last in/first out order after you prioritize by sender or subject.
- On first reading, do something with your email even if it is only to group "Read Later" emails into a separate folder rather than leaving them to clutter your inbox.
- Deal with all non-urgent messages at one time rather than taking time out to deal with each message as it comes up during the day.[21]
- Use the archive function to inspect emails not previously saved and either delete or archive them. Set up the function to delete nonessential email older than a specific number of days.
- Delete chain letters and spam. Unsubscribe to mailing lists that you rarely read.
- Set up other email accounts, outside of your office email, for ordering online and for personal correspondence.
- Use filters to prioritize your mail into separate folders. Some suggestions for ways to organize your mail include:
  - Separate internal office mail from outside mail that does not include your company's domain name.
  - Separate mass mailings from those addressed directly to you. Route messages in which you were cc'd or bcc'd to a mass mailing folder.
  - Create a separate folder for family email after you are sure your company's electronic policy allows you to receive personal email at work.
- If you are going to be away from the office and your email account for an extended period, activate an automatic "Out-of-the-Office" reply. That way, anyone who needs your answer right away will know not to expect it. This feature will help you keep messages from building up in your inbox while you are away.[22]

Following E-Policies in the Workplace

To reduce potential liability associated with e-messages, organizations develop **e-policies** that address matters ranging from corporate-level wheeling and dealing to employee jokes. Email with its shoot-from-the-hip qualities of speed and mass distribution exponentially increases this potential for liability. International Data Corp. estimates that millions of U.S. workers send billions of email messages daily.[23] Not all these messages are related to work. According to an Elron Software survey, many employees with Internet access report receiving inappropriate emails at work, including emails with racist, sexist, pornographic, or inflammatory content. Whether solicited or not, these types of messages in employees' electronic mailboxes can spell e-disaster for the company and the sender.[24]

e-policies
Guidelines for using electronic communication.

Evidence of executives and companies stung by their own emails abound. Probably the most well-known example is Microsoft's Bill Gates who was stung along with other Microsoft top executives in the first phase of his company's antitrust trial when the government's lead attorney used emails from as far back as 1992 to back claims of Microsoft's corporate aggressiveness. Other examples that proved costly to corporations include the 1995 Chevron Corp. case in which sexist emails, like "25 Reasons Beer Is Better Than Women," circulated by male employees cost Chevron $2.2 million to settle a sexual harassment lawsuit brought by female employees.[25]

In another case involving *The New York Times*' shared services center in Virginia, nearly two dozen employees were fired and 20 more were reprimanded for violating the company's email policy. They had sent and received emails with sexual images and offensive jokes.[26] In this case, one of the fired offenders had just received a promotion while another had just been named "Employee of the Quarter." These kinds of incidents cost any company negative publicity, embarrassment, loss of credibility, and inevitably, loss of business.[27]

Another example was a lawsuit against American Home Products Corp. over a rare, but often fatal lung condition some consumers developed after taking the company's diet pills. Insensitive employee emails contributed to the company's decision to settle the case for more than $3.75 billion dollars, one of the largest settlements ever for a drug company.[28] One email printed in the *Wall Street Journal* expressed an employee's concern over spending her career paying off "fat people who are a little afraid of some silly lung problem."[29] This example shows how important it is to review and revise your emails for potentially inflammatory content before sending.

Still another example involves unauthorized use of a company's computer system for a mass emailing. At Lockheed Martin, an employee sent 60,000 coworkers an email about a national day of prayer and requested an electronic receipt from all 60,000. The resulting overload crashed Martin's system for six hours while a Microsoft rescue squad was flown in to repair the damage and to build in protection.[30] The company lost hundreds of thousands of dollars, all resulting from one employee's actions. The employee was fired for sabotage.

The list of corporate disasters goes on. Companies recognize that emails can cost them dearly. If one badly-worded, thoughtless email gets in the hands of defense attorneys in a lawsuit, the company may have to pay up to the tune of a six-to-ten-figure settlement.

With these kinds of productivity, profit, and public relations disasters resulting from email and Internet use, it is not surprising that employers are implementing e-policies. Whether a company consists of two part-time employees or 2,000, the best protection against lawsuits resulting from employees' access to email, IM, and the Internet is a comprehensive, written e-policy, one that clearly defines what is and is not acceptable use of the organization's computers.[31] According to Nancy Flynn, Internet guru and policy consultant, the best policies are straightforward, simple, and accessible.[32] These e-policies inform employees of their electronic rights and responsibilities as well as tell them what they can expect in terms of monitoring. While employees believe they have a right to privacy where their email is concerned, the federal Electronic Communications Privacy Act (ECPA) states that an employer-provided computer system is the property of the employer, and the employer has the right to monitor all email traffic and Internet activity on the system.

According to the 2000 American Management Association Survey of Workplace Monitoring and Surveillance, the number of major U.S. firms checking employee email and Internet activity is up to 38.1 percent, more than double since 1997. More than half of participating companies monitor Internet access for inappropriate uses, ranging from playing games to viewing pornography or shopping.[33] Similar surveys today report more than 80 percent of companies monitor employee electronic activity. Monitoring programs can determine if email is being used solely for business purposes.

Some companies monitor every keystroke made on every keyboard by every employee. Electronic monitors screen for words that alert managers to inappropriate content and some monitors can be configured to spot "trigger" words that indicate everything from profanity and sexually explicit or racially offensive language to the exchange of sensitive information, such as trade secrets and proprietary information.[34]

Although sanctioned by the ECPA, monitoring has its downsides. For example, employees may resent being unable to send an occasional personal email to check on their kids or to make dinner reservations online without violating company policies. The feeling of being watched can hurt morale and eventually may cause good workers to quit, which means lost money for the company.[35] Some employees are striking back at e-policies, claiming protection under state statutes that include a right to privacy.

Finally, even monitoring programs cannot ferret out the hostile tone or the badly-worded document open to misinterpretation, especially when being dissected by opposing lawyers. The haste and recklessness with which we tend to write email, along with its high volume, exacerbate an old problem: mediocre writing skills. Only careful attention to employees' writing habits can halt the production of bad email documents.[36]

Policies that govern email use often incorporate electronic writing policies, electronic etiquette (netiquette) guidelines, company-use-only guidelines, and retention policies that outline how long emails are saved before they are automatically deleted from mail queues and mail host backups.[37] Some companies purge all email older than 30 days and do not keep backup tapes of email. Some experts advise companies to retain no email since a lawsuit could require a review of all backed-up email.[38] However, some industries such as financial firms are governed by federal securities laws that require them to retain emails relating to the brokerage firm's overall business for three years.[39]

What is clear is that more organizations and employers are implementing e-policies. Therefore, as an employee, you need to familiarize yourself with your employer's e-policies and follow them closely. Since every document you write, including email or IMs, is not only a reflection of your professionalism but also a reflection of your organization's credibility, you need to reflect corporate goals by writing clear, clean emails that observe company policy. In addition, before firing off that first email at your job, take time to learn your organization's email culture.

Getting Familiar with the Email Culture of Your Organization

Corporate email culture varies greatly.[40] Some companies use email as their main form of communication; others use it only occasionally. Some companies do not mind if you send personal messages email; in others, it is a dismissible offense. Some top executives welcome emails from staff; others follow a strict hierarchy of who can message whom. Therefore, when you move to a new job, take time to observe the email culture for answers to the following questions so you avoid email mistakes or, worse yet, the email career killer (ECK). A good example of an ECK is sending your new boss a jokey "let's get acquainted" email, only to discover his or her dislike for using email except for formal communications.

- Notice the tone of the emails among coworkers. Is it formal or informal? Do they use email for socializing, joking, and gossiping, or is it reserved for formal, job-related messages?
- What is the email chain of command? Can staff skip over their supervisors and email suggestions or questions to upper management?
- How are urgent or sensitive messages usually communicated?
- Do all employees have access to email? Do they check it regularly?
- How are staff-wide announcements made, via email? bulletin board? memo? intranet?
- Are personal announcements, such as "My house is on the market" or "I just found the cutest puppy; any takers?" acceptable over companywide email?
- What are the company's written policies on email, IM, and Internet use? Is any noncompany-related browsing allowed? To what extent are email, IM, and Internet use monitored?[41]

Become familiar with your employer's e-culture as soon as possible. If you are uncertain about where the organization stands on such matters, do not guess. Ask your immediate supervisor or an HR representative.

Writing Effective Email Messages

The fiction that quick, poorly-written business messages are acceptable is fostered, in part, by the medium itself. Unlike writing memos and letters where we tend to be more guarded and take time to write thoughtfully, email brings out the worst of our bad writing habits. As Gregory Maciag, president and CEO of ACORD, the nonprofit industry standards association, points out, the medium can hamper communication: "In place of thoughtful content, we send and receive short bursts of often grammatically and emotionally challenged communiqués that we sometimes regret."[42] As writing consultant Dianna Booher laments: "They log on; they draft; they send."[43]

To avoid creating confusion, misunderstandings, and hurt feelings with your emails and to ensure that you get the busy reader's attention, think and plan before you draft an email message, then revise it before you hit Send.

Drafting Effective Email

Competition for the electronic readers' attention grows daily with tens of millions of email users online. The challenge for those who send emails is to get their targeted readers' attention. To help you get your reader's attention, follow the guidelines discussed here, beginning with the all-important subject line.

Subject Line

Some busy executives get hundreds of incoming emails a day. To get through them quickly, they look at the subject line. If that grabs their attention, they scan the first screen. Messages that do not get the reader's attention at the subject line run the risk of never being read or of being deleted. To avoid "email triage":[44]

- Always include a clear, informative subject line. It should communicate the topic of the message, and like the subject line of a memo, it should be specific and brief. A subject line such as "Staff meeting changed to 3 p.m." provides the necessary information. A subject line such as "Meeting," "Information," "Guidelines," or "Hey" convey nothing. If you use "URGENT" or "!" too often, you reduce the urgency of the message in the reader's mind, and he or she eventually ignores it.[45]
- Be brief. Write the important points of your topic in the first half of your subject line. In an inbox, only the first 25–35 characters of the subject line usually appear. In a string (thread) of exchanged emails, change the subject line if the subject changes. That way, if you have to refer to an old email, you won't have to reread every message with the same "Re."[46]

Body

The first thing to appear in the body of your email may be your **emailhead**. Emailheads appear routinely when email is used to transmit formal contracts, proposals, offers, and other business transactions. Companies that provide eletterheads, like Dynamic Email Stationery from StationeryCentral.com, offer graphically enriched email that reflects your corporate identity.[47]

emailhead
Email letterhead.

When you use an emailhead, be sure you have a clear purpose for including it, such as making it clear that the message is from your company. Otherwise, the reader may object to the wasted lines.

Always include a salutation. Including a greeting at the start of your message personalizes it, plus it establishes your role in the message's history.[48] If you normally address a person by his or her title or if you do not know the person well, include his or her title when addressing them: "Ms. Jones," "Dr. Smith," "Prof. James," etc.

If you are unsure how to address a person, err on the side of formality (this cannot be stressed enough). Being too formal will not offend even the most informal of people and will satisfy the most traditional formalists. However, if you are too informal at the outset, you run the risk of ruining your credibility with that person or organization. You can always change the salutation in subsequent messages if the recipient indicates that informality is fine.

Although business correspondence greetings like "Dear Sirs" are outdated forms in the United States, greetings are more formal in other countries, such as Japan and Germany.

Keep the body of your message short, no more than 25 lines or one screen's worth of words.[49] For longer documents, send an attachment. Some companies do not accept attachments for fear of viruses, so check with your recipient first. If you must attach a document to an email, use clear headings in that document to break up the text and allow for skimming.

Remember there is a live person on the other end of your communication, so get right to the point in the first or second sentence and support your main idea with details in the next paragraph. Begin your email with your main idea or request. Most important, keep the body of the message brief and to the point. Focus on developing one topic only, and make responding easy. For example, phrase a message so that your reader can respond with a quick "yes" or "no."[50]

In summary, make your message reader oriented to help your recipient grasp your message quickly:

- Begin with your bottom-line, main idea, or a precise overview of the situation. If you want the reader to take action, begin by making your request. Include the requested action in the subject line for emphasis. If the message is for the reader's information only and needs no follow-up, put "FYI" in the subject line.
- Keep messages under 25 lines long, and use short sentences.
- Use white space before and after your main idea to highlight it.
- Use short sentences and short paragraphs that cover one idea. Separate short paragraphs with white space.
- Use bulleted or numbered lists to help readers quickly differentiate multiple points or directions.[51]

Signature Block

Always close with something, even if it is only your name. Simple closings, though, like "Regards" or "Best wishes," add a touch of warmth to this otherwise cold medium. Add credibility to your message by adding a signature block after your name. For consistency, create a signature file (.sig) containing, at a minimum, your name, title, and address. You could also include phone and fax numbers; your web address or website; an advertising message, slogan, or quote; your business philosophy; or ASCII art created from text and symbols. However, many organizations have eliminated these slogans and quotes from employees' signature files in an effort to steer clear of potential litigation. The best advice is to know and follow your organization's policy. Avoid duplicating material in the signature that is already in your emailhead, if you use one.[52]

Revising Email Messages

Every document you write is a reflection of your professionalism and your organization's credibility. Good email is businesslike, well written, and free of grammatical errors and typos. Before you hit Send, always take a minute to proofread. And remember, whatever you send could get forwarded to the one person you do not want to see the message. Message privacy is not guaranteed in cyberspace.

Proofread Your Email Carefully

Do not let poor spelling and typos detract from the credibility of your message. Use the grammar and spell checker, but then reread the message yourself before you hit Send. You cannot depend on a grammar or spell checker to catch every error.

Contrary to popular belief, businesspeople do pay attention to typos, no matter what the medium. Typos and grammatical errors undermine your credibility and the credibility of your email message, and they subject the message to misinterpretation. Sloppy writing shows a lack of respect for your reader. Remember that some decision makers go out of their way to catch spelling or grammatical errors in business documents. Catching careless coworkers' errors in office email is a common pastime in many organizations.[53] To avoid being the joke of the day, proofread. Take the following example:

> *"If emale is writon with speeling mestakes and gramitckal errors, you mite git the meening, however, the messige is not as affective, or smoothly redable."*

After proofreading, always ask yourself, Would I want to see this message in a *New York Times* cover story or taped to the office refrigerator? If not, do not send it.

e-tone

Refers to the miscommunication that occurs when the writer has one tone of voice when writing the email, but the recipient reads it in a totally different tone.

Humanize Your Emails

Use a tone appropriate to your audience. Email is an impersonal medium that is fertile ground for misunderstandings and hurt feelings. Temper your messages with politeness and objectivity. Strive for a professional, yet conversational tone. Use personal pronouns ("I," "you," "me)" to humanize the connection.

To humanize your email, avoid the "**e-tone**." Nancy Friedman, a consultant and trainer, invented the term "**e-tone**" to refer to the miscommunication that occurs when you have one tone of voice in mind as you write email, but your recipient reads it with a totally different tone. Friedman urges email writers to use words that express feelings—"please," "thank you," "I'm happy to report," or "sorry to say"—to tell them how you feel.[55] Friedman contends that even sensitive topics, such as apologies, can be addressed in email if done properly.

Although not widely used in business documents, emoticons (smileys) can give your email humor and temper the e-tone. However, overuse of punctuation marks and emoticons can also cause misunderstandings. For example, a subject line in an email to a professor that says "Grades???????" might sound as if you are unhappy with your grade, when you may just want to find out what your grade is. Similarly, overuse of the exclamation point can offend people because it can make you sound pushy or overexcited!!!!!!!!

Proofread to Guard against the "ECK"

An ECK (Email Career Killer) could be the too-cute reply to your boss or the off-color joke sent to your buddy that offends whoever sees it or whomever it might be forwarded to. ECKs stem from emails greatest assets: speed and ease. To avoid them, carefully check every message for tone and errors, and look twice and three times at the names listed in the TO, CC, and BCC blocks to ensure that nothing you have written could be misinterpreted or cause a problem for anyone.[56]

Be sure the email you are about to send does not contain a reckless or emotional outburst. For example, do not send an email in which you stridently complain about your boss to a coworker because the offensive email could get forwarded, accidentally or on purpose, to your maligned boss.

netiquette

Guidelines for acceptable behavior when using electronic communication via email, instant messaging, text messaging, chat rooms, and discussion forums.

Observing Email Netiquette

We are all encouraged to follow the rules of netiquette (electronic etiquette) when developing and sending email messages. **Netiquette** refers to etiquette rules governing electronic content and use. Netiquette rules apply to all electronic communications—blogs, social media, email, text messages, etc.—business or personal.[57] These rules of polite behavior have sprung up alongside the etiquette that governs our off-line behavior and have quickly become a universally understood behavioral standard that transcends cultures, businesses, and geographical boundaries.[58]

Avoid Flaming

Avoid publicly criticizing people in email or discussion groups using inappropriate language. An email **flame** is a hostile, blunt, rude, insensitive, or obscene email. Flames are immediate, heated reactions and have no place in a business environment. If you are upset or angry, cool down and rewrite your hastily-written, angry message before it damages you and your organization.[59] Remember that any email you send—whether strictly business, gossip, complaints, or personal issues—could wind up in your boss's inbox.

flame
A hostile, blunt, rude, insensitive, or obscene email.

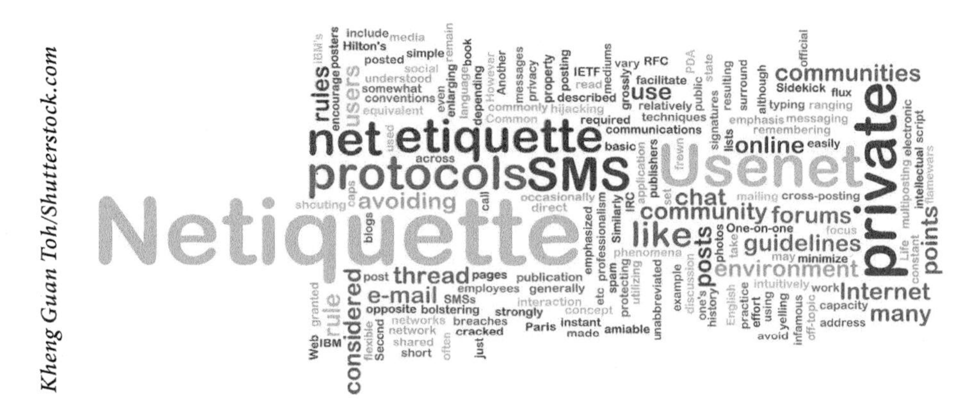

Kheng Guan Toh/Shutterstock.com

Avoid Shouting

Shouting is using all CAPS in your message. An email with the line THE MEETING WILL BEGIN AT 3PM will be interpreted as demanding and obnoxious. Conversely, do not write in all lower case letters. Stick to standard capitalization in email.

shouting
shouting Using all CAPITAL LETTERS in an email.

Avoid Spamming

Spamming refers to posting junk or unsolicited email posts to a large number of mailing lists.

spamming
spamming Posting junk email or unsolicited posts to a large number of email lists.

Avoid Acronyms and Abbreviations in Email

Do not use acronyms and abbreviations unless you are emailing good friends. If you do use them, always explain what they mean. It seems that with increasing volumes of email coming across computer screens daily, the substance of each communication drops, so that the average email message now looks something like this: "OMG did u c K's latest x LOL!!!!!!!!!!" or "I'm on my way. I'll be there soon," has now degenerated to "im on my way!!!!!!!!! Ill be thr son!!!!!"[60] Remember that email not only reflects your level of professionalism, but also lives on in backups that could come back to haunt you.

Watch What You Forward

Forward is perhaps the most dangerous command on your email program, second only to Reply All. While forwarding is a time-saving way to share information, used without thinking, it can turn into an ECK. Think twice before Forwarding email. Although sometimes necessary for business reasons, most people do it more often than they need to.[61] For example, refrain from passing on chain letters, jokes, rumors, and wacky stories unless you know your recipient shares your love of this Internet flotsam and jetsam.[62] Many email users resent having jokes and stories fill their inboxes. Respect your reader's time and ask before you forward.

Do not forward all or parts of messages without the consent of the sender as well as the intended recipient. Remember that anything a person writes is copyrighted the minute the author writes it, whether it is an article from the *Wall Street Journal* or the musings of your best friend and coworker.[63] Would you send a photocopy of a handwritten letter to someone else? Phillip Zimmermann, creator of Pretty Good Privacy encryption, says the same thought and respect should go into forwarding emails.[64]

Ask Before You Send an Attachment

Some organizations prohibit email attachments due to hidden viruses. Before you send an attachment, ask if the reader would prefer receiving the material as an attachment or in the text of the message.[65] In addition, do not simply send an attachment without explaining what it is in the body of the email. "See attached" or a blank email with an attachment raises suspicions that it might be a virus. In such a situation, the recipient will most likely delete the message without opening the attachment.

If You Need an Immediate Response or Action to Your Email, Use the "Receipt Notification" Option

Receipt Notification lets you know when the reader opens the message. However, some readers resent the use of receipts, saying that receipts imply a lack of trust on the sender's part. Your better option might be to phone the recipient, letting them know that a pressing email is on its way and that you would appreciate a quick response.[66]

If You Want Your Reply to Go Only to the Sender, Hit "Reply," Not "Reply All."

If you hit "*Reply All*," your message will go to all the recipients on the list. Getting messages not meant for them irritates many listserv members. They might send you a flame for this netiquette breach email. If you want your reply to go only to the sender, type in that person's address. Be considerate of people's privacy.

Only CC People Who Need to Read the Email

Many managers and executives complain about being copied on messages they do not have to read. Unless your boss has requested it or it is standard practice for your work-group, keep message CC's to a minimum.

Beware of the blind copy (BCC). BCCing means you can sneak a copy of a message to someone without the main recipient knowing, since the BCC'ed person does not appear in the main recipient's header. You have no assurance that the person BCC'ed will keep the "message" "secret." [67]

Before Emailing Over Your Boss's Head to Upper Management, Know What Is Customary in Your Office

In most cases, your boss wants to be kept in the loop when you are emailing up the chain of command, so let him or her know about ideas, requests, or questions that you plan to send to upper-management. Tell your boss before you email up the command chain, rather than in by CCing him or her.[68]

Do Not Use Company Email to Circulate Personal Requests

Otherwise, use bulletin boards or a web message board to post personal requests.[69]

International Emails Pose Language, Culture, Time Challenges

Before writing and sending an email message internationally, think about your reader's communication needs (see Figure 6.2).

- Before you begin writing, determine your reader's needs. You may have to translate your message into your reader's language.
- Be careful with dates and times. Most Europeans would interpret 3/5/14 as May 3, 2014 (with the month appearing in the middle), rather than March 5, 2014. The Japanese, in contrast, sometimes use a year/month/day format. To avoid misunderstanding, write out the name of the month as in March 5, 2014 or 5 March 2014.
- Most countries use a 24-hour system, so be sure to use that time format when setting up videoconferences, conference calls, and IM meetings: "The conference call will begin at 13:00 on 3 March 2014."
- Since most countries use more formal written communications than United States does, be formal when you email internationally. Address people by their surnames and titles, and use a formal tone throughout your message. To help you sort out titles used in different countries, like "*Monsieur*" or "*Madame*" or "*Herr*," check a reference work, like Peter Post and Peggy Post's *The Etiquette Advantage in Business.*[70]
- Use specific language and avoid acronyms, abbreviations, business or technical jargon, and humor. They do not translate well.
- Before using monetary denominations, state the currency (e.g., US$10,000).
- Give country codes for phone numbers. The U.S. country code is "1 (e.g., 001-608-123-4567)."
- Use generic names rather than brand names (e.g., photocopy rather than Xerox).
- Be specific when you mention geographical locations: Use New York rather than East Coast.
- When indicating time, be sure to indicate which time zone you mean. For example, "I'll call you at 6 p.m." could mean your time or theirs.

FIGURE 6.2. International Email Etiquette

Source: Adapted from Samantha Miller, Email Etiquette *(New York: Warner Books, 2001) and Nancy Flynn,* The ePolicy Handbook *(New York: American Management Association, 2001).*

Summary: Section 1—Writing Email Messages

- Businesses and organizations increasingly use email to send routine and sometimes even sensitive messages to employees, clients, and business contacts. Email recipients scrutinize what employees are writing, looking for typos.
- Along with business-related email, workers are also sending many nonwork-related emails, all of which are backed-up regularly. For employers, this increased use of email causes liability concerns.

- When deciding whether to send an email, consider how formal or informal the situation is and your need for speed and mass distribution. Know the tips on how to manage high volumes of email to avoid getting overwhelmed.
- Employers are developing e-policies to protect themselves from lawsuits. Know your company's e-policies and why they were designed.
- Become familiar with email your organization's email culture before you send emails.
- Know how to write effective emails to communicate your thoughts clearly and to avoid misunderstandings. Rather than drafting and sending a message, take time to plan, draft, and revise before you send it email.
- Abide by the rules of email etiquette (netiquette) to avoid offending a recipient.
- Before you send messages internationally, familiarize yourself with the rules of international email etiquette.

Writing Effective Instant Messages

Even though email is more widely used in U.S. organizations, instant messaging is also used extensively.

Major Uses of IMs in Business

Instant messaging (IM) is a form of online chat. While IM was once used mainly by night-owl teens chatting with their friends, it is now a standard business communication medium. The two defining characteristics of IM, presence awareness and near real-time operation, make it a compelling alternative to email and voicemail. According to a recent report, some 60 percent of business phone calls never reach their intended recipients.[71] These deficiencies of traditional forms of business communication help explain the explosive growth of business IMs.

Combining the real-time benefits of using the phone with the convenience of email, IM offers a variety of advantages to the corporate communicator.[72] IM's popularity as a business communication tool can be explained by its quickness, flexibility, and versatility.

IM Is Quick

A sender can detect whether a user is online, send an IM, and institute a back-and-forth conversation, virtually, in real time.[73] Unlike email that can remain unanswered in someone's overloaded inbox for days, IM can detect someone's presence online, which is good for an for immediate response. IM also eliminates long email threads.[74]

IM Is Flexible

Many people can be in on the same conversation. In that way, work groups can use it to get tasks done quickly.

Along with making team communication easier, IM allows users to have more than one message thread going at a time. IMs are so easy to handle that you can be on the phone and still respond to them. In terms of security, with commercial-grade IM software, users have the option of archiving IMs for legal or management reasons or purging them to avoid being susceptible to court-ordered discovery processes.[75]

IM Is Versatile

IM facilitates communication among geographically distributed workgroups; it improves communication with business partners and suppliers; it quickens response time between customer service and support departments and customers; and it facilitates cross–business unit communication.[77] Users can even send files via most IM applications when a report, contract, or invoice needs to be quickly reviewed or approved.[78] As a collaborative tool, IM easily enables team members to meet in a dynamic space, share files, set up whiteboards, and discuss changes. The space disappears when users are finished. Collaboration is much easier than in the pre-IM day.[79]

The buddy list, an IM staple first developed by AOL, has become "*presence management*" in business contexts, where detecting who is online to answer a question or to buy a product in real time means increased productivity and profits.[80] **Presence management** refers to being able to determine if others are online and available. Companies like AT&T, IBM, Boeing, the U.S. Army and Navy, National Cancer Institute, and a host of other high-tech, financial, and retail companies use IM to assemble virtual teams from locations around the world. IMs allow employees to communicate with coworkers, clients, customers and other business contacts from their virtual offices.[81] Retailers like Landsend.com and 1-800-flowers.com use IM to answer customer questions when they arise. It is faster than email and cheaper because customer reps can reply more quickly.[82] HP, Gateway, and Mail Boxes Etc. also use chat in their sales and service.

presence management

Being able to determine if others are online and available.

A good example of a business that thrives virtually by using IM is the CPA firm of Carolyn Sechler. She heads a 14-member virtual office workforce that serves 300 clients, primarily nonprofit organizations and technology entrepreneurs, in several states and countries.[83] From her home office, Sechler works with CPAs from Alabama to British Columbia. She meets her core team of four every two weeks using her IM service, ICQ. She believes in making people feel comfortable: "At 9 a.m. we all tune in—and we can archive the chats. What's the point of making people go anywhere when they can be comfortable?"

IM is the tool that makes her virtual business possible, a business that has grown 10–15 percent every year. Sechler says that she has used ICQ since its inception. She keeps a "buddy list" of team members up and running on her computer so that they can communicate throughout the day. She can exchange quick messages or files with any of them by IM—a faster service than email—and they can have "impromptu conversations" that bring together four, five, or six members into one chat area. According to Sechler, being accessible on-line strengthens ties with clients, circumvents crises, and lets the firm find out early about new consulting opportunities.[84]

How to Use IM

Instant messaging works this way: A small piece of client software is loaded onto a PC, smartphone, or other device and maintains a constant connection to a central hosting service. Anyone who is logged onto the service is flagged as online. The software includes a "buddy list" that enables the user to store the nicknames (everyone has a nickname on IM) of clients, coworkers, and other business contacts. When a buddy is online, the name or icon lights up. The user clicks on the icon to send the buddy an IM or a file attachment.[85] After that, messages fly back and forth, all in the same window that scrolls up as the conversation continues. Unlike email, which can take time to reach its destination, IMs reach their destination instantly even if the person is continents away.

You can also control when others can send you a message. Privacy options include saying that you are away at lunch and that you don't want to be bothered. You can even block the fact that you are online.

IM is great for those messages that are too brief to pick up the phone or too urgent to try to play phone tag. In many cases, IM has replaced phone use for short transactions.[86] Since it does not require your full attention, you can even talk on the phone while messaging other people.

Another useful feature allows you to invite several people into the same session. Unless you need the archiving function to save the message, these messages usually disappear when you log off.

Writing Effective Text Messages

Texting is a popular way to communicate in the workplace, thanks in great part to the proliferation of affordable, dependable smartphones and tablets working in sync with wifi networks.

Business Uses of Text Messages

Many of the same rules and uses for IM apply to text messaging. A **text message** is a type of IM. Text messaging has gained wide popularity in the United States and elsewhere

Today's mobile communication technologies allow drivers to send and receive text messages, tweets, and phone calls while operating vehicles. Unfortunately, many drivers who chose to do so have caused car, truck, motorcycle, bus, and train accidents, resulting in injuries and fatalities. Some of these drivers lived to tell their side of the story in court, while others were not so lucky. Texting, tweeting, and placing phone calls while driving defies both logic and responsible citizenship. It is basically selfish, narcissistic behavior. However, this has not stopped all drivers from taking chances with their own fate as well as the fate of others! These activities have prompted several U.S. municipalities and states to consider or pass laws making it illegal for drivers to text, tweet, and participate in phone conversations while driving. Several countries outside of the United States are considering or have already passed similar laws. The U.S. government is also considering such measures.

Drivers in some countries have made great headway in reducing instances of simultaneous driving and texting, tweeting, and talking on cell phones. For example, in the United Kingdom texting while driving is socially unacceptable behavior. Thus, UK citizens are policing themselves on this matter. There appears to be a similar trend in Los Angeles. These are positive signs.

©2013 by karen roach, Shutterstock, Inc.

Hopefully you are not risking your safety and the safety of others' by practicing such dangerous behaviors. And hopefully, the defensive driver in you is constantly on the lookout for other drivers who are not as considerate.

While driving recently, I observed an additional texting-related distraction. It was a humorous, yet serious bumper sticker. It read, "Honk If You Love Jesus. Text While Driving If You Want To Meet Him!" No matter how tasteless you might find this bumper sticker, it is one more reminder to those who read it to be responsible, careful drivers.

Drivers, Beware. Or Is It Beware of Drivers?

since its inception. The first text message (Merry Christmas) was sent on December 3, 1992. Although usually sent from one mobile phone to another, text messaging is often integrated into IM software so that messages can be sent via an IM, but received on a mobile phone. Text messages are no longer limited to text. Video clips and pictures, which are often captured with a mobile phone, are easily integrated into these messages.

Text messaging is popular in the U.S. business community for many reasons. First of all, text messages are relatively inexpensive compared to phone messages. While some communications are involved and do not comfortably lend themselves to text messaging, some companies find they can save more money than phone calls. Another text message advantage is its ability to communicate in real time. When emailing or IMing, if the recipient is away from their computer or if their phone lacks Internet capability, they cannot be reached. Text messaging has the mobility of phone calls as well as email functions, such as one message going to several people at once. It should be no surprise that text messaging is used extensively in industries, such as real estate, construction, and transportation. Each of these industries has people working in the field who are not connected to a computer all the time. They are able to receive alerts, updates, and quick messages throughout the day without having to rack up phone bills.

Of course text messages are not limited to business use. In the United States the most common use of text messaging is to communicate with consumers through retail or reality TV shows. It is not uncommon to see advertisements soliciting ring tones by texting "win" to a certain phone number. Nor is it uncommon to have the next American Idol decided by voting via text message. While these are the first uses of texting, more are coming. Text message marketing is a foreseeable field, whether it be advertisements or product notifications. Further, text messaging can notify buyers of order confirmations, shipping, back orders, etc.

Instant Messaging and Text Messaging Etiquette

Many companies do not allow employees to install IM services because those employers believe IMs distract employees from their work. For instance, you might be in the middle of an important project and suddenly a message from your spouse pops up with a reminder to take the dog in for a flea bath. Unwanted interruptions can be avoided by using the privacy functions on your IM service or by having two IM services, one for work and the other for family and friends. Finally, be sure to practice good messaging etiquette by following these suggestions:

- If someone is marked as unavailable or if you have received an away message from that person, refrain from messaging until they return. Although you may be in a chatty mood, a coworker or family member may be busy and unable to respond promptly. If the recipient uses the same IM service for work and personal messages, the person cannot shut it down since it is a work tool, like the phone. If they are at work and available, be courteous and ask them if they have time to chat, "Got a minute?" If they do not, do not be offended. Ask them to message you when they are free. If you are on the receiving end, don't be afraid to let people know you are busy. Learn to say "no" or ask them to message back later.[87]
- When responding to someone's message, it is a good idea to type your answer, send it, and then wait for the recipient's reply before sending another message. That way, things will not get confusing because you each respond to one idea at a time.
- Know when to stop. Do not let a thread go on and on. A simple "got to go" or "bye" or "ttyl" (talk to you later) should be sufficient to end the conversation.

- Check your grammar. If you are communicating at work, take a few seconds to reread what you have written before you send it. More and more people are chastising those who misspell the same words consistently. Misspelling can ruin your credibility.
- Do not use IMs for long messages. Betsy Waldinger, vice president at Chicago-based OptionsXpress Inc., spends a lot of her day working on an online customer service chat system. She recommends either sending short messages or breaking up long ones over many screens. Typing long messages keeps the person at the other end waiting while you type. IM should not replace email. Use IM for quick-hit messages that require fast responses. Use email for longer discussions.[88]
- If you work in an office where IM is part of the culture, send a message before you drop in on someone. This gives the other person a chance to say whether the visit would be convenient.
- Log off IM when you are not using it. Otherwise, you may come back to find messages waiting and senders wondering why you are ignoring them.
- Do not hide online. Some managers use the ability to be invisible online as a way to watch employees. This is a quick way to discourage workers from using IM. Being on or off IM is not a good way to determine whether your remote employees are working. Be courteous, and if you are on, be visible. Use the service busy icons, like "Do Not Disturb" or "Busy" to indicate your availability.[89]
- Use proper punctuation and capitalization as though you were typing an email or letter.
- Use two carriage returns to indicate that you are done and the other person may start typing.[90]
- Remember that chat is an interruption to the other person. Use only when appropriate.
- Be careful if you have more than one chat session going at once. This can be dangerous unless you are paying attention.
- Like email, set up different accounts for family and friends and for work-related IMs.

The need for speed when instant messaging has prompted the creation of acronyms that express nearly every sentiment. Figure 6.3 contains some examples of common IM and texting acronyms. In addition, visit buzzWhack.com online for the latest acronyms.

When writing IMs and text messages, be cognizant of using standard acronyms and abbreviations. Not all your communication partners may understand what you are saying. I noticed a drink coaster recently in a local restaurant. It had a reminder printed on it that applies to using acronyms and abbreviations in IMs and text messages. It read, "2 much txting mks u 1 bad splr." No matter whether you are writing business reports, letters, memos, emails, IMs, text messages, blogs, website content, or tweets, misspellings have the potential to wreck your credibility.

IM	instant message
BRB	be right back
CID	consider it done
OMW	on my way
G1	good one
VM	voicemail
W8	wait

FIGURE 6.3. Instant Messaging and Texting Acronyms

Summary: Section 2—Instant Messaging and Text Messaging

- Presence awareness and real-time operation make IMing and text messaging compelling alternatives to email and voicemail in the workplace.
- IM's flexibility is also another feature that attracts workers who use it. For example, since many people in remote locations can be in on the same conversation thread, team communication is easier and faster than in the past.
- IM is versatile. It facilitates communication among remote work groups, it can improve communication between business partners and suppliers, and it quickens response time between customer service and customers.
- Some firms see IM's downsides as stumbling blocks to its use. For example, security and interoperability are two issues getting attention now by commercial-grade IM services. In addition, some managers believe that IMs are a distraction that interferes with employee productivity.
- If you use IM in your workplace, follow the rules of messaging etiquette.
- Text messaging is a viable alternative to email and phone calls. Text messaging can cut costs and keep people in touch no matter where they are. The uses continue to grow for company-to-consumer relations.

Writing Effective Business Blogs

Blog is short for "web log," a type of user-generated website that acts as a communicating and networking medium for the masses. It is a type of self-publishing via the Internet done through easy-to-use Internet applications and websites. Many websites that host blogs are free, so they are easy to start and require little technical knowledge to operate.

Blogs are built for flexibility and designed to be short and updated frequently, unlike long essays or prose. Although blogs have been championed by social networks for personal use, they have made a significant impact on the business world, the journalism world, and

blogosphere
The universe of blogs.

academia, where they are used in the classroom. External blogs that the public has access to are considered to be in the **blogosphere**, which is the universe of blogs.

Business world blogs can be broken down into several categories. The main type of business blog is the corporate blog. Corporate blogs can be used as e-newsletters, viral marketing campaigns, and an open channel between businesses and consumers. For making sure readers are current on a blog post, frequent blog readers use a tool called RSS. RSS stands for "really simple syndication," and it is a way to notify readers that their favorite blogs have been updated. The ease of use and relatively low cost make blogs an excellent medium for businesses, which is why they are growing so quickly in corporate settings.

Corporate Blogging

Since blogs are usually updated at least once a week, blog managers make sure the corporate website is constantly freshened with new information. When a blog becomes popular, it guarantees that consumers or even frequent bloggers will view a company site again and again. These blogs differ from the traditional website in several ways. They are

- interactive
- written in a conversational tone
- frequently updated
- can be used to express personal opinions, not only company policy[91]

The most popular type of blog is the external blog, which can be seen by anyone with an Internet connection. Internal blogs are also growing in popularity. *Internal blogs* are established by companies so that only employees or a select group of people can see them. These are usually password protected, and sometimes are on a local intranet. Internal blogs are a good starting place for companies that are testing the waters of blogging. Internal blogs can be used as testing grounds for speed, content, frequency, and authorship before a company enters the blogosphere. Although they can be a warm up for external blogs, internal blogs also have a place once a company is comfortable with the medium.

BLOG

dslaven/Shutterstock.com

Shel Holtz's point about personal blogs coincides with another use of blogs, knowledge management, which are sometimes referred to as K-Blogs or knowledge blogs. **K-Blogs** refer to when an expert keeps a topic-specific blog that allows users to tap into his or her expertise when needed. New tools and applications are being designed specifically for K-blogs to facilitate references to documents, journals, essays, and emails.

K-Blog

Refers to when an expert keeps a topic-specific blog that allows users to tap into his or her expertise when needed.

As time-ordered business records, blogs have many uses (see Figure 6.4). They are great for project members or leaders who can track projects and ideas without adding an extra archiving step. New project members can easily get an overview of the project and how it is developing. Possibly the most valuable use is for legal discrepancies. If a decision or action was subject to questions, the business would have clear and thorough documentation.

One big selling point for internal blogs is that they can act as email killers. Imagine this, Jill wants her department to know cake and sandwiches will be in the break-room at 2 p.m. Since everyone in her department has something to say, they all hit Reply to All and send everyone in the department responses like "Can't wait!" or "See you there!" Now everyone's inbox is full of the original memo and fifteen responses. If Jill had posted a blog with the memo everyone could have logged on in the morning, seen that cake and sandwiches would be in the break-room, and left their comments on the blog page, leaving all the inboxes free of clutter.

Communications and technology expert Shel Holtz provides us with the following list of ways you can use an internal blog:

Alerts. Don't you hate getting those emails that let you know when the server's going to be down? People who need to know can subscribe to the list server status Weblog. Instead of having to send out those emails, IT can simply request that employees subscribe to their blog.

Projects. Companies have terrible institutional memories when it comes to projects. Anybody who needs to delve into a project's records to find out how a decision was reached a year ago is probably out of luck. Project teams can set up a group blog to maintain an ongoing record of decisions and actions. Project leaders can also maintain a blog to announce to the rest of the company the current status of the project.

Departmental. Departments can maintain blogs to let the rest of the company know of current offerings or achievements. Imagine the marketing department being able to submit a simple post to its own blog announcing the availability of new marketing brochures or other collateral.

News. Employees can contribute industry or company news to a group blog or cover news they have learned in their own personal blogs.

continued...

Brainstorming. Employees in a department or on a team can brainstorm about strategy, process, and other topics over blogs.

Customers. Employees can share the substance of customer visits or phone calls.

Personal blogs. Even though it sounds like a timewaster, a personal blog can prove valuable in the organization. Consider an engineer who reads a lot and attends meetings of his professional association. He updates his blog with summaries of the articles he's read in journals (with inks to the journal's website) and notes he took at the meeting. Employees who find value in this information (other engineers) will read the blog; those who don't care probably aren't missing anything if they don't. And if the engineer posts an article or two that has nothing to do with work, well who said work can never be fun?

CEO blogs. What a great way for the CEO to get closer to employees. Imagine a new CEO hosting a blog called "My First Hundred Days" in which he writes about his experiences daily and lets employees comment in order to help him get acclimated.

FIGURE 6.4. What to Do with Your Internal Blog

Source: Debbie Weil, **The Corporate Blogging Book: Absolutely Everything You Need to Know About Blogging to Get It Right.** *(New York: Penguin Group, 2006).*

External Blogs

While internal blogs are useful, to reach their full potential, corporate blogs need to be part of the blogosphere for everyone to read. One of the biggest distinctions in external corporate blogs is who writes them. According to Chris Anderson, author of *The Longest Tail: Why the Future Is Selling Less of More*, the best business blogs are not written by top managers or marketers, but by midlevel employees.[91] They are often more entertaining, more honest, and less likely to repeat everything in the latest ad campaign or mission statement. They can also be more informative about the company's inner workings. Consulting company Accenture directs new recruits to its Careers page, which displays blogs written by employees in various departments to give recruits a taste of the day-to-day activities of a new management consultant. Or, they can read the rants of a recruiter so they know what not to do. Everyone wins in these situations because recruiters no longer have to read vapid cover letters, and prospective employees may discover, for instance, that they do not care for the extensive travel required of management consultants.

For organizations, one big draw of blogs is the feedback they receive from consumers. Because bloggers comment on blog posts in the relative anonymity of the Internet, consumers are less reluctant to tell companies exactly what they think. This uncovers valuable information, formerly only available through expensive polls and surveys. It is also instant, and sometimes preemptive. Bloggers can tell a company what they think of a product before it hits stores shelves, for example. That is the real key to successful blogs; a conversational style that solicits and receives feedback.

The most-often cited use of external blogs is for marketing. Instant publishing is a powerful tool, as it can potentially be read by the very demographic the company in targeting. With a real-time, searchable, easy-to-use public relations tool like blogs, some have asked if blogs will replace traditional marketing communications. While it is true that blogs are one of the latest waves of Internet marketing, they are not meant to be a one-stop shop for companies. They are an additional avenue, that when combined with TV and print advertising creates a more robust marketing campaign.

Much like K-Blogs, external blogs establish thought leaders. **Thought leadership blogs** are used to establish experts and respected leaders in narrow fields. These blogs are usually updated frequently; usually several times a week if not daily. CEO and senior executive blogs usually fall into this category, along with top consultants.

thought-leadership blogs
Used to establish experts and respected leaders in narrow fields.

One type of blog referred to as *customer evangelists* has become a powerful resource for companies, and they do not even pay for it. Some company brands have a powerful and passionate consumer following, so these companies benefit from consumers who do the blogging for them. **Customer evangelists** are loyal consumers who start blogs devoted to their favorite product or company. These customer-written blogs act as a meeting place for passionate consumers to discuss their favorite products. They are often seen as a fair judge of products, because they are independently written. While these blogs are a great asset for the companies and products they are devoted to, they have drawbacks. Because the company has no say in the content, if the company makes a decision that their blogging customers dislike, there is a chance the blogs can turn negative. It takes a different kind of public relations effort to manage customer evangelist's blogs.

customer evangelists
Loyal customers who start blogs devoted to their favorite product or company.

Whether blogs are written by evangelists, midlevel managers, or CEOs, they all have the power to impact product or service sales. Therefore, it is more important than ever to craft coherent messages through writing style, content, and type of blog. While many companies may know what an ideal blog should be, they do not always know how to create one. With this gap between the perceived power of business blogs and how to create them has emerged a new profession—the **professional blogger**. This is someone (in-house or consultant) whose job is to create, write, edit, and promote a corporate blog as well as stay up to date with other corporate blogs.

Blog Writing Tips

The main concern for many businesses when starting a blog is: What do we write about? The answer to this important question determines the blog's success or failure.

First, the blog should have a narrow focus, but broad and complex enough that you do not run out of things to say. If a blogger jumps from topic to topic, he or she risks losing readers. Bloggers who focus on a particular subject tend to do well because they grow a faithful fan base. That said, to keep content fresh and the writing style entertaining, controversial content works well. Taking risks on controversial topics often increases the number of links to your post, which increases the number of readers. At the same time, an author who wants to be controversial must to be ready for the consequences. A controversial post can elicit comments both attacking and defending it. Editing these comments can be a daunting task. Ultimately, the decision whether or not to make a blog controversial depends on the company's goals and objectives for it. If the goal is merely to garner a large number of hits and readers, then controversy is the way to go. But if an organizational blog exists to portray a company in the blogosphere, then it is inappropriate to include controversial material.

In any case, controversial topics must remain balanced and in check. Even if the organization desires high readership, one slip in an entry or an unchecked comment could result in costly lawsuits and the loss of jobs. Nancy Flynn, author of *Blog Rules*, recommends having a lawyer review official corporate postings and the incoming comments.[92] Of course, for a small business without staff lawyers, the guideline is to be careful and follow good blogging etiquette.

After a topic has been established, the trick to getting readers to return is the writing style. The appropriate tone for a blog is tricky because if it is too dry or resembles a press piece, no one will read it. But if it is too informal, it might affect the organization's professional image. The true winners are the blogs that maintain this balance yet feel like a conversation. They are informal as if you were writing an email to a friend, but professional enough not to taint the corporate image. Debbie Weil, author of *The Corporate Blogging Book*, says she thinks of a blog as if she were sending an email to the world or, as she puts it, the "cc:world." She says much of her blog content comes from emails from readers, and that it is like writing a letter.[93]

One feature that makes blogs a more powerful tool than a website is the frequency of updates. So it would only make sense that a good writing practice for blogs is to maintain the frequency. There are no hard and fast rules, but a general guideline is that good corporate blogs are updated at least once a week. Often corporate bloggers give themselves deadlines so that every week on a certain day at a certain time they update the blog. This way readers know when to expect posts, and it adds to the conversational atmosphere. For busy business executives there is often an extended time between posts, which readers may understand, although they expect an explanation. Keeping readers informed of a break in posts shows respect to the readers.

Blogging Guidelines

In the exciting world of corporate blogging it is important to establish guidelines so the organization's employees know what is acceptable and what is not. How restrictive the policy is depends on the company. On both the company side and the employee side the threat of lawsuits posed by blogging is real. Everything from defamation to intellectual property infringement to trade secrets can be litigated, plus a whole lot more. In several cases employees have lost their jobs due to personal or company blog postings, so it is in everyone's best interest to abide by a common set of guidelines. Employment attorney Mara Levin has crafted blogging guidelines for several private companies. Here are some guidelines:

- Do not defame or discuss your colleagues and their behavior
- Do not write anything defamatory
- Do not write personal blogs on company time
- Identify your blog as a personal blog and state that the views are your own (e.g., include a disclaimer)
- Do not reveal confidential information
- Do not reveal trade secrets"[94]

Although it is important to have rules in place to protect companies and their employees, some would argue it is more effective to offer advice on how to blog. That is what IBM's blogging policy does. There is a difference between a policy and a code of ethics. A policy covers what a person can and cannot say for legal or company reasons; whereas, a code of ethics gives instructions on how to act in the blogosphere. The code of ethics works in tandem with blogging etiquette.[95]

hedonometer

A computerized sensor that surveys the web to measure the collective happiness of millions of bloggers.

In closing, here is a blog-related term to add to your already-extensive vocabulary. The term **hedonometer** refers to a computerized sensor that surveys the web to measure the collective happiness of millions of bloggers.

An example of IBM's corporate blogging policy is located at: https://www.ibm.com/blogs/zz/en/guidelines.html

Writing for Websites

website

A set of interconnected web pages starting with a homepage.

A **website** is a set of interconnected web pages that build from a homepage. Websites are commonplace in today's businesses and are used predominately for sharing information with and selling goods and services to customers.

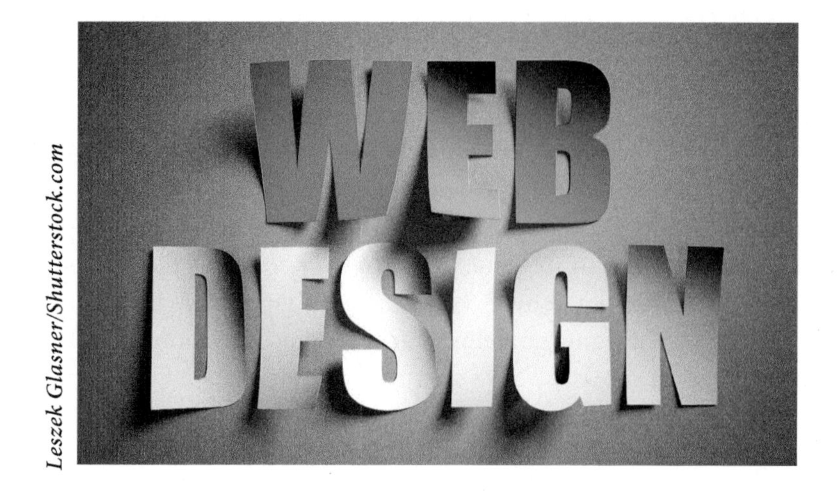

Leszek Glasner/Shutterstock.com

What follows are some suggestions to keep in mind when developing written materials for business websites.

Most People Who Browse the Internet Skim

This means they do not read content in its entirety. Nor are they big on scrolling. What does this mean to those who write web content? Write concise, skimmable text. Grab your readers' attention quickly. Make your main points on the first screen. If you don't, your reader will likely be off to one of the millions of other sites, just clicks away.

Write Concisely

Web readers do not want to wade through excess verbage, so write concisely. However, do not do so at the expense of clarity. Include the level of detail needed to achieve this goal also.

Write with Design in Mind

Include keywords, subheads, bulleted lists, and short paragraphs. Use clear, informational headlines. Then provide objective, supporting details. Experts say that your web page is too text heavy if you can place your open palm over a block of text on your website without touching a graphic image.

Write Comprehensively

While it seems like a contradiction, web writing also demands comprehensiveness. Web browsers will stay around if they like what they see.

Use Links to Interior Pages Where You Can Tell the Rest of Your Story

Use links to keep them hooked. However, keep links to a minimum and always provide a link back to your home page.

Avoid Puffed, Exaggerated "Marketese" and Stuffy, Bureaucratic Prose

Write in a conversational voice, do not talk down to web readers, and do not exaggerate your product or service.

Use the Web's Interactivity

Build your website so your readers can react to your site through email feedback or discussion boards, and use drawings, graphics, animation, audio, or video where they enhance your words.

Your website must serve the business community's needs. Knowing how to write effective website content will serve you well in this environment.

Writing Effectively at Social Network Sites

Social networking refers to online media tools that allow users to form communities, communicate with each other, and produce content. Online **social networking sites**, also referred to as social media sites, allow users to increase personal and professional networks and find others with whom they share interests. The sites help users connect with other people, and keep up communication with friends and associates with whom it may be otherwise hard to stay current. Boyd and Ellison define social network sites as "web-based services that allow individuals to (1) construct a public or semi-public profile within a bounded system, (2) articulate a list of other users with whom they share a connection, and (3) view and traverse their list of connections and those made by others within the system."[96]

social networking
Refers to online media tools that allow users to form communities, communicate with each other, and produce content.

social networking sites
Online media tools allowing users to form communities, communicate with each other, and produce content. Also called *social media sites.*

Social network sites are hosted on a number of popular websites including Facebook, LinkedIn, and Twitter. These sites usually require users to create a profile that features demographic information, interests, preferences, pictures, and updates about current events or happenings in that individual's life. The sites also provide tools for users to create and facilitate content. Users can upload pictures and videos, create blogs, review media, and share opinions and comments. Users can also group themselves depending on geographic location, interests, heritage, professional and social associations, or topics of interest. These groups make possible communication and management of subgroups within larger social networks.

White art/Shutterstock.com

Social Networking Sites

Facebook

facebook
The most popular social networking site.

Facebook is currently the largest social network site, with approximately 850 million users worldwide. Originally this site was only accessible to college students, but has since opened up so that anyone can create a free account. Users are loosely grouped by "networks" such as school affiliations or geographical locations. Users can join multiple networks and can also join or start other groups based on any criteria they chose. Facebook uses the "wall" onto which users can post comments or links to their friends. They also use "status updates" so users can post what they are doing or thinking and can also post media links.

Interestingly, a Pew Internet and American Life Project survey reported that several college students said they regularly shared their Facebook passwords with friends to force themselves to study for exams.[97] They instruct their friends to change their password, which temporarily locks them out of their Facebook account and they can study. Aside from the potential study benefits, for obvious reasons one would certainly want to know that the person to whom he or she turns his or her account is an honest, long-term friend.

MySpace

MySpace
A social networking site especially popular with musicians.

MySpace was at one time the largest social network site. It currently has approximately 33 million users. MySpace maintains a special niche for musicians. Bands create their own pages and upload music. Then they use these pages to keep fans informed of shows and information about the band.

LinkedIn

LinkedIn is a specialty social network site that caters to business professionals. Users are able to upload résumés and present a professional online presence rather than the social one portrayed on other social networking sites. LinkedIn has approximately 150 million users worldwide, nearly half of them in the United States.

LinkedIn
A specialty social networking site catering to business professionals.

Nancy Flynn, author of the *Social Media Handbook*, sums up social media etiquette as follows: Adhere to the rules of social media etiquette. Be polite, polished, and professional. Write, post, and publish content that is 100 percent appropriate, civil, and compliant.[98] Figure 6.5 contains additional social media etiquette advice.

Twitter

Twitter is one of the fastest-growing social network sites on the Internet, with approximately 300 million users worldwide sending 400 million tweets per day. Twitter differs from other social network sites in users do not set up an information page. Twitter is a series of miniblogs ("tweets"), similar to the Status Updates on Facebook. These tweets are usually one to two sentences and can contain links, to a maximum of 140 characters.

Twitter
An online social networking and microblogging service enabling users to send and read text-based messages of up to 140 characters.

- Fill out your online profiles completely with information about you and your business.
- Use a different profile or account for your personal connections.
- Create a section on your main profile detailing whom you want to befriend.
- Offer information of value.
- Don't approach strangers and ask them to be friends with you just so you can then try to sell them your products or services.
- Pick a screen name that represents you and your company well.
- Don't mail requests for birthdays, invitations to play games, or other time-wasters for those using the site.
- Don't put anything on the Internet that you don't want your future boss, current clients, or potential clients to read.
- Check out the people who want to follow you or be your friend.
- If someone does not want to be your friend, accept their decision gracefully.
- Never post when you are overly tired, jet lagged, intoxicated, angry, or upset. You do not want to post argumentative or inappropriate texts.
- Compose your posts, updates, or tweets in a word processing document so you can check grammar and spelling before you send them.

FIGURE 6.5. Social Media Etiquette

Source: Lydia Ramsey, Top 12 Rules of Social Media Etiquette. www.pewinternet.org/Reports/2012/Facebook-users/Summary.aspx

Users can subscribe to other users' tweets and receive their messages via their account or through text messages. With the rapid growth of Twitter, businesses have been discovering new and creative way to leverage Twitter to communicate with customers. To date, companies use Twitter predominately as a customer service and public relations tool.

Annette Shaff/Shutterstock.com

premature tweeting
Posting messages on Twitter before the facts are properly ascertained.

If you suffer from a bit of vocabulary deficit, add the term premature tweeting to your word bank. **Premature tweeting** refers to posting messages on Twitter before the facts are properly verified. Obviously, confirming your facts before tweeting is both professional and courteous. The same advice applies when posting entries on Facebook.

- **Immediacy**. Real-time flow of comments and adaptability to mobile handsets makes tweeting more immediate than blogging.
- **Brevity**. Limiting messages to 140 characters makes them easier to produce and digest.
- **Pull and Push**. The ability of users to choose whose tweets they follow makes it less random than email.
- **Searchability**. Messages can be searched, making the content more accessible than the comments on a social network.
- **Mixing the Public and the Personal**. A user's personal contacts are on equal footing with public figures' contacts.
- **Retweeting**. By copying and retransmitting messages, users can turn the network into a giant echo chamber.

FIGURE 6.6. Six Secrets to Online Success with Twitter

Source: Richard Waters, "Sweet to Tweet." Financial Times *(February 27, 2009).*

Social networking sites are useful to businesses in a variety of ways. One of the most well-researched uses is online advertising that targets customers.

Much like blogs, businesses can use external social networking sites, and they can create their own internal networks. Internal networks are useful for collaboration, problem solving, and building relationships within a company, without some of the legal issues that arise when employees use external network sites.

In growing numbers small businesses owners and operators are discovering how social networking sites can promote their businesses. While some small business owners write, blog, and post their own Twitter and Facebook entries, others do not have the time to do so. This situation has fueled a number of support businesses that specialize in providing advertising and public relations services to businesses for their social networking sites. Among these support businesses are 3 Green Angels, Everywhere LLC, Red Square Agency Inc., and ThinkInk LLC.[99]

"5 Benefits of Social Media Business Owners Need to Understand" @ http://www.inc.com/peter-roesler/5-benefits-of-social-media-business-owners-need-to-understand.html

Choosing the Right Communication Medium

Imagine you are a businessperson who wants to communicate with existing and/or potential customers or clients. Given the many electronic and non-electronic options available to you, how do you select the right medium? Do you choose email, IM, text messaging, tweeting, website, or blogging? Do you place your message on a social network site such as LinkedIn or Facebook? Do you choose to place a phone call or have a face-to-face meeting? Or, do you saturate the market, so to speak, by transmitting your message over all available media options, in the hopes that current and potential customers or clients will spot you? With so many choices, making the right media choice can certainly be overwhelming!

Jim Blasingame, one of the world's foremost experts on small business and entrepreneurship, offers some practical advice to help you make good media choices. He encourages businesspeople to use the following two approaches when making communication media choices: (1) ask yourself which communication medium best suits the circumstance and (2) ask your customers and clients which communication medium they prefer.[100] Regarding Blasingame's second approach, remind yourself that not all of your clients or customers will share your interest in and your ability to use each electronic communication tool that you are comfortable using. Nor do they all have access to the technology you prefer to use. For example, most people around the globe lack access to the Internet by circumstance or choice, including many in the United States. Blasingame's advice mirrors some of the basic communication suggestions presented earlier in the book. In chapter 1 you were urged to reflect on the circumstances before selecting a medium. For example, messages about routine matters can typically be communicated effectively via email, while messages involving important and/or controversial matters are best communicated face-to-face. Furthermore, you learned earlier that the better you know our communication partner, the more likely you are to achieve effective communication.

Summary: Section 3— Business Blogs, Websites, Social Network Sites, and Choosing the Right Communication Medium

- The main type of business blog is the corporate blog.
- Web content that is readable, accurate, and user friendly is what keeps users coming back to a website.
- Web content planning demands that you analyze your audience, develop a purpose and objectives, and organize the site.
- Navigational aids such as links help users find information quickly on your site and show where they are in the site's structure.
- Test your content for authority, accuracy, objectivity, and currency.
- To name links, use specific, informative words or phrases.
- When drafting web content, be sure that it is clear, concise, and credible.
- Elements that enhance scanning are headings, bolding text, bulleted lists, graphics, captions, and clear topic sentences.
- When editing and proofreading web content, check for consistency of style and usage. After you run the spellchecker, proofread the text again for spelling and other errors.
- Know that social networking sites have several business applications, and sites are growing in popularity within organizations.
- When making a communication media choice, be it electronic or otherwise, consider which one best suits the situation and which one suits your communication partner prefers.

End Notes

[1] Nancy Flynn, *The ePolicy Handbook* (New York: American Management Association, 2001), 1.

[2] Ibid., 49.

[3] Samantha Miller, *Email Etiquette* (New York: Warner Books, 2001), 93.

[4] Flynn, 51.

[5] Ibid, 49.

[6] Ibid, 50.

[7] Dianna Booher, *E-Writing: 21st Century Tools for Effective Communication* (New York: Pocket Books, 2001), 1.

[8] Ibid., 2.

[9] Ibid, 12.

[10] Miller, 103.

[11] Ibid., 105.

[12] Ibid., 103.

[13] Ibid., 102.

[14] Ibid.

[15] Flynn, 101.

[16] Booher, 13.

[17] Flynn, 89.

[18] Ibid.

[19] Miller, 33.

[20] Miller, 22; Booher, 33.

[21] Miller, 31; Booher, 36.

[22] Booher, 41.

[23] Flynn, 3.

[24] Ibid., 4.

[25] Ibid., 7.

[26] Ibid.

[27] Ibid.

[28] Ibid., 50.

[29] Ibid.

[30] Ibid., 9.

[31] Ibid., 81.

[32] Ibid.

[33] Ibid., 34.

[34] Michael R. Overly, *E-policy: how to develop computer, E-policy, and Internet guidelines to protect your company and its assets* (New York: American Management Association, 1999), 27.

[35] Alan Cohen, "Worker Watchers Want to Know What Your Employees Are Doing Online," *Fortune* (June 1, 2001): 70–81, http://infoweb7.newsbank.com.

[36] Valli Baldassano, "Bad Documents Can Kill You," *Across the Board* 38 (September/October 2001): 46–51.

[37] Flynn, 54.

[38] Ibid., 52.

[39] "Some Wall Street Firms Did Not Retain Required Emails–NYT," (May 7, 2002), http://www.reuters.com.

[40] Miller, 94–95.

[41] Ibid., 95.

[42] Gregory A. Maciag, "Email Might Be the Killer Application, but Poorly Managed, It Could Bury You," *National Underwriter* 106 (March 18, 2002): 33–35.

[43] Booher, 9.

[44] David Angell and Brent Heslop, *The Elements of Email Style* (New York: Addison-Wesley, 2000), 18–19.

[45] Ibid., 19–20.

[46] Ibid., 19.

[47] "Promote Corporate Identity Through Email," *Business Forms, Labels & Systems* 40 (March 20, 2002): 18.

[48] Angell and Heslop, 21–22.

[49] Ibid., 20.

[50] Ibid., 24.

[51] Ibid., 15–32.

[52] Ibid., 117.

[53] Flynn, 146.

[54] Dawn Rosenberg McKay, "Email Etiquette," *Online Netiquette Uncensored: Courtesy #3*, http://www.onlinenetiquette.com.

[55] David Stauffer, "Can I Apologize by Email?" *Harvard Management Communication Newsletter* (November 1999): 3.

[56] Ibid., 4.

[57] Nancy Flynn, *The Social Media Handbook: Policies and Best Practices to Effectively Manage Your Organization's Social Media Presence, Posts, and Potential Risks* (Pfeiffer, 2012), 327.

[58] Flynn, 97.

[59] Ibid.

[60] David Spohn, "Email Rules of Engagement" (May 18, 2002), http://webworst.about.com.

[61] Miller, 43.

[62] Ibid.

[63] Ibid., 51.

[64] Brian Sullivan, "Netiquette," *Computerworld* 36 (March 4, 2002): 48.

[65] Flynn, 99.

[66] Ibid., 100.

[67] Ibid., 99.

[68] Ibid.

[69] Ibid., 107.

[70] Peter Post and Peggy Post. *The Etiquette Advantage in Business* (New York: Harper Resource, 1999): 64-65.

[71] Danny Bradbury, "A Pressing Message," *Communications International* (January 2002): 40–43; Matthew Schwartz, "The Instant Messaging Debate," *Computerworld* 36 (January 7, 2002): 40–41.

[72] Mandy Andress, "Instant Messaging," *InfoWorld* 24 (January7, 2002): 36.

[73] Suzanne Gaspar, "RUOK w IM?" *Network World* 19 (February 25, 2002): 40.

[74] Matt Cain, "META Report: The Future of Instant Messaging," *Instant Messaging Planet* (April 29, 2002).

[75] Ibid.

[76] Cain.

[77] Andress, 36.

[78] Brad Grimes, "Peer-to-Peer Gets Down to Business," *PC World* 19 (May 2001): 150–154.

[79] David LaGesse, "Instant Message Phenom Is, Like, Way Beyond Email," *U.S. News & World Report* 130 (March 5, 2001): 54–56.

[80] Laura Schneider, "Is There Room in Your Cubicle for Instant Messaging?" *What You Need to Know About,* http://chatting.about.com.

[81] Jim Sterne, "People Who Need People," *Inc.* 22 (2000): 131–132.

[82] Michael Hayes, "What We Sell Is Between Our Ears," *Journal of Accountancy* 191 (June 2001): 57–63.

[83] Ibid.

[84] Bradbury, 41.

[85] Lagesse, 56.

[86] Laura Schneider, "Instant Messaging: Annoyance or Necessity?" *What You Need to Know About,* http://netconf...about.com.

[87] Brian Sullivan, "Netiquette," *Computerworld* 36 (March 4, 2002): 48.

[88] Jeff Zbar, "The Basics of Instant Messaging Etiquette," *Net.Worker News* at www.nwfusion.com accessed on May 13, 2002.

[89] Ibid.

[90] "Netiquette Guidelines," http://www.dtcc.edu.

[91] Chris Anderson, *The Long Tail: Why the Future of Business Is Selling Less of More* (New York: Hyperion, 2006).

[92] Nancy Flynn, *Blog Rules: A Business Guide to Managing Policy, Public Relations, & Legal Issues* (New York: American Management Association, 2006).

[93] Weil, 105.

[94] Ibid., 43–44.

[95] Ibid., 45.

[96] Danah Boyd and Nicole B. Ellison, "Social Network Sites: Definition, History, and Scholarship," *Journal of Computer-Mediated Communication*. 13 (2008): 210–230.

[97] www.pewinternet.org/Reports/2012/Facebook-users/Summary.aspx

[98] Flynn, 327.

[99] "Social Networking Sites Account for More than 20 Percent of All U.S. Online Display Ad Impressions" (September 1, 2009), http;//www.reuters.com.

[100] Ibid.

[101] Sarah E. Needleman, "Firms Get a Hand With Twitter, Facebook," *The Wall Street Journal*, October 1, 2009.

[102] Jim Blasingame, "High Tech or High Touch: It's Really Not Complicated," *The Wall Street Journal*, October 6, 2009.

CHAPTER 7

COLLABORATION

Teams improve communication by sharing information within an organization. Because of their versatility teams can produce large amounts of work, including written documents, that individuals would have difficulty accomplishing on their own. Teams can increase productivity when writing collaboratively by brainstorming ideas, dividing responsibilities, working together to gather information, and editing each other's work. Effective teams can solve problems creatively and efficiently. The better equipped you are to analyze what is happening in and around your team, the more successful and satisfying your group writing project will be.

Developing Team Skills

During the first stage of team development members of the group need to get to know each other. Trading contact information and discussing the skills that each member feels comfortable contributing to the group, as well as being honest about fears and weaknesses, will help individual members know what to expect from each other.

Conflict

As teams develop they should expect conflict to arise. How a team manages conflict determines the quality of its performance. Teams may experience two kinds of conflict: cognitive conflict and affective conflict. Cognitive conflict revolves around project-related issues and is a necessary and healthy aspect of a well-functioning group. Creative thinking and stimulating discussion should be the result of cognitive conflict. Affective conflict centers around feelings and personalities rather than issues—on people rather than the matter at hand. Affective conflict tends to be emotional and may erupt into personal criticism, which is destructive to trust, a necessary component for the team's success. *Groupthink*, a term coined by Irving Janis, may be equally destructive. Groupthink occurs when team members are overly eager to agree with each other and fall victim to faulty decision making processes. Teams suffering from groupthink may fail to examine all alternatives, self-censor thoughts and ideas, and refuse to question the decisions of a strong leader. Effective teams encourage open discussion, focus discussion on issues, and evaluate many alternatives carefully.

Communication

Although the end goal of the group project may be a written document or documents, clear verbal communication is also necessary. The best teams speak clearly, contribute ideas freely, and encourage feedback from one another. Constructive disagreement is encouraged. Teams whose members enjoy working together take their tasks seriously but are also able to laugh at themselves and inject humor into their interactions. Two of the most important tools for effective communication are the abilities to listen and to read body language.

Even though communication in groups is often informal, conscious efforts to listen will increase the team's ability to work well together. To increase listening skills:

- Stop talking and focus internally on the speaker (block out competing thoughts)
- Maintain an open mind
- Don't interrupt; lean forward and maintain eye contact with the speaker
- Don't fidget or try to complete other tasks while listening
- Ask clarifying questions
- Rephrase and summarize what the speaker has said in your own words; ask the speaker if your summary is a correct interpretation of what he/she intended to communicate
- Listen between the lines (observe nonverbal cues)

Nonverbal communication, or body language, often tells the listener more than the speaker's words. Just as you use nonverbal language to communicate your interest in someone else's words (by leaning forward, making eye contact), it is important to pay attention to the signals others are giving you. Leaning forward, opening hands, sitting on the edge of the chair, placing hands behind one's back, and sitting in a relaxed position show cooperation and feelings of confidence, while fidgeting, arm crossing, eye rubbing, and throat clearing may indicate defensiveness or nervousness. Hunching over, pen chewing, hand wringing, and neck rubbing are signs of insecurity or frustration. However, as people are individuals, they each may have their own ways of communicating different emotions. The goal is to pay attention to the clues that the speaker is giving, nonverbally, and if you detect cues that seem to contradict the speaker's words, politely seek additional clues by asking questions:

- "Please tell me more about..."
- "Do you mean that..."
- "I'm not sure I understand..."

Asking questions of this nature reassures the speaker that you are interested in understanding his/her real feelings and message, which will go a long way toward establishing trust and open communication.

Compromise

When team members cannot agree, negotiation and compromise may be necessary. It is important to remember, however, that when a successful negotiation occurs both parties win. During negotiation listen carefully, be empathetic, explain your position logically and clearly, avoid expressing hostility or engaging in personal attacks, be willing to compromise and expect compromise from others, and be willing to follow through with whatever you have agreed on.

Team Leadership and Roles

Unless your team has been assigned a leader, you will have to organize your group on your own. Effective teams utilize the idea of roles within the group to maintain order and produce work efficiently. Sometimes a leader emerges naturally and members of the group volunteer for different positions. Positions that are important in all teams, whether the end product is a written or oral report, include:

- Leader (to plan and conduct meetings)
- Recorder (to keep a record of group decisions)
- Evaluator (to determine whether the group is on target and is meeting its goals)

Teams also must decide whether they will be governed by consensus, where everyone must agree, or by majority rule.

The role of the leader should be to facilitate meetings—not to do all the talking or make all the decisions. The purpose of meetings is to exchange views, and the leader's role is to encourage all members to give their opinions, ideas, and responses. The leader can also encourage the group to avoid digressions and to adhere to a schedule. Finally, when the group seems to have reached a decision, the leader can summarize the group's position and check to see whether everyone understands and agrees. No one should leave a meeting without full understanding of what was accomplished. An effective leader may ask each person to recap briefly the content of the meeting and decisions that have been reached in order to clarify that each person understands.

An important item to discuss when a team first begins to meet is how to deal with group members who fail to fulfill their duties or pull their share of the load. The leader of the team, along with the evaluator, can work together to keep members of the group on task. Including a brief report (along with the final written report) which discusses how each member of the group contributed to the overall project will be incentive for individual members to fulfill their assigned responsibilities. Or, a chart may be used in which each member of the team rates the other members from 1 to 5 on how they performed in the following areas:

- Contributed his or her fair share to all phases of the project
- Participated actively in meetings
- Was dependable, prompt, and courteous as a group member
- Overall rating for this person's contributions to the group

Organizing and Managing the Project

The most effective way to organize a project is to make a project plan. The team begins by discussing how to break the project into parts and assign responsibilities to each member, how to collect the most accurate and useful information, and who will be responsible for gathering what information. Once the project has been divided into discrete parts, the team establishes deadlines for completing tasks and for writing and editing the document.

The complete project plan should involve a list of the **tasks** that must be accomplished, **who** will be responsible for completing them, and a **schedule** or timeline for the process, including intermediate and final deadlines. A good place to begin is to create a PERT or Gantt chart for the project. PERT charts, short for Program Evaluation and Review Technique, map out various paths of work for the project and include discrete project phase start dates and end dates.

PERT chart

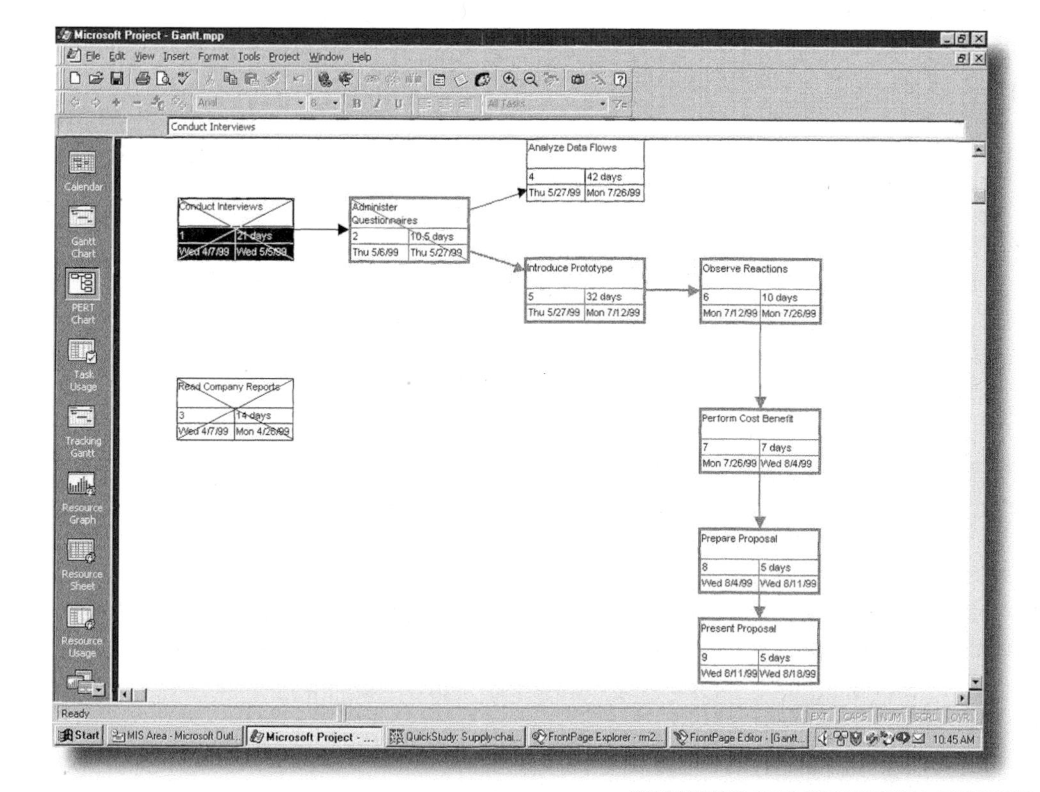

Gantt charts show various overlapping tasks in a project as well as indicating the duration of each task. Using a Gantt chart enables team members to see where they are in the process, who has successfully completed tasks, how much time is left to complete those tasks remaining, and where adjustments need to be made in time allotments.

Gantt chart

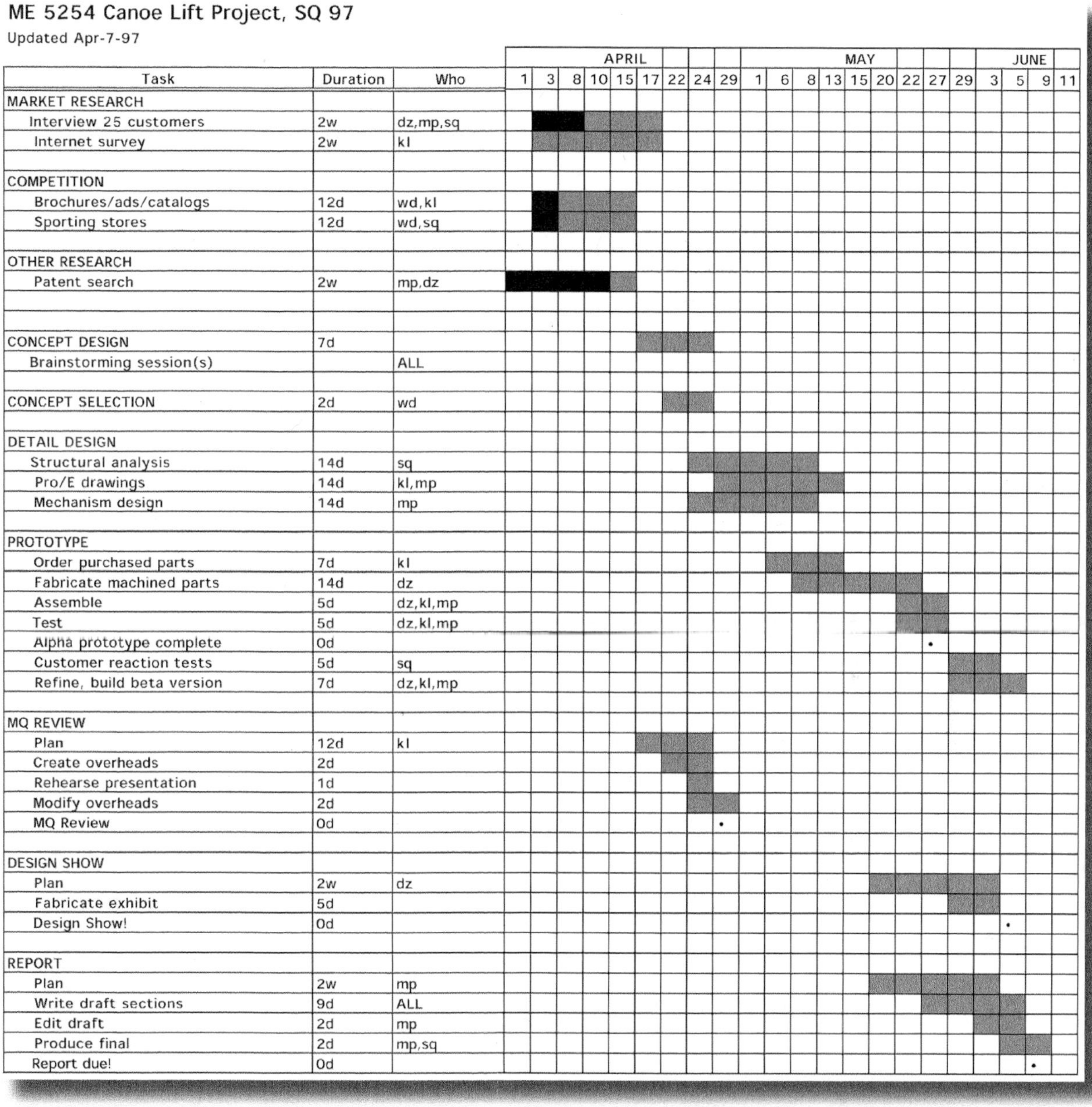

ME 5254 Canoe Lift Project, SQ 97

Updated Apr-7-97

APRIL: 1, 3, 8, 10, 15, 17, 22, 24, 29 — MAY: 1, 6, 8, 13, 15, 20, 22, 27, 29 — JUNE: 3, 5, 9, 11

Task	Duration	Who
MARKET RESEARCH		
Interview 25 customers	2w	dz,mp,sq
Internet survey	2w	kl
COMPETITION		
Brochures/ads/catalogs	12d	wd,kl
Sporting stores	12d	wd,sq
OTHER RESEARCH		
Patent search	2w	mp,dz
CONCEPT DESIGN	7d	
Brainstorming session(s)		ALL
CONCEPT SELECTION	2d	wd
DETAIL DESIGN		
Structural analysis	14d	sq
Pro/E drawings	14d	kl,mp
Mechanism design	14d	mp
PROTOTYPE		
Order purchased parts	7d	kl
Fabricate machined parts	14d	dz
Assemble	5d	dz,kl,mp
Test	5d	dz,kl,mp
Alpha prototype complete	0d	
Customer reaction tests	5d	sq
Refine, build beta version	7d	dz,kl,mp
MQ REVIEW		
Plan	12d	kl
Create overheads	2d	
Rehearse presentation	1d	
Modify overheads	2d	
MQ Review	0d	
DESIGN SHOW		
Plan	2w	dz
Fabricate exhibit	5d	
Design Show!	0d	
REPORT		
Plan	2w	mp
Write draft sections	9d	ALL
Edit draft	2d	mp
Produce final	2d	mp,sq
Report due!	0d	

Projects don't always go exactly as they are planned; however, a project plan enables all members to see how things are progressing and to make changes when necessary. Having a schedule and deadlines for a project is essential to a team's ability to complete the project successfully.

Document Production

The advantages of teamwork can be easily sabotaged without careful consideration about how to maximize the range of ideas and strengths that members bring to a group project. After developing team skills and planning the broad framework of your project the labor intensive tasks of drafting, revising, and editing present new challenges.

While the principle of writing collaboratively may seem simple, many find that the actual process of collaborating is contrary to the majority of the educational experience, which requires one to function as an individual author. To reap the benefits of teamwork it is vital that your group consider exactly *how* it will write together.

Plan Early

Half of the battle of creating a successful group project is a commitment to leave nothing to chance. One way to ensure members are willing to make such a commitment is to involve *all* group members in the initial planning, regardless of the role they will play in the project (leader, recorder, evaluator, etc.).

In addition to creating a project plan sets up a schedule for key deadlines and the division of project tasks, it is important to establish frequent group meetings to serve as checkpoints throughout the project. One of the most frustrating pitfalls of writing collaboratively is for one group member to invest a substantial amount of time and energy on a project task that is not essential or becomes obsolete. Frequent checkpoints allow the group to continually assess its progress and refine or alter goals if necessary.

Three helpful strategies beyond project scheduling for maximizing your group's document production time are:

1. Use a style guide
2. Create a storyboard
3. Use technology

Use a Style Guide

One of the most difficult challenges in any collaborative writing project is maintaining consistency throughout the document, especially when parts written by various team members are brought together into a master document. Groups can reduce their workload later in the project by using a style guide to help ensure individual writers are using a uniform style from the beginning of the drafting process.

Style guides dictate document design aspects such as the formatting of page layout, headings, body text, and graphics. Style guides also set up guidelines on mechanics such as what kind of abbreviation to use, whether to spell out numbers, what rules for comma usage to follow, and how to treat nonstandard spellings such as "email" vs. "email" or "web site" vs. "website."

General style guides such as the *Chicago Manual of Style*, the *CBE Guide* (for writers in the biological sciences), or the *APA Style Manual* (for writers in the social sciences) are all good choices. However, your group may want to sit down at an early meeting and create its own particular style guide. Certainly it will be helpful to make early decisions about how to handle the jargon or industry specific language that may surround your project topic.

Create a Storyboard

Based on the way Hollywood screenwriters depict a movie's story prior to filming, storyboarding as a practice of document production enables team writers to draft separate portions of a document so that they fit together seamlessly and require a minimum amount of revision. The storyboarding technique works like this: after creating a comprehensive document outline in the planning phase of the project, the group assigns sections to individual members. Before the next meeting each member creates a storyboard for his/her section(s). Typical storyboards address the following:

1. Summarize the particular document section in one complete sentence.
2. List the key points that will be developed in the document section.
3. Note if a visual should be included, and, if so, the purpose, type, and title of the graphic aid.

When the group meets again, members put all the separate storyboards together. Groups often tape their storyboards on a wall or make photocopies of all the separate storyboards prior to the meeting. Bringing the storyboards together enables the group to form a comprehensive vision of the final document and to see how each section works together. This process often reveals overlapping information, inconsistencies, and omissions, but most importantly, after team revision, your group's storyboard serves as a detailed guide for drafting!

Use Technology

Because collaborative writing projects involve individual work as well as group work, it is vital that groups communicate regularly. Technology offers business writers effective and easy ways to enhance collaborative writing tasks.

- **Email:** Email is a perfect vehicle for keeping all group members abreast of progress, planning or changing meetings, asking each other questions, transmitting drafts back and forth, etc. Although email is by nature a more casual form of communication, it is important that all group members maintain their professionalism.

- **Word Processing:** Microsoft Word offers several tools for drafting and revising documents. The most useful collaborative tool is the **comment** function. Selecting the **View>Toolbars>Reviewing** tool enables writers and reviewers to add comments to a document—small post-it style notes or editor's remarks. To do this highlight the text you wish to comment on, select **Insert>Comment** (there is also a yellow, post-it-like "comment" button), and then type your comment (see Figure 7.1). The comments appear in different formats depending on what version of MS Word you're using.

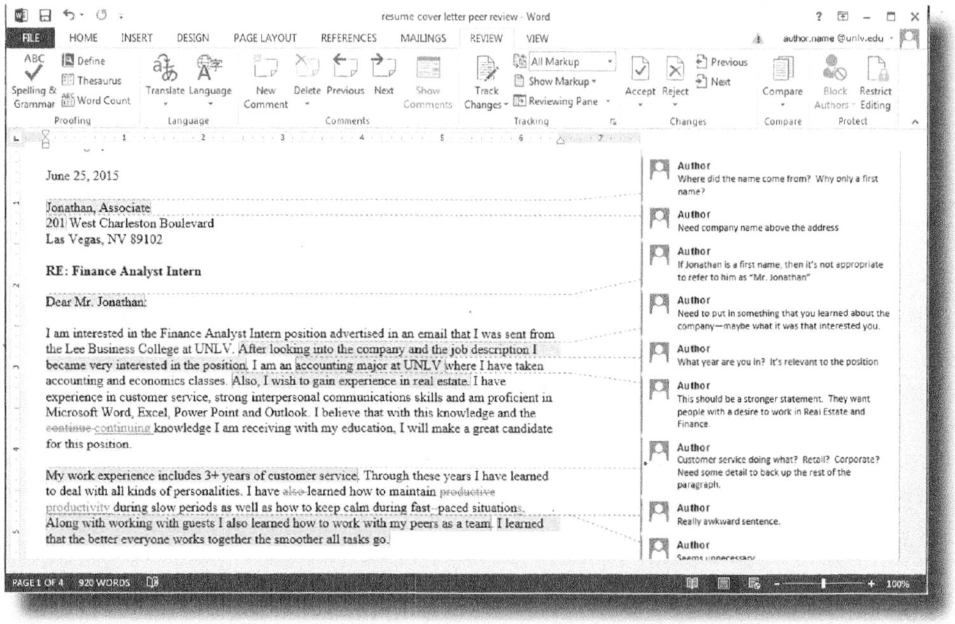

FIGURE 7.1. The MS Word Review tools make it easy to collaborate via email

Other collaboration aids include the **track changes** tool, which creates a record of all changes made to a document. While this feature is a bit cumbersome for basic student group writing tasks, some organizations require it. The **highlighter** tool can be used to mark typos or color-code types of mistakes.

Don't overlook basic file-management issues when working collaboratively. Files should be named in such a way as to be easily recognizable and prevent accidental overwrites. Successive versions are usually numbered (e.g., report_draft_1, report_draft_2, etc.). Everyone should keep backup files of all document versions in case any one person should lose a file by accident or computer malfunction.

- **Google Drive and Docs:** Google Drive is a free web-based file storage space available to users with Google accounts. Google Docs is a web-based word processing application that allows users to collaborate via the Internet without using MS Word or email (the Google Apps for Work tools also includes Sheets, for spreadsheets, and Slides, for presentations). The comment and editing tools are very similar to MS Word. One advantage of Google Docs is that it offers real time collaboration tools where one version of the document is continuously updated as various users make changes, which reduces errors in file version management. Changes in traditional MS Word are based on the separate versions of the document shared among users via email or other means, and sometimes individuals working in teams can mistakenly use or submit older versions of a document (Microsoft now offers a similar real time online collaboration service with Office 365). MS Office applications tend to have more features, but Google Apps for Work work well for planning and prewriting tasks and most basic documents.

For online tutorials on how to use **Google Drive and Docs**, see:

- "Google Drive Tutorial 2014 - Quick Start" by Anson Alexander at https://www.youtube.com/watch?v=i125gM-UAoY
- "Google Drive and Docs: Sharing and Collaborating" by GCFLearnFree .org at https://www.youtube.com/embed/Dsoa9skxVuk
- "Collaborative Editing | Docs, Sheets, Slides | The Apps Show" by Google for Work at https://www.youtube.com/watch?v=ekn-jXNQcqc

- **Computer Conferencing:** Business meetings are increasingly held online using web conferencing software, also called groupware. Computer conferencing requires members to use applications such as Skype for Business, Webex.com, or GoToMeeting.com and depending on the service can include videoconferencing as well as desktop sharing, whiteboard, and chat tools. This method of communication is particularly helpful to groups whose members are working in different cities or countries. Groupware enables members to work almost as if they were sitting in the same room because team members can share documents, video, and other collaboration tools in real time.

For a side-by-side comparison of **web conferencing** services, see "The Best Videoconferencing Services of 2015" by Molly McLaughlin at http://www.pcmag.com/article2/0,2817,2388678,00.asp

Revision and Editing Stages

By the revision and editing phases of a collaborative project all group members have invested considerable effort into drafting their assigned sections of the document. Group members may feel a certain sense of ownership to their individual work that can create a

barrier for effective revision. Paul Anderson, author of *Technical Communication*, offers two strategies for effective team writing:

1. **Be considerate when discussing drafts:** Acknowledge that group members have invested individual skill and creativity in producing their portion of the draft. One way to show consideration is to pose your suggestion as an option. You could say, "Here's another way you could communicate your point..." Another strategy is to focus on the positive reasons for choosing one option over another. As always, coupling suggestions for change with praise for strong features of a draft usually helps the writer member realize that the suggestions are not criticisms of that person's overall writing ability but specific instances where the writing is less effective. It is also important when your portion is up for critique that you set aside your ego and present yourself as open to constructive criticism. Oftentimes group members will fail to make important revision suggestions for fear of hurting a group member's feelings. In this situation no one wins.
2. **Treat drafts as team property:** In order to revise fully and adequately, group members must (1) let go of personal ownership in their portions of the draft and (2) take ownership of parts written by others. The document is the product of the entire team, thus everyone is responsible for ensuring that each section is effective and coherently fits with the rest of the document. This does not mean that individual group members should passively yield to any and all revision suggestions, but that all group members should engage in a reasoned discussion about the best way to write all portions of the *team's* communication. One way to decrease personal ownership is to alternate responsibility for portions of the document in subsequent drafts. For example, the team member who was responsible for writing the background section of the first draft would be responsible for the recommendations section in the second draft.

Revise and Edit Thoroughly

Groups are often lured away from effective revision and editing practices by the seduction of a completed first draft. Don't be duped. All effective, professional communication needs to undergo a thorough revision process. Ideally your group should solicit feedback after the first draft has been completed from (1) your instructor or supervisor, (2) peers or coworkers, and (3) your internal team members. Carefully consider the feedback and revision suggestions your group receives and decide how to revise together.

Revision applies to global considerations related to the rhetorical context of the document. Before worrying about checking for surface issues like grammar, punctuation, and spelling, the team should focus on big-picture document issues such as focus, key points, organization, and content. The group should check:

- **Plans vs. draft:** Does each section adhere to plans established in initial outlines and storyboards? Was anything unintentionally left out? Did new content work its way in, and, if so, is the new content relevant and useful?

- **Reader's point of view:** Does the document answer all the major questions readers would have such as purpose of the document, problem being addressed, research conducted to address the problem, presentation and interpretation of research findings, and cogency of conclusions and recommendations being made? Does information provided, claims made, or actions requested take into account the needs, values, and background of the reader?
- **Flow:** Does each section flow coherently into the next in a way that readers can easily follow? Clear headings, introductions/openings, transitional phrases, and topic sentences all serve as cues that help readers comprehend the structure of your document.

Editing, a microscopic investigation of the surface features your document should happen after all major revisions are complete. As the last stage in your collaborative project before submitting a final draft, editing is the way to catch seemingly minor errors that have escaped everyone's attention up until this point. Professional communication professor James Porter advises that editing should be done by more than one person in more than one setting. He suggests that editing tasks should be divided among group members in the following manner:

- **Document design**—Are there consistent margins, pagination, position of text, spacing, indentation, column alignment, layout of pages?
- **Names, titles, formalities**—Are all important names and nouns correctly spelled, capitalized, and with proper titles acknowledged?
- **Numbers**—Are numbers consistently written, using either words or numerals? Are numerical totals, data, and arithmetic correct?
- **Grammar, word usage, punctuation, and spelling**—Are all periods, commas, semi-colons, colons, and apostrophes in the right place? Has the computerized spell-checker missed any usage mistakes (e.g., too, two, to) or other mistakes (e.g. a singular word spelled correctly but that should be plural).
- **Visual aids**—Are all tables and figures consistently designed and labeled?
- **Macro view**—Are all parts of the document together and in order?

After all the hard work and time you have invested in creating the document, in the editing phase make certain that every aspect of the document is representative of the group's commitment to excellence. The beauty and perhaps the pressure of collaborative writing is that there is always more than your own individual professionalism on the line.

CHAPTER

8

REPORTS

The following provides an overview of reports as a type of professional writing. It includes a discussion of report organization and design, including page layout, style sheets, and graphics.

Reports as a Type of Professional Writing

All reports address a particular need by conveying in-depth information, often aimed at persuading readers to act on that information. Reports are written for many purposes, including:

- **Proposals.** Proposals written by private companies for government contracts can be hundreds of pages long, aimed at persuading the government agency the company is capable of providing the desired service in a cost-effective manner. Such proposals include detailed discussion of the company's qualifications, feasibility of the work, necessary methods and materials, production schedules, and budgets.
- **Business plans.** A type of proposal, business plans document the scope of a business venture and persuade investors of its feasibility. Such plans include a description of the proposed product or service, technical explanations of technologies used, market analyses, production requirements, facilities and personnel needed, projected revenues, funding requirements, legal issues, qualifications of start-up personnel, and investment potentials.
- **Feasibility and recommendation reports.** These reports provide in-depth analysis helping readers to make informed decisions. Feasibility reports consider

if a proposed action is practically, economically, and technologically possible. Recommendation reports compare two or more options against similar criteria and advocate for a particular option.

- **Primary research reports.** These reports detail the results and implications of research conducted in experiments, surveys, and other field work. Research and development companies produce primary research reports as the first step in considering practical applications of advances in science and technology.
- **Technical-background reports.** Often called *white papers*, these reports provide background on a given topic that readers need to understand better, such as new technologies, market projections, or other business trends. Technical-background reports typically argue a particular position or solution, but in an objective, fact-based manner.
- **Usability reports.** A combination of primary research and recommendation reports, usability reports detail the results of user testing conducted to ascertain the user-friendliness of a prototype or working website.

A key feature of reports is that they are written for different kinds of readers and **different types of reading**. Studies show, for example, that executive decision makers spend most of their time reading the executive summary, the detailed synopsis appearing at the beginning of the report. Decision makers may not ever read the entire report. However, some may decide to look more closely at parts of the report. For example, a manager may want a closer look at the recommendations, so after reading this executive summary she will scan the table of contents and then skip to the recommendations section. Others read the whole report more closely. Someone might be charged with carefully studying the main points of the report and formulating an opinion about whether to follow the report's recommended action. In this case, the person may read more closely from front to back, but he or she may also decide to skip from section to section. In fact, most readers do not read reports linearly, or from front to back. Instead, they skip around from the opening to the conclusion to the appendices, etc. Many of the **purposefully redundant** elements are designed to facilitate these different ways of reading complex persuasive documents (e.g., the report is summarized more or less in several places: the transmittal letter, the executive summary, the opening, and conclusion). Each report section also includes clear summaries that inform the reader of the purpose of that particular section.

Each section of the report contributes to its overall persuasiveness. A report's background persuades the reader that there is indeed a problem or need that calls for action. The report methodology establishes that the writers have carefully thought about how best to study that problem. The findings and recommendation sections aim to convince readers that the writers carefully collected and analyzed information in light of the problem or need being addressed. Appendices help document careful research and provide details that ordinarily would disrupt the flow of the main report discussion.

You can review the following links to get a sense of the variation among report writing practices:

- California Energy Commission Reports
  http://www.energy.ca.gov/research/new_reports.html
- OneStart Portal Usability Report
  http://www.indiana.edu/~usable/reports/test3_report.pdf
- The National Commission on Writing's report "Writing: A Ticket to Work ... Or a Ticket Out: A Survey of Business Leaders" (2004) http://www.college-board.com/prod_downloads/writingcom/writing-ticket-to-work.pdf

Report Format

Reports vary in length and format. Most reports are the result of considerable investment of time and resources, so careful attention is paid to dressing up the report in the trappings of commercial publishing. That is, the reports will include elements that one typically finds in published books, such as a cover, table of contents, and pleasing page design. Some reports, however, are distributed in memo format and therefore dispense with more formal elements like a cover. Except for memo (or email) reports, most reports contain the following elements:

Front Matter	Transmittal Letter Cover (or Title Page) Executive Abstract Table of Contents List of Figures
Body	Overview/Introduction Background Recommendations Method Findings Closing
Back Matter	References Appendices

Front Matter

Front matter includes the transmittal letter, cover, executive summary, table of contents, and list of figures.

- **Transmittal Letter.** The letter of transmittal is a cover letter that accompanies the delivery of documents to external audiences. The letter should follow standard letter format and be limited to one page. The letter should state the purpose of the report, highlight pertinent recommendations, and briefly describe research methods. Include a courteous closing that expresses willingness to work further with the reader and offers to answer any questions. This letter can either be bound with the report (inside the cover) or delivered with the report as a separate document.
- **Cover.** The cover (or title page) includes the title of the report, date, names of the people writing the report, and name of the recipient. It might include the address of the client company. If it is a report for school, the cover could also include your instructor's name and course information. To add to the report's overall persuasiveness some attention should be paid to the cover's visual appearance. You can find sample report cover and page designs on the Internet.
- **Executive Summary.** After reading the executive summary, the reader should know the gist of the whole report, including background, methods, findings, and recommendations. This condensed version of the final report summary is for decision-making audiences who lack the time to read the entire report closely. It should be written in lay terms, designed for managers who may not have the technical expertise of other readers. Because the reader is most interested in what your final recommendations are, be sure to emphasize this part of your executive summary with subheadings and lists. You may wish to omit a summary of the findings, mentioning relevant findings only as you present each recommendation in some detail. The executive summary should not exceed 2 pages and can be placed either in the front matter (before the table of contents, paginated with lower-case Roman numerals) or in the body of the report (as page 1).
- **Table of Contents.** The table of contents provides readers with an overview of the organization of the report and a way to find the information that they want quickly and efficiently. List all major headings and all subheadings. Use double spacing, indentation to indicate subheadings, and leader dots. Technical reports often include a decimal numbering system for easy reference (e.g., 1.0, 1.1, 1.1.1, 1.1.2, etc.).
- **List of Figures.** The list of figures is a separate page after the table of contents that lists the titles of all visuals with page references. Tables are listed separately from figures. Consult models and style guides for specific ways to format the table of contents and list of figures.

Managerial Organization

Unlike traditional research report organization which is arranged in linear or chronological order, the managerial style organization arranges the more important recommendations before the research methods and findings sections.

Body

The body, or main part of the report includes the following sections: introduction, background, recommendation, methods, findings, and closing. This arrangement, which presents the recommendations early in the report, is known as **managerial organization** because it emphasizes suggestions for action over methods and data. The managerial organization is recommended for most reports other than primary research reports, which should follow the traditional scientific introduction, methods, results, discussion arrangement.

- **Introduction.** Review the gist of the report, focusing on the statement of the problem, pertinent background/history of the project, results, and recommendations. Include a summary of the report's major sections.
- **Background.** You should include the background information from your Design Plan, including a description of the report's purposes and audiences, your preliminary evaluation, and testing goals. The background section should be formatted with a first-level (L1) heading, while the preliminary evaluation and testing goals sections should be subsections formatted with second-level (L2) headings. In usability reports, you may want to add a screenshot of your site's home page in the background section.
- **Recommendations.** Make several recommendations to improve the issue based on the trends you noted in the data. Be sure to support your recommendations with references to your data (e.g., "As the results indicated...," or "Because of this, we recommend...."). Provide suggestions on how to fix the problems you identify. Arrange this section with specific subheadings that summarize each recommendation (e.g., "Recommendation #1: Sell Unnecessary Assets"). Consider adding visuals to help clarify recommendations. One simple visual is a figure that lists the recommendations.
- **Method.** Describe whatever research methods were used to address the question or problem motivating the report. Primary research reports provide a narration of the research procedures used.
- **Findings.** This section presents the results of any primary research conducted.
- **Conclusion.** End the report with a persuasive appeal that suggests the benefits of accepting the report's claims. If applicable, discuss any limitations of the current study or anything that could use further analysis or research.

Back Matter

References and appendices appear at the end of the report.

- **References.** If you paraphrase or quote any outside sources to support your work, you must document these sources with in-text citations and a list of references. If you have fewer than 3 citations, cover them fully in a parenthetical reference in the text or with a footnote. If you have more than 3 sources, use a list of references. Be sure all references follow an accepted style manual. Chicago style and APA are the most common. (See the style guide section below for links to information on documenting sources.)
- **Appendices.** Appendices are for supplemental information that ordinarily would interrupt the flow of the main discussion of the report. Be sure to reference appendices in your discussion. For example, when you describe your test methods include a reference to your data sheets (e.g., "see Appendix A: User Test Protocol") and include a copy as an appendix. Avoid putting graphics in appendices, since they typically help clarify key points. For each appendix make a divider page with the appendix title or put the appendix title at the top of the first page of the appendix.

Report Design

Business and technical reports pay careful attention to visual presentation. Aside from aiding readability, careful presentation demonstrates the investment of organizational resources into a document that will be read by many different people within and outside an organization.

Page Numbering

Paginating reports follows traditional publishing practice:

- All pages are numbered except for the front and back covers and the transmittal letter. Page numbers are not always displayed.
- All front matter pages are numbered with lowercase Roman numerals (e.g., i, ii, iii, iv).
- The report body begins on page 1 and is labeled with Arabic numerals (e.g., 1, 2, 3, 4).
- The executive summary is either included in the front matter before the table of contents (labeled with lower-case Roman numerals) or as the first section of the body (labeled as page 1 with Arabic numerals).
- Page numbers are not displayed on the transmittal letter, cover, first page of the table of contents, page 1 of the introduction, and appendix divider pages (if used).
- You can place page numbers either centered or to the right of either top or bottom margins. You can also include the report title and, if applicable, a company name in headers or footers.

Page Layout

When multiple writers contribute to a document the best way to ensure consistency is to use a style sheet. Style sheets dictate font styles, heading formats, and page layout (see Figure 8.1). You can download a Word file of a sample style sheet for use with your final report, or you can follow similar guidelines, examine some sample reports, and design your own.

Graphics

Whenever you want to emphasize a point, catch your reader's attention, or clarify a complex idea, add a visual graphic. Next to section headings, captions on graphics are the second most read parts of a report. This underscores the importance of adding visuals to important documents like reports. The following discusses types of visuals and conventions for integrating visuals within the prose discussion of the report.

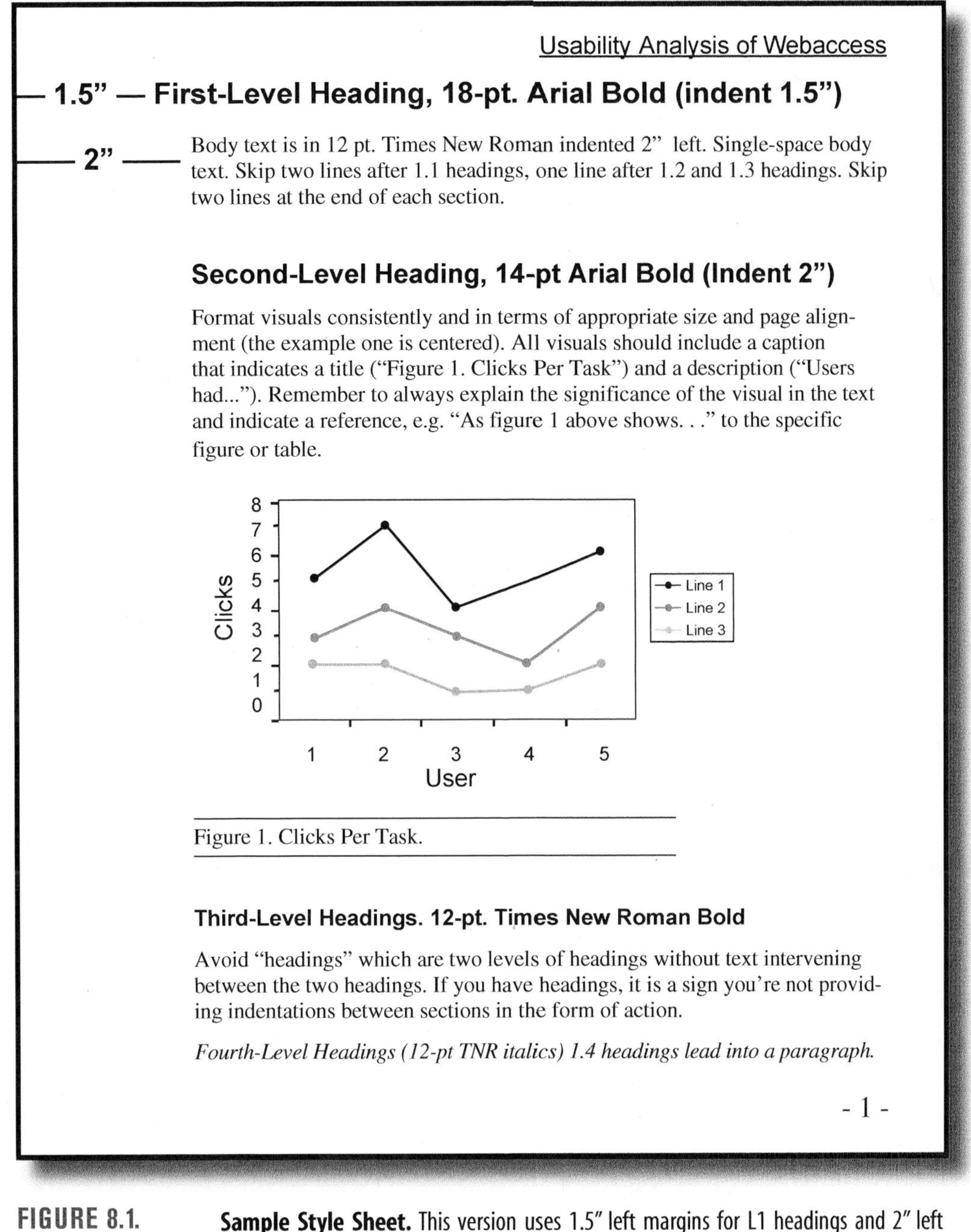

Usability Analysis of Webaccess

— 1.5" — **First-Level Heading, 18-pt. Arial Bold (indent 1.5")**

— 2" — Body text is in 12 pt. Times New Roman indented 2" left. Single-space body text. Skip two lines after 1.1 headings, one line after 1.2 and 1.3 headings. Skip two lines at the end of each section.

Second-Level Heading, 14-pt Arial Bold (Indent 2")

Format visuals consistently and in terms of appropriate size and page alignment (the example one is centered). All visuals should include a caption that indicates a title ("Figure 1. Clicks Per Task") and a description ("Users had..."). Remember to always explain the significance of the visual in the text and indicate a reference, e.g. "As figure 1 above shows. . ." to the specific figure or table.

Figure 1. Clicks Per Task.

Third-Level Headings. 12-pt. Times New Roman Bold

Avoid "headings" which are two levels of headings without text intervening between the two headings. If you have headings, it is a sign you're not providing indentations between sections in the form of action.

Fourth-Level Headings (12-pt TNR italics) 1.4 headings lead into a paragraph.

- 1 -

FIGURE 8.1. **Sample Style Sheet.** This version uses 1.5" left margins for L1 headings and 2" left margins for body text. This creates visually pleasing white space and allows for binding.

Tables

Tables are good for showing precise data, comparison, and juxtaposition of abstract ideas. The most common type of table presents data in rows and columns like an Excel spreadsheet. Another form of table is called a matrix, which presents text in tabular form. For example, Figure 8.2 is a matrix table comparing several computers according to various specifications.

Table 1. Best Value Desktop PCs. ***Dell offers the best all-around value. For a few more dollars, Gateway offers more speed and storage. Sony and iBuyPower: too pricey.***

Computer	Processor	Hard Drive	RAM	DVD	Price	Comment
Dell Dimension 4600	2.8 GHz Pentium 4 CPU	120GB hard drive	128MB NVidin GeForce FX 5200 graphics	16X DVD-ROM and 48X CD-RW drives	$1129	Best price, comes with an LCD screen, good speakers. Performance is good.
Gateway 510XL	3-GHz Pentium 4 CPU	160GB hard drive	128MB NVidin GeForce FX 5200 G graphics	8X DVD±RW and 16X DVD-ROM drives, media-card reader	$1400	Big hard drive, fast processor. Gateway warranty is a hidden extra cost
Sony VAIO PCV-RS520	3-GHz Pentium 4 CPU	160GB hard drive	128MB TI Radeon 9200 graphics	8X DVD±RW and 16X DVD-ROM drives, media-card reader	$1529	Very nice design, excellent performance. Keyboard is sub-par and sound system is weak
iBuyPower Back To School	2.2 GHz Athlon 64 3400+ CPU	80GB hard drive	256MB ATI Radeon 9800 XT graphics	8X DVD±RW drive, media-card reader	$1694	Weakest performance and hard disk. Best RAM, but cost is high.

FIGURE 8.2. **Sample Matrix Table.** Some cells have precise data, but other cells include words, allowing for easy comparison of different criteria. Notice that captions are located above table graphics. **Source:** PCWorld.com. Top 15 Desktop PCs. August 2004.

Charts

Some distinguish between *graphs*, which show values plotted on an x and y coordinate system, and *charts*, which do not; but the terms are often used interchangeably. Charts include the following:

- **Bar graphs**—show comparison and contrast.

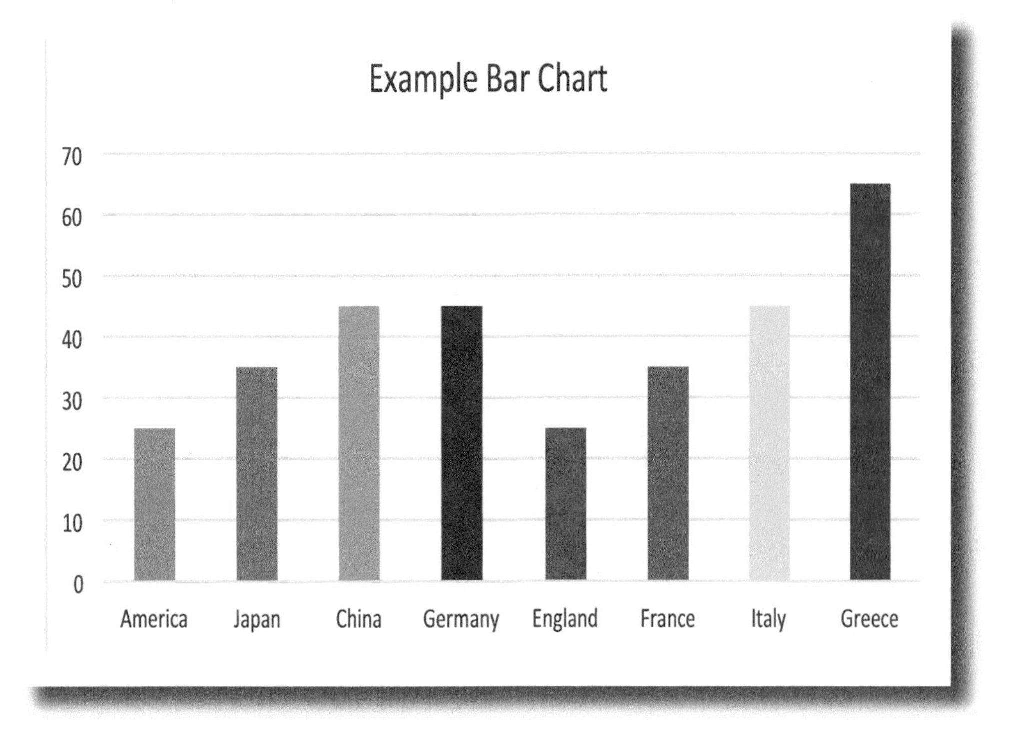

- **Line graphs**—show trends.

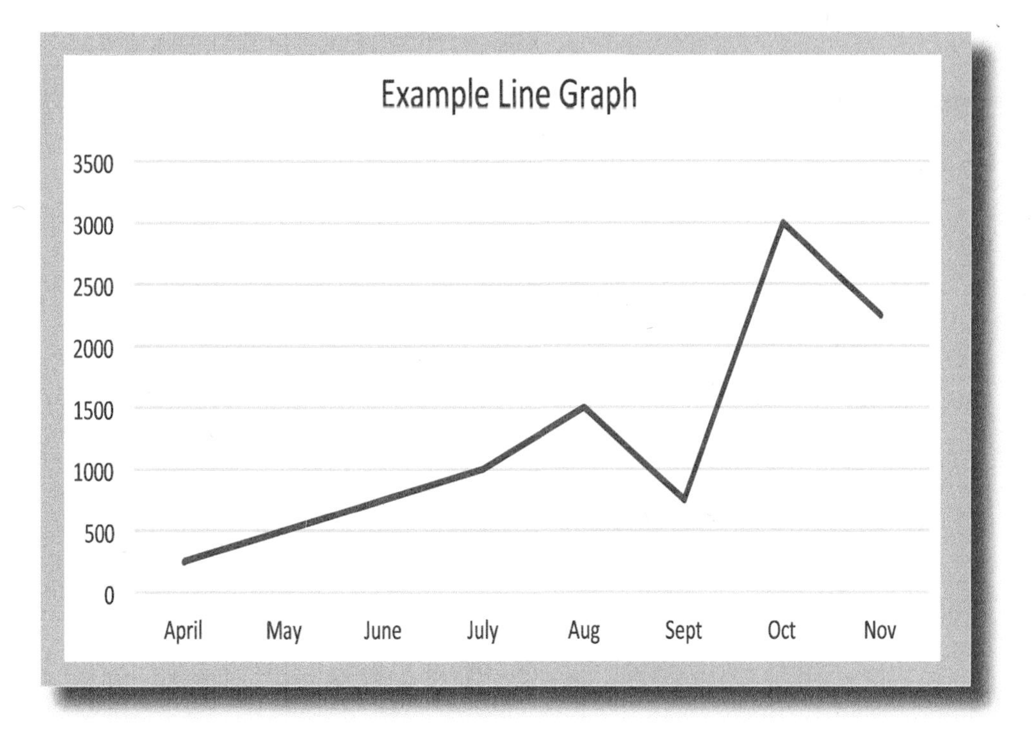

- **Pie charts**—show percentages, or parts of a whole.

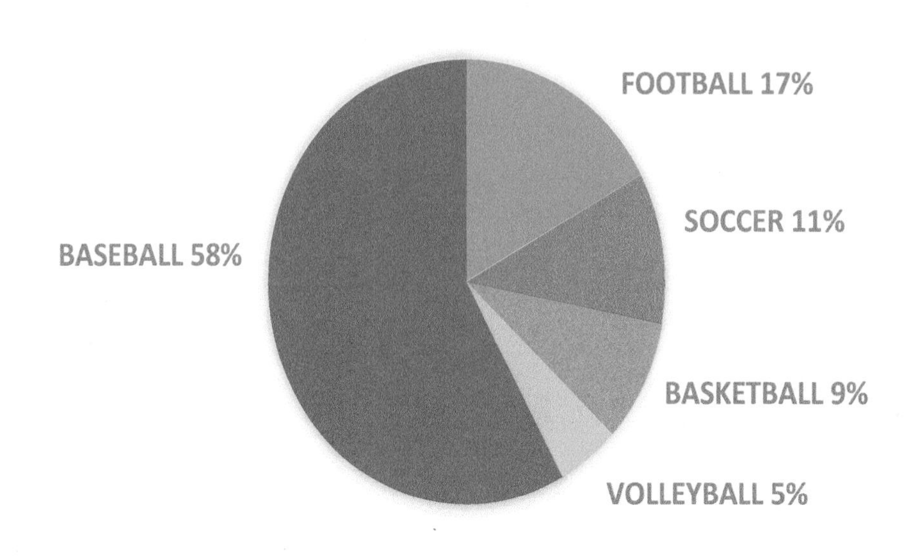

- **Organizational charts**—show structure or hierarchy.

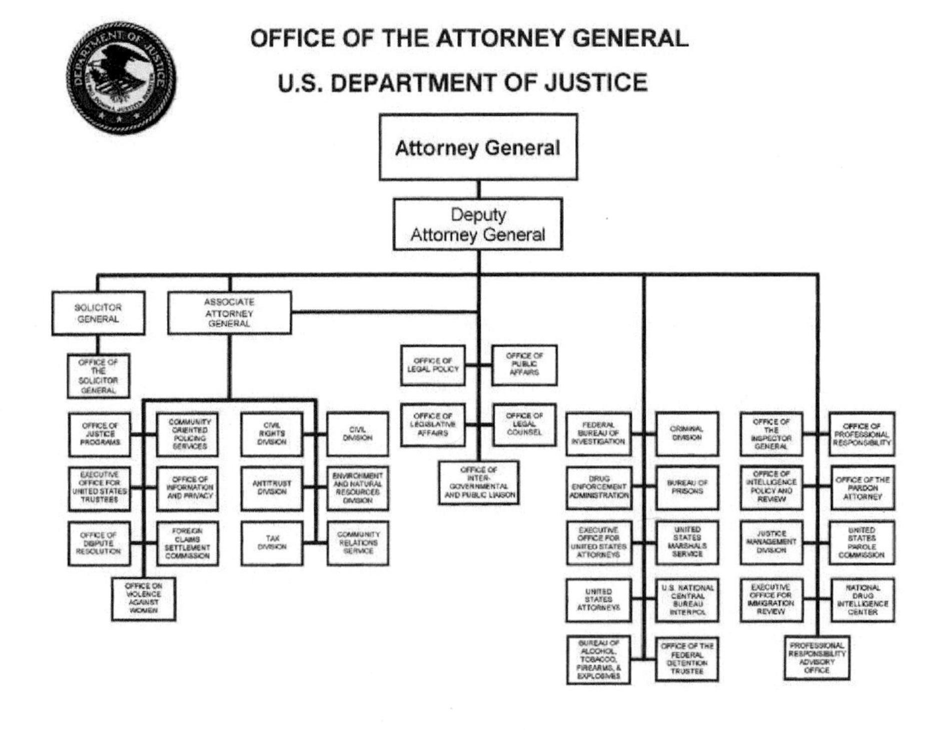

Approved by: ALBERTO R. [illegible]ZALES, Attorney [illegible]ral Date: 8-8-05

- **Flow charts**—show steps in a process.

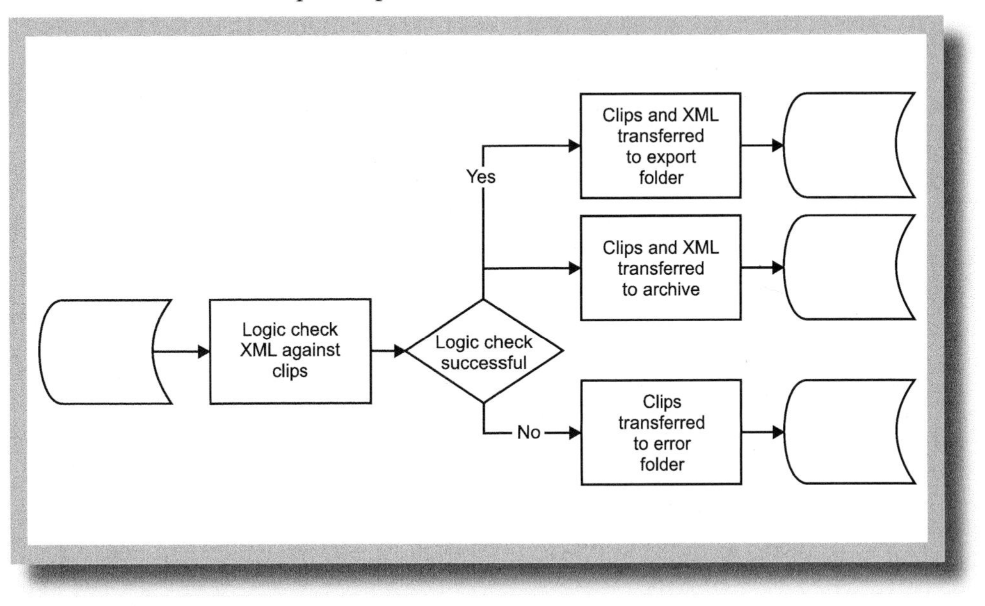

- **Pictogram**—a cross between a bar graph and a diagram, show comparisons using symbols.

Pictogram

Sales as of March 31, 2007

CITY

Boston

Dallas

Los Angeles

Orlando

Seattle

St Louis

* Each Computer equals 100000 units

Keep in mind the different emphases of each chart type when deciding which to use. Which chart would be good for showing how users rated the difficulty of a task on a scale of 1 to 5? Which chart would help show the different levels of Internet use among users tested?

Photographs, Illustrations, Diagrams, and Screenshots

Sometimes an exact representation is called for. Instructions often contain illustrations and diagrams (illustrations with labels) to help readers follow the written instructions and visually understand special relationships among parts (e.g., an arrow showing how to insert part A into part B). Diagrams can also be used to clarify the relationship of abstract concepts.

A screenshot, or screen capture, is an image of whatever is displayed on a computer monitor at the time of the screen capture (see Figure 8.3).

FIGURE 8.3. **Sample Screenshot.** To make a screenshot, select the "Print Scrn" key on the keyboard, then paste the image into a Word file. Callouts (the red highlighting and explanatory text) help make a point.

Screenshots are helpful in usability reports when discussing faults with or making recommendations to improve an existing website interface. To create a screenshot of your website, open the desired web page in a browser and click the PRINT SCREEN key. Then go to a Word file and click PASTE. You can resize and crop the image using Word's drawing tools. Screen captures can be turned into useful diagrams with the addition of *callouts*, which are arrows and explanatory text added to a screenshot. This can be done with screen capture software or with Microsoft Word's drawing tools.

For more help with screenshots, see:

- "How to a take a screenshot on a PC with Windows 7, 8, or 10" by Brandon Widder at http://www.digitaltrends.com/computing/how-to-take-a-screenshot-on-pc/

Integrating Graphics within the Text

Visuals don't speak for themselves. This means that you must not leave interpretation of the visual up to the reader. You must reference the visual in the text and explain its *purpose* (what is it?) and *significance* (what is its main point?) to the reader. Create your graphics before you write your text and then, keeping in mind what is depicted in the graphic, write the explanation of the graphic. Use the same concepts and language depicted in the graphic for your explanation. *The visual always appears immediately following the paragraph where it is first referenced in the text.* For instance, you might write: "As Figure 1 shows, profit has decreased steadily since the merger," or, "Profit has decreased steadily since last quarter (see Figure 8.4)." Then you would insert the visual in the space immediately following this paragraph.

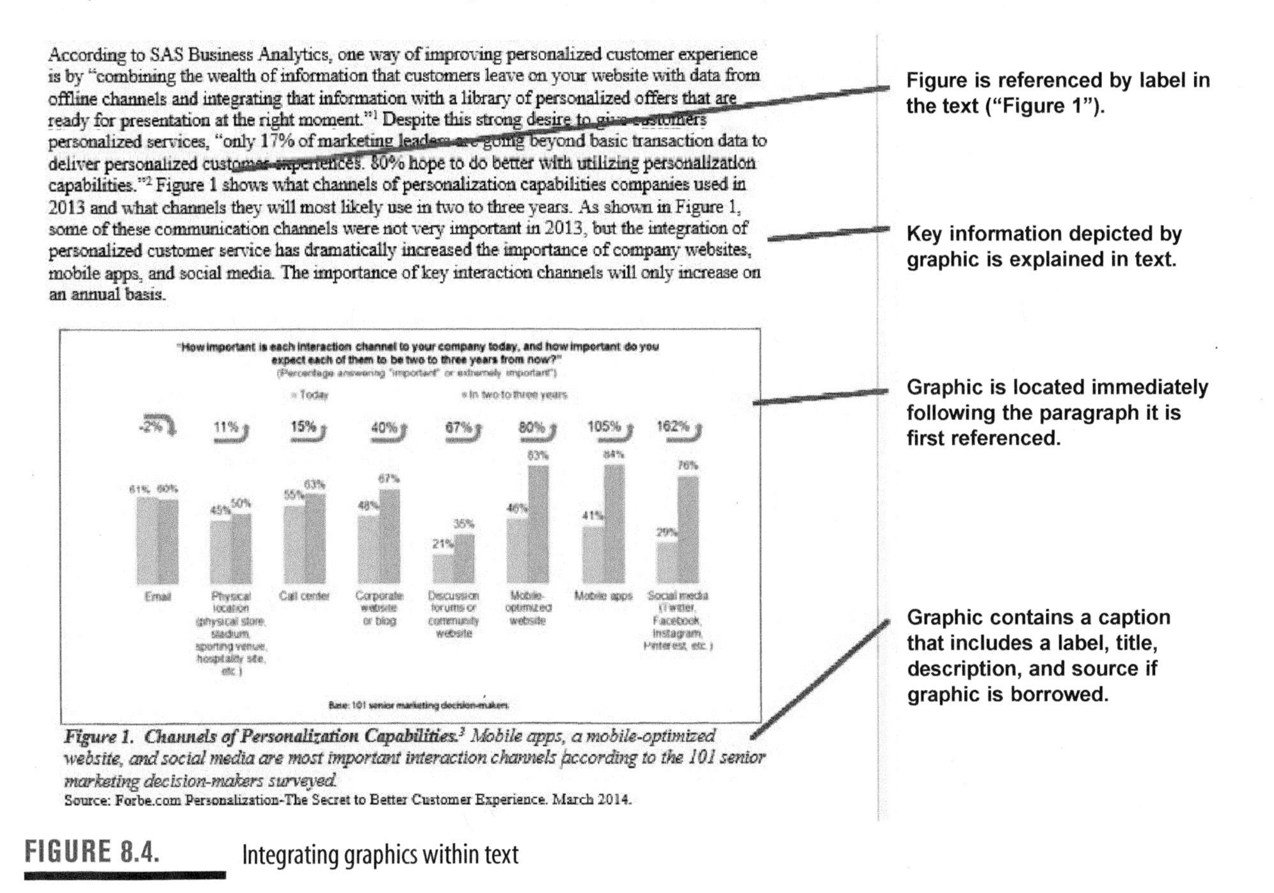

According to SAS Business Analytics, one way of improving personalized customer experience is by "combining the wealth of information that customers leave on your website with data from offline channels and integrating that information with a library of personalized offers that are ready for presentation at the right moment."[1] Despite this strong desire to give customers personalized services, "only 17% of marketing leaders are going beyond basic transaction data to deliver personalized customer experiences. 80% hope to do better with utilizing personalization capabilities."[2] Figure 1 shows what channels of personalization capabilities companies used in 2013 and what channels they will most likely use in two to three years. As shown in Figure 1, some of these communication channels were not very important in 2013, but the integration of personalized customer service has dramatically increased the importance of company websites, mobile apps, and social media. The importance of key interaction channels will only increase on an annual basis.

Figure 1. Channels of Personalization Capabilities.[3] *Mobile apps, a mobile-optimized website, and social media are most important interaction channels according to the 101 senior marketing decision-makers surveyed.*
Source: Forbe.com Personalization-The Secret to Better Customer Experience. March 2014.

FIGURE 8.4. Integrating graphics within text

All graphics include a caption that includes the label, title, and description. Notice that for tables the caption is placed above that table. For figures, the caption is placed below the graphic.

Label ("Table 1").

Title ("Best Value Desktop PC").

Description ("Dell offers...").

If borrowing information for graphic, indicate source.

Table 1. Best Value Desktop PCs. *Dell offers the best all-around value. For a few more dollars, Gateway offers more speed and storage. Sony and iBuyPower: too pricey.*

Computer	Processor	Hard Drive	RAM	DVD	Price	Comment
Dell Dimension 4600	2.8-GHZ Pentium 4 CPU	120GB hard drive	128MB NVidia GeForce FX 5200 Graphics	16XDVD-ROM and 48X CD-RW drives	$1129	This well-appointed Dell offers fine performance for its configuration; and LCD and good speakers sweeeten the deal.
Gateway 510XL	3-GHz Pentium 4 CPU	160GB hard drive	128MB NVidia GeForce FX 5200G Graphics	8X DVD±RW and 16X DVD-ROM drives, media-card reader	$1400	Nicely configured machine can handle basic text and spreadsheet documents, as well as multimedia presentations
Sony VAIO PCV-RS520	3-GHz Pentium 4 CPU	160GB hard drive	128MB ATI Radeon 9200 graphics	8X DVD±RW and 16X DVD-ROM drives, media-card Reader	$1599	System with stylish design and competitive performance are undermined by inferior keyboard and tinny sound system.
iBuyPower Back To School	2.2-GHz Athlom 64 3400 + CPU	80GB hard drive	256GB ATI Radeon 9800 XT graphics	8X DVD±RW drive, media-card reader	$1694	This system's high-flying performance makes it an appealing choice; outlandish front panel might give even Gamers pause, though.

Source: PCWorld.com Top 15 Desktop PCs. August 2004.

FIGURE 8.5. Integrating graphics within text

Since the best use of graphics is to clarify key points, always locate your visual as close to the relevant discussion as possible. While unimportant graphics can be located in appendices, this is not the best use of the persuasive power of visuals.

The design of graphics should be kept simple, uncluttered, and easy to read. Avoid overly complex three-dimensional visuals for this reason. If a reader cannot determine the point of the graphic in five to ten seconds, the graphic is not clear enough. Make sure to consistently re-size all visuals to fit within the margins of the page (e.g., all visuals 4 x 4" and center aligned).

Each chart or illustration is identified as a "Figure" and labeled consecutively. Each table is identified as a "Table" and labeled consecutively but separate from figures.

All figures and tables should be labeled with a caption that includes a number and a descriptive title that conveys the main point of the visual. You can also add a one or two sentence interpretation to the caption that further explains the graphic for readers. Place captions for figures below the graphic, and place captions for tables above the graphic (see Figure 8.5).

If you borrow information to make a visual, or borrow someone else's visual, identify the source in small print below the visual.

Adopt a Style Guide

Lastly, if you were writing a report for a large company, you would likely adhere to a particular style guide. Style guides dictate rules for proper punctuation, spelling, number usage, and page formatting. Since standards vary from company to company and field to field, it is important that writing teams decide ahead of time issues like how to spell non-standard words. For example, how will your team spell W-E-B-S-I-T-E? According to Associated Press style, it should be spelled "website." However, it used to be spelled "Web site" and some conservative publications still use the older spelling. The AP spelling is more recognized, but either version is acceptable, as long as the word is spelled consistently within the document. Style guides help resolve questions of style and mechanics during all phases of report production (drafting, revising, and editing).

ONLINE

Here are some examples of style guides:

- Microsoft Manual of Style https://ptgmedia.pearsoncmg.com/images/9780735648715/samplepages/9780735648715.pdf
- "Developing a Departmental Style Guide" by Jean Hollis Weber (2011) http://techwhirl.com/developing-a-departmental-style-guide/
- Resources for Documenting Sources by Purdue OWL (Good list of style guides arranged by field) https://owl.english.purdue.edu/owl/resource/585/02/

CHAPTER

9

RESUMES

The resume is still the gold standard when it comes to the job application process. Online profiles like those found in LinkedIn are felt by many employers to be a useful adjunct to the resume, but 95 percent of employers still strongly prefer the resume as the main screening tool.[1] Social profiles are often not as complete as resumes and can't be printed as easily as resumes when it comes time for interviews.

This chapter emphasizes that resumes are rhetorical strategies that change for each job. That means that when it comes time to apply to a specific job, you should carefully consider the particular employer and prepare a resume specifically for that one audience. Students often think they only need one resume and that's what they'll submit for any job. This is not a good strategy. If a student is training to be an accountant, plays sports, and likes to scuba dive, for example, they might have three different resumes for accounting-related work, sports camp work, and scuba diving instruction. All three resumes would emphasize different skills, and it would be difficult to fit all the skills for those three areas on a single one-page resume. You should not expect to have just one resume for all occasions. There is a place for generic resumes such as career fairs and online job-hunting databases, but when it comes time to apply to a specific job advertisement (or "job ad" for short), an effective writer understands how to adjust his or her resume for the particular occasion. You might start from a generic or preexisting resume, but you should *always carefully tailor it based on the reader's expectations* according to your understanding of the job requirements and that company.

Purposes

The resume introduces you to an employer in a short, quick way. It doesn't get you the job, but it can convince the employer that you're among the most qualified applicants and should be contacted for an interview. The resume then serves as a point of reference during the actual interview. Your resume, therefore, should

- Be easy to read
- Establish your ethos as a qualified professional
- Provoke a desire for more information

Your immediate goal is to convince employers to spend more time on your resume. (Your ultimate goal is to convince them to call you for an interview.) To succeed at this, you need to be familiar with what most experienced resume readers have come to expect from resumes. These are the rules, or *conventions*, of resume writing. You also need to be able to strategically utilize **page design principles** to create an initial and lasting impression with your resume. Most importantly, you need to be keenly aware of what that particular employer is looking for in an ideal candidate. Generic resumes cannot compete against resumes written by applicants who took the time to carefully tailor their job documents to the specific interests of the given audience.

Together with the cover letter, the resume is typically the first example the employer sees of your ability to communicate in writing—clearly, concisely, and neatly. The resume is an **outline** of your relevant skills. You cannot put your life's accomplishments into one or two pages, not now and especially not after you've been a working professional for several years. Your aim is to provide only the details that will convince an employer that you are qualified for the position. It is like a one-minute elevator pitch. This usually means including details of your education, relevant work experience, and anything else that shows you're qualified and capable. But you need to carefully and strategically think about what details should be included, what should be emphasized—and what should be excluded.

Format

Some basic format considerations include the following:

- **Length.** Generally speaking, readers expect the resume to be one or two pages long. Entry-level resumes—from freshly minted college grads— typically do not exceed one page. Only if you have extensive relevant accomplishments and skills from your college activities can you consider two pages. The length of experienced professional resumes (with 5 or more years' experience) varies according to the field, but most businesses still expect to see one- or two-page resumes.

- **Submission Format.** Most employers prefer electronic resumes submitted as an email attachment or through a form on the company website. Usually, a job ad will specify the submission format. MS Word documents are still the most common, but PDF versions are acceptable if specifically requested by the employer. If you have doubts about how to submit your resume, contact the employer.
- **Font.** Limit the font styles to no more than two. Keep heading and body fonts to between 10 and 12 points per inch. Since you want your name to stand out, you can get away with 14, 16, or maybe even 18 point font there. Keep in mind the look, feel, and impression that font styles make. Be conservative when selecting fonts. Your resume is not the time to dazzle with Matisse ITC—unless you judge it to be appropriate given your audience.
- **Balance.** Be sure your resume, as a visual unit, looks balanced across the four quadrants formed by the vertical and horizontal axes. Is your resume top-heavy, with too much text toward the top and very little toward the bottom? Is it left-heavy, with too much text toward the left margin? See the Visual Design section for more on this.
- **White Space.** This is a design term that refers to spaces on a page without text. You should always have a visually pleasing 1" border of white space around the resume. While you can adjust this a little (e.g., 0.8" bottom margin), do not try to cram everything onto a page with 0.2" margins.
- **Chunking.** Break up or "chunk" your resume by adding white space after headings, between sections, and between items in sections. Is your resume so densely packed with single-spaced lines of text that it is difficult to read? If so, you should break up dense chunks with white space. Space consistently between and within sections. The spaces separating sections are usually larger than the spaces separating items within sections (Hint: Use the Page Layout > Paragraph Spacing > Before and After tools to consistently adjust the spacing of headings and other content. There should be more space between sections than after headings.).
- **Conventional Correctness.** Both the cover letter and resume should be free of mechanical mistakes and presentational errors: no misspellings or typos. You should, however, understand that as an outline your resume does not have to adhere to the same rules for grammar and punctuation as prose documents (see the Style section).
- **Headings and Lists.** Because your resume is an outline you should use headings and bullet lists to help guide the reader quickly through your document.

Organization

The main parts or sections of a resume include contact information, opening, education, and relevant work experience or skills. Other relevant details such as computer skills, languages, honors and awards, or activities are often included but are dependent on the particular situation. The basic rule of thumb when organizing information in your resume is to **put the most important information toward the top** of the resume. Western readers are conditioned to read left to right, top to bottom. Don't bury key details related to the position at the bottom of the resume. At the same time, resume readers have come to expect to see certain sections, like those listed above, so you must carefully work within those expectations but still try to take advantage of strategic organization to emphasize key information.

The opening stragey of this chronological resume includes a qualifications summary with a "profile" heading.

The chronological resume emphasizes work history to present relevant skills. Skills are arranged by company name.

Notice the visual hierarchy, or clear levels of information, created by formatting fonts differently so that the reader can quickly tell which fonts are section headings, which fonts are company/school name, and which fonts are job titles/major.

Skill lists are in parallel grammatical form, or each item beginning with a verb in the same tense (present tense for current jobs, past tense for older jobs).

This writer lists her education after her work expereince because her experience is more immediately relevant despite her pending degree.

Dates are listed in a consistent vertical column on the right margin.

Jane Smith
janegsimth@gmail.com • (702) 688-1459 • http://linkedin.com/in/janeasmith

Profile

- Dynamic, well-versed hospitality professional with 4 years experience at a AAA Five-Diamond resort
- Currently manage front office operation at a 7,095 all-suite property
- Highly skilled in guest recovery and specialize in development and training staff to provide exceptional guest service
- Proven ability to lead a team in a fast paced, high stress environment

Work Experience

The Venetian Resort Hotel Casino, Las Vegas, Nevada — October 2014-Present
Front Desk Manager
- Oversee front line staff of over 70 employees at a 3,095 all suite property
- Ensure daily operations of the check-in/check-out process runs efficiently
- Mentor, coach, and guide employees in team atmosphere
- Assist in rectifying guest complaints and issues as they arise
- Ensure staff is providing AAA Five-Diamond service

Lane Property Group, Las Vegas, Nevada — October 2013-Present
Realtor
- Promote sales of properties through advertisements, open houses, and participation in multiple listing service
- Advise clients on market conditions, prices, mortgages, legal requirements and related matters
- Act as an intermediary in negotiations between buyers and sellers, representing one or the other
- Confer with escrow companies, lenders, and home inspectors to ensure terms and conditions
- Coordinate property closings, overseeing signing of documents and disbursement of funds

The MGM Resort Hotel Casino, Las Vegas, Nevada — August 2011-October 2014
Front Office Lead
- Communicated effectively with focused attention to guest needs and satisfaction while utilizing the LEARN recovery model when handling guest challenges
- Assisted in the command center with emphasis on reports and internal calls from guests and other departments within the hotel by adjusting charges to ensure accuracy of guest accounts
- Used up-selling and cross-selling techniques to increase guest satisfaction and revenue
- Maintained confidentiality of guest information and pertinent hotel information

Education

University of Nevada Las Vegas — Expected: May 2016
BS Business Administration
- Major: Management
- GPA: 3.5

Phi Delta Theta Fraternity, Nevada Beta Chapter
Vice President — 2013
Recruitment Chairman — 2012

FIGURE 9.1. **Chronological Resume Sample.** This writer is applying for an assistant hotel manager job that requires "minimum 1 year guest service experience, minimum 1 year management experience, excellent customer service skills, ability to lead a team, knowledge of Opera PMS system, and associates or bachelor's degree preferred."

There are some basic variations for organizing and presenting your information. Employment experts sometimes refer to these as resume **formats** or **styles**. These formats include the **chronological**, the **functional**, and the **combination** chronological/functional.

Chronological

This format is the most common and preferred by employers. It organizes your work experience and education information in a straightforward reverse-chronological order (i.e., list more recent experiences first) (see Figure 9.1 on page 140). For entry-level candidates, this usually means education comes before work experience because their education is the most recent, most important relevant experience. For professionals with more time on the job whose work experience becomes more immediately relevant, education is generally moved to the bottom.

This format is best if you have a strong history of relevant school and work experience. You basically organize your resume by where you worked or company name. The writer develops his or her relevant qualifications within the listing of jobs held. This style is less effective if you have unrelated work experience.

Functional

This format has fallen out of favor with most employers for reasons explained below. A functional resume emphasizes relevant skills, skills that are not necessarily related to jobs that you've held (see Figure 9.2). Instead of organizing information by company name or job title, the functional resume is organized by skills or competencies related to the position. Instead of a Work Experience section, functional resumes include a Professional Summary, Professional Experience, or Relevant Skills section. For example, if an internship asks for organizational, leadership, and communication skills, you might consider developing evidence of your skills in each of these three areas.

Functional resumes emphasize relevant skills over work history, and are typically recommended for people without a lot of relevant work experience, who are changing jobs, or seeking a promotion. It appears the trend is against functional resumes. If you google "should I use a functional resume," some of the first results will be "Four Reasons Not to Use a Functional Format" and "Why You Should NOT Use a Functional Resume Format."[2, 3] According to these sources, employers don't like reading functional resumes because they make it harder to connect your skills to where you worked, can confuse resume screening software programs, and raise a red flag as if you are hiding something. However, despite their disadvantages, if you read "The Demise of the Functional Format," Katharine Hansen provides a long list of people who might still consider a functional resume:

- *Those with very diverse experiences that don't add up to a clear-cut career path*
- *College students with minimal experience and/or experience unrelated to their chosen career field*

- *Career-changers who wish to enter a field very different from what all their previous experience points to*
- *Those with gaps in their work history, such as homemakers who took time to raise the family and now wish to return to the workplace*
  - *For them, a chronological format can draw undue attention to those gaps, while a functional resume enables them to portray transferable skills attained through such activities as domestic management and volunteer work.*
- *Military transitioners entering a different field from the work they did in the military*
- *Job-seekers whose predominant or most relevant experience has been unpaid, such as volunteer work or college activities (coursework, class projects, extracurricular organizations, and sports)*
- *Those who performed very similar activities throughout their past jobs who want to avoid repeating those activities in a chronological job listing*
- *Job-seekers looking for a position for which a chronological listing would make them look "overqualified"*
- *Older workers seeking to deemphasize a lengthy job history*[A]

If you read this list carefully, most people in college still fall under one or more of these categories. So the question becomes, if employers prefer chronological resumes, how can one write a convincing entry-level accounting chronological resume, for example, if his or her last three jobs were Starbucks barista, hotel cabana host, and retail salesperson? One answer is that if employers prefer chronological resumes, you better do all you can *while you are still in college* to get career-related experience in the form of internships and part-time work that will fit neatly into a chronological resume. If you don't, and insist on working unrelated survival jobs, you risk putting yourself at a rhetorical disadvantage on the job market come graduation.

However, if you fall under any of the categories above, what other strategies can you employ to make your resume more effective? Hansen suggests using at least a **combination format** that includes dates of employment.

Combination

A combination resume uses elements of both the chronological and functional styles. Usually, a combination resume begins with a summary of relevant qualifications followed by a work history section (see Figure 9.3). Opening with a clear summary of your qualifications has many rhetorical advantages because it can grab the quick-reading employer's attention even faster. A combination resume also still includes a work history, which satisfies employers and resume reading software. Unrelated experience and gaps will still be apparent to employers; however, they tend to expect some irregularities and you mainly need to be prepared to convincingly explain them in a job interview.

The combination format is preferable for most occasions because, as previously mentioned (and explained in more detail in the Content section), including a clear opening that

The opening stragey of this functional resume includes a profile statement.

A funcational resume can still have an education section. This student is applying for an internship so his academic background is particularly relevant and thus is placed toward the top of the resume.

The functional resume emphasizes work-related skills arranged by type of skill or competency (not place of employment).

Headings should highlight keywords (and thus relevant skills) listed in the job ad.

Relevant work history can still be listed in a functional resume but notice that no skills are developed under the job, only place of employment is listed. The skills are developed in the functional section.

Derek Hart
6494 W. Teton Blvd.
Las Vegas, NV 89129
(702) 326-2108
dhart@unlv.nevada.edu

A highly motivated student with strong communication skills, ability to multitask, stay organized, and 3+ years of experience in the customer service industry.

EDUCATION

University of Nevada, Las Vegas Expected: May 2016
Bachelor of Science in Business Administration, Finance
- GPA: 3.0

College of Southern Nevada May 2013
Associate of Business
- GPA: 3.44

FINANCIAL SKILLS

Communication
- Prepared and delivered speeches for oral communications course
- Facilitated multiple group training sessions
- Communicated with employees and managers via e-mail daily

Multitasking/Organization
- Tasked with preparing and organizing multiple training sessions simultaneously
- Held multiple positions with multiple job functions

Data Skills
- Analyzed data to pitch buy/sell orders for Rebel Investment Group course
- Evaluated large sets of metrics data to determine areas for improvement

Computer
- Created income statements and balance sheets for financial accounting course
- Used Microsoft Excel at work and school
- Created fixed income spreadsheets for Investments course

WORK EXPERIENCE

Apple Inc., Las Vegas, NV 2012 – Present
Specialist, Mentor, and In-Store Guest Trainer

The Betty R. and Robert S. Winthrop Foundation 200p – Present
Board Treasurer

FIGURE 9.2. **Sample Functional Resume.** The writer is applying to a financial analyst internship. The qualifications listed in the job ad include "excellent verbal and written communication skills, highly organized with proven multi-tasking ability, excellent problem solving and analytical skills, ability to create spreadsheet models, MS Excel required, Database and/or Visual Basic a plus."

summarizes your qualifications can leave a strong first impression rather than the reader having to determine your qualifications by carefully reading through an entire resume that contains no opening.

Content

Content refers to what details you choose to include and how you choose to represent them. The resume usually provides the following information: **contact information**, an **opening**, **education**, work **experience**, and **other relevant details**.

Contact Information

Include your name, telephone number, email address, and online content. When deciding what contact information to include, list only the information that will make it easy for the employer to contact you, so use only one phone and email address preferably. Make sure your email address is professional, and you don't need to write "phone" or "email" next to the actual information because it is obvious.

Experts disagree about including a physical address, but the trend is toward leaving it off your resume. For one, few employers correspond with applicants via traditional mail anymore. Also, recruiters warn against various abuses, such as the possibility of discrimination based on the perceived length of your commute, identity theft, even sexual harassment. Including the cities and states of your employers should be enough to indicate your location, and if you are willing to relocate for a position, you can indicate that in your resume opening and letter. Including just your city and state in the contact information is another option.

You can add a graphic element such as a horizontal line to visually separate your contact information from the body of your resume. Some resumes include a simple image or design for visual appeal, but resumes designed for resume screening software called applicant tracking systems (ATSs) should avoid graphics.

You can also use the same design and layout of your contact information on your resume and cover letter to give both documents a more professional and consistent visual design.

Referencing Your Online Content. Should you include website addresses (URLs) of any of your social media sites or other online content such as a personal website or blog as part of your contact information? You will have to decide if they are relevant or appropriate. Having a LinkedIn or About.Me account to promote your professional image makes sense, but do your Facebook, Twitter, Instagram, Tumblr, or blog pages contain any content that employers might find questionable? You must assume employers will search to find your online content regardless of whether it is listed on your resume. Some employers are concerned about how

The opening stragey of this combination resume includes a headline, branding statement, and summary of qualifications.

The summary of qualifications is a highlight of the candidate's main qualifications drawn from the rest of the resume and that corresponds to the key qualifications listed in the job ad.

This student's work experience is not directly related to the casino marketing position he is applying to, but he has chosen to include his work experience to satisfy employers' desire to see this information. What could this writer have done in school to bolster his marketing skills?

Here are some keywords related to the marketing position advertisement including "utiilze social media" and "promote."

The student also includes computer and internet skills listed in the job ad.

James Xiu
Las Vegas, NV
(702) 364-4995
xiu11@unlv.nevada.edu
http://linkedin.com/in/jtxiu

CUSTOMER SERVICE AND MARKETING
Committed to providing exceptional customer service by contributing restaurant supervisory experience and knowledge of diverse cultures to assist management for marketing purposes.

SUMMMARY OF QUALIFICATIONS

- Two years' experience in customer service
- Strong interpersonal communication skills
- Proficient in MS Office and social media

EDUCATION

University of Nevada, Las Vegas Expected: July 2016
Bachelor of Arts Interdisciplinary Studies
- Emphasis in Asian studies
- 3.0 GPA

WORK EXPERIENCE

Supervisor Mar. 2013 - Present
Lee's Sandwiches
- Oversee productivity and performance of up to 8 personnel
- Attend to customers' needs
- Manage monetary transactions and inventory control

EFL Tutor Sept. 2014 – Dec. 2014
Volunteer
- Planned, prepared, and instructed lessons to children and adults in Taipei, Taiwan
- Evaluated students' class work and progress
- Maintained an upbeat and relaxed environment for students

ACTIVITIES

President Aug. 2010 – Sept. 2013
Asian Film and Drama Club
- Utilized social media and other techniques to attract member base
- Interacted with several film corporations for film rights purposes

Student Government Campaigner Sept. 2012 – Sept. 2013
Consolidated Students of the University of Nevada (CSUN)
- Helped promote candidates by word-of-mouth and distributing flyers
- Communicated with students to gather their concerns

COMPUTER AND SOCIAL MEDIA SKILLS

MS Word	MS Excel	MS PowerPoint	Twitter
Facebook	Instagram	LinkedIn	Pinterest
Tumbler	Google Plus	Vine	Yelp

FIGURE 9.3. **Sample Combination Resume.** This writer is applying for a casino marketing coordinator position that requires "ability to communicate in English with guests and co-workers in a professional manner; maintain dialogue with all cultures and ethnicities while upholding a warm, positive and friendly persona; minimum of 1 year work experience in hospitality or service environment; and proficiency with MS Office."

their potential employees' social media content could negatively affect the company's image. On a positive side, the personality or work-related thinking you demonstrate in your online content might actually sway employers based on its unique qualities.

So, for these reasons, some experts recommend that you always include links to social media content in your resume and that instead of trying to create separate personal and professional online presences that you instead carefully manage, or curate, your online presence as one coherent personal–professional image or brand. To simplify things for the employer and limit the amount of links you present, you can use sites such as LinkedIn, About.Me, or Google Plus as a single landing page URL that contains further links to the rest of your online content.[5]

If you are going to include links to your social media and other websites, be sure to have additional content not available in your resume and cover letter, such as a professional photograph, list of references, quotes or testimonials from references, and samples of your work.

Having your own website is another way to organize your online image and showcase your technical knowledge.

Resume Opening

resume opening
Following the contact information, the resume opening attempts to summarize as succinctly as possible how you meet the qualifications of the job and/or what type of employee you are.

Traditionally, after the contact information a resume would open with an objective statement stating the applicant's goals either short-term, such as for the specific job, or long-term, such as one's 5- or 10-year plan. However, most employers don't like to see objective statements because they are either so vaguely written they are meaningless or because they are too self-centered and employers are more focused on learning what you can do for them.[6, 7] In "The Resume Objective: Why it Needs to go the way of the Dodo," Louise Fletcher writes that instead of an objective statement, a resume should have an **opening** that "truly communicates how and why you will add value" by using strategies such as a headline and resume summary that grabs attention.[8]

In other words, the alternative to omitting an objective statement should not be leaving the most important top portion of the resume blank. In another article by Hansen titled "Your Job-Search Resume Needs a Focal Point," she suggests making the beginning of the resume as effective as possible by using some combination of a **headline**, **branding statement**, **qualification summary**, **professional profile**, or **keywords** to convince the busy reader to spend more time reading the rest of your resume.[9]

- **Headline.** A headline is a **phrase** (never a complete sentence) appearing in the opening that concisely states your value as a candidate. It is emphasized with all capital letters or bold text (hence, the term headline) and can precede other opening strategies noted below such as a branding statement or professional summary. As with any of the opening focal-point strategies, try to use keywords from the job ad to connect with your reader and emphasize your fit to the position (see Figure 9.4).

SALES PROFESSIONAL

Top-Producing Sales Professional

Honor Roll Student with Tutoring Experience in Numerous Subjects

Senior Accounting Major with Two Years Accounting Experience

Cook with Extensive Fine Dining Experience

IT Professional with Five Years Experience in Software Support

Technology-Proficient Worker with Administrative Experience

Detail-Oriented History Student with Curatorial Experience

FIGURE 9.4. Example resume headlines[10]

- **Branding Statement.** According to Hansen, a branding statement "defines who you are, your promise of value, and why you should be sought out. A branding statement is a punchy 'ad-like' statement that tells immediately what you can bring to an employer."[11] As the examples in Figure 9.5 show, the branding statement is a one to three lined statement or list that can include the following:
    - Relevant qualifications
    - Personality or passion
    - Distinctive qualities or characteristics
    - Significant or relevant accomplishments

A highly motivated student with strong communication skills, ability to multitask, stay organized, and 3+ years customer service experience.

Specialize in raising the bar, creating strategy, managing risk, and improving the quality and caliber of operations.

Vibrant and creative college student with an insatiable thirst for knowledge. Works efficiently in a fast-paced environment and possesses strong leadership skills. Analytical and versatile thinker; effective in developing and carrying out ideas.

Positioned to contribute accounting principles and auditing experience to MGM's diverse, innovative environment for industry standard assurance, consistent and accurate reporting, and company profit expansion

Bringing a wealth and diversity of skills in management, education, and human relations and prepared to apply knowledge of medicine, science, and nutrition to enhance quality of life by promoting pharmaceutical products and services.

Dedicated nursing professional committed to excellence in patient care and poised to deliver unsurpassed, individualized nursing care in an acute care setting

The following examples combine a headline with a branding statement:

SENIOR-LEVEL SALES and MARKETING LEADER
Positioned to contribute progressive, innovative sales and marketing methodologies with energy, enthusiasm, and professionalism to your organization in a leadership role.

TAX ACCOUNTANT
Offering accounting experience and specific expertise in tax research, strategy, and planning.

FIGURE 9.5. Example resume branding statements[12]

- **Qualifications Summary.** A qualifications summary is a four to five item bullet list of your relevant qualifications specific to the particular job ad. A well-written qualifications summary will closely resemble, without appearing to copy, the main qualifications listed in the job ad (see Figure 9.6). This is a particularly effective strategy because the resume reader usually has a specific list of criteria in mind based on the job ad and the qualifications summary can quickly establish you possess them. When you write your list of qualifications, arrange them in order of importance. You can begin your summary with a headline, or with the heading "Qualifications Summary."

Qualifications Summary
- Majoring in business management at UNLV
- Over 5 years administrative experience
- Strong written and verbal communication skills
- Computer skills include Word, Excel, PowerPoint, Cisco Systems
- Type 60+ WPM

SUMMARY OF QUALIFICATIONS
- Accurately create financial reports, budgets, and cost analysis
- Successfully research and reconcile all guest billing issues
- Report any variances and discrepancies to senior management
- Proficient in Excel,Access, LMS, CMS, and Infinium

FIGURE 9.6. Example summaries of qualifications

- **Professional Profile:** Very close to a qualifications summary, a professional profile tends to be a summary of your employment history and skills relevant to the job. It is somewhat of a hybrid between a branding statement and qualifications summary. Profiles tend to be short statements, while longer ones should be put into bullet lists for better readability (see Figure 9.7).
- **Keyword Summary.** Keywords are specific terms drawn from the job ad that recruiters have designated as meaningful searchable words in computer applications called **applicant tracking systems** (ATS), or databases that collect information from applicants and manage the hiring process, including scheduling interviews, checking references, and completing hiring paperwork.

You can add a section of keywords and use heading such as "Key Skills," "Core Competencies, or "Areas of Expertise." Employer opinions vary on the value of keyword lists (see Figure 9.8). Some ATS software will rank keywords higher if they are in context, or embedded within other words, so be sure to use keywords in other places of your resume, as well, such as your work history skill lists.

You can also open with a more succinct list of keywords appearing in the headline positon, accompanied by other opening strategies (see Figure 9.9).

In the end, it is your choice whether you want to leave the most important, first-impression forming space at the top of the resume completely blank or include an opening that summarizes why an employer should hire you.

University graduate with educational and practical experience in financial reporting for a Fortune 500 gaming and hospitality company.

A dynamic and well-versed hospitality professional with 4 years of experience at a AAA Five-Diamond resort. Currently manage the front office operation at a 7,095 all-suite property. Highly skilled in guest recovery and specialize in the development and training of staff on providing exceptional guest service. Proven ability to lead a team in a fast paced high stress environment.

Detail-oriented history student at UNLV with experience in preservation and museum work. Praised for ability to give well-organized and informative museum tours. Award-winning customer service and communication skills.

Editor and Writer
Award-winning editor and technical writer with five years of experience. Successfully implement current web design technology to develop and maintain sites for start-up IT companies.

Bilingual Health Care Provider
Nursing graduate fluent in Spanish with experience in rural health care. Successfully established multiple clinics with nonprofit health care groups, providing service for thousands of rural women, children, and infants.

Non-Profit Fundraiser
- Detail-oriented fundraiser with three years of experience
- Plan and execute events for nonprofit organizations
- Strong interpersonal skills with training in conflict mediation
- Proficient in current web design technology

FIGURE 9.7. Example resume profiles[13]

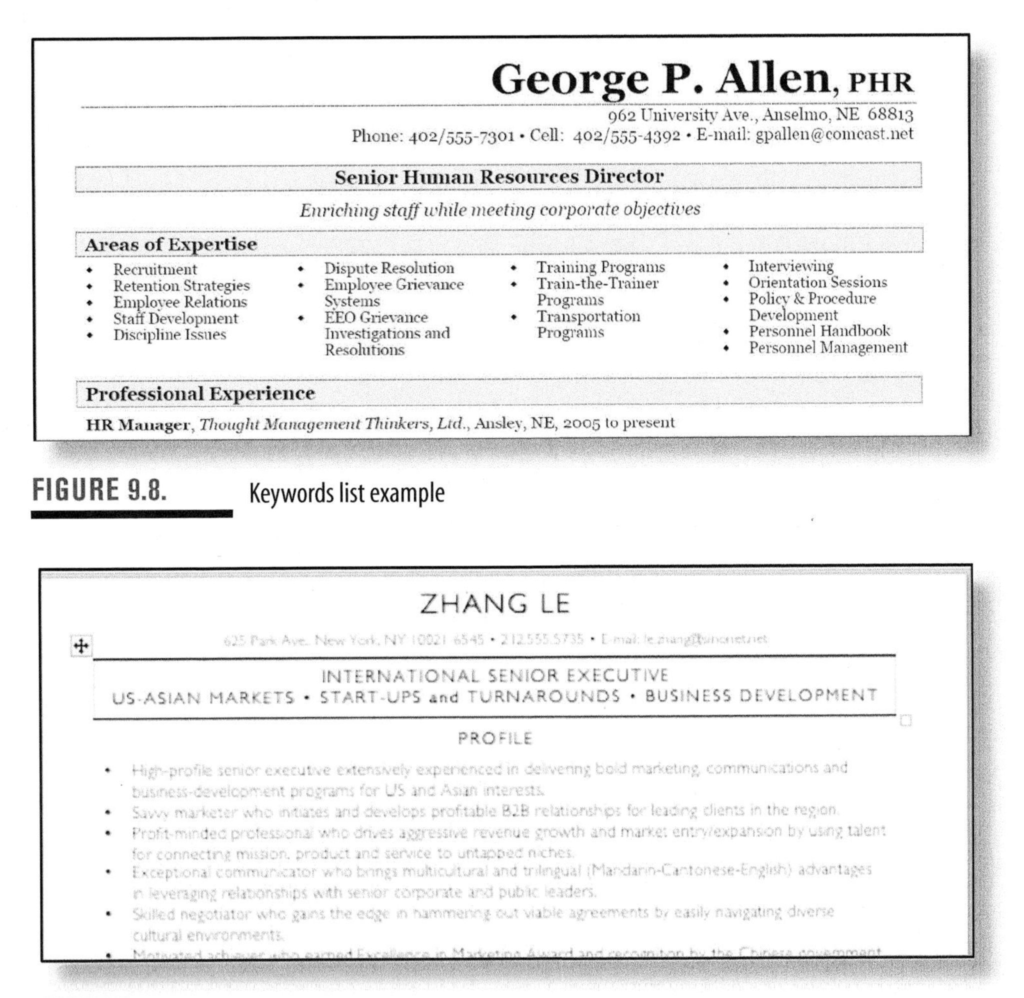

George P. Allen, PHR

962 University Ave., Anselmo, NE 68813
Phone: 402/555-7301 • Cell: 402/555-4392 • E-mail: gpallen@comcast.net

Senior Human Resources Director

Enriching staff while meeting corporate objectives

Areas of Expertise

- Recruitment
- Retention Strategies
- Employee Relations
- Staff Development
- Discipline Issues
- Dispute Resolution
- Employee Grievance Systems
- EEO Grievance Investigations and Resolutions
- Training Programs
- Train-the-Trainer Programs
- Transportation Programs
- Interviewing
- Orientation Sessions
- Policy & Procedure Development
- Personnel Handbook
- Personnel Management

Professional Experience

HR Manager, *Thought Management Thinkers, Ltd.*, Ansley, NE, 2005 to present

FIGURE 9.8. Keywords list example

ZHANG LE

625 Park Ave., New York, NY 10021-6545 • 212.555.5735 • E-mail: [illegible]

INTERNATIONAL SENIOR EXECUTIVE
US-ASIAN MARKETS • START-UPS and TURNAROUNDS • BUSINESS DEVELOPMENT

PROFILE

- High-profile senior executive extensively experienced in delivering bold marketing, communications and business-development programs for US and Asian interests.
- Savvy marketer who initiates and develops profitable B2B relationships for leading clients in the region.
- Profit-minded professional who drives aggressive revenue growth and market entry/expansion by using talent for connecting mission, product and service to untapped niches.
- Exceptional communicator who brings multicultural and trilingual (Mandarin-Cantonese-English) advantages in leveraging relationships with senior corporate and public leaders.
- Skilled negotiator who gains the edge in hammering out viable agreements by easily navigating diverse cultural environments.
- Motivated achiever who earned Excellence in Marketing Award and recognition by the Chinese government

FIGURE 9.9. Keyword headline

Education

Include name of college, degree, major, and date of graduation. You only list the date the degree is received or expected, not the dates of attendance. Format the fonts and levels of information consistently with the experience section (i.e., school is equivalent to company name, major is equal to job title).

You are technically only to list degrees and certifications already earned or expected to obtain. You do not list schools you merely attended and transferred from, for example. It is convention not to list your high school diploma on your resume, unless you had some particularly noteworthy achievement such as class valedictorian. Arrange multiple degrees in reverse-chronological order.

You don't have to include your GPA, but employers will assume that if your GPA is missing that it must be low, so include even "average" GPAs. You may also list significant courses or academic honors in this section, though employers know they can request a transcript if they want to know exactly what courses you've taken, so only list courses if you need to fill space until you get more substantial internship or work experience.

For entry-level candidates, education tends to be one of the most important qualifications, so it is listed toward the top of the resume. Working professionals with several years of experience tend to move their education to the bottom of the resume.

Experience

Include your name of company, job title, and dates of employment, generally in that order. Provide detailed bulleted action lists that evidence your relevant qualifications, skills, and accomplishments. Be very specific and concrete. You should consider which skills from your past experience are most suited to the job you're applying for. Instead of a generic list of eight skills, choose the four or five that best match the employer's criteria and keywords from the job ad.

Don't just cut and paste your job duties, instead list particular accomplishments/achievements. Use industry jargon and numbers/statistics to list specific projects you worked on and productive results. Do not overdevelop the obvious. If your job is pizza delivery person, you don't need to list delivered pizzas as a skill.

Make sure your lists are parallel in structure. If you start your job descriptions with a past tense verb, then you should start all job descriptions with a similar past tense verb (e.g., *created, sold, supervised*). Remember, you can develop your experience chronologically or functionally, or combined.

Generally, employers are most interested in what you have done lately. That means your relevant experience from the last 5 years is most important, and you usually go back no more than 10 years. Employers will notice gaps in your work history, but if you are prepared to explain them in an interview, you can have some gaps in your experience chronology.

Other Relevant Information

Along with your education and experience, you should consider what other curricular or extracurricular skills, activities, or accomplishments help establish your qualifications and ethos. It is often a good idea to include a separate computer skills section. Employers want to know how you can contribute to the particular position you're applying to. They're very interested in your career-related skills, academic- and community-oriented activities, and special aptitudes. Being involved in a range of activities helps show that you can balance a busy schedule and are not one-dimensional. Employers are less interested in your hobbies or outside interests, however.

If you're involved in college athletics or social organizations (e.g., Greek fraternities or sororities), make sure you don't overemphasize this. Executive positions held in social organizations, such as being president or treasurer, might help show relevant leadership or financial experience, but don't list every position you've held in your sorority for the last 4 years. Employers generally want to know what you've done lately, so if you're a junior or senior in college, this usually means that your high school accomplishments aren't that valuable.

References

It's taken for granted that you have a list of references prepared if employers request it. You shouldn't list references and contact information in your resume. You don't even need to have this section, unless you maybe need to fill some empty white space on your page (e.g., "References—Available upon request").

Resume Style

Style refers to the accepted ways to present resume content. Remember, above all, that the resume is an **outline**. As such, you do not need to conform to the same grammatical and mechanical principles as prose documents. The following guidelines help you present your information clearly and quickly:

- Do not use complete sentences and avoid the first-person "I." Short, concise fragments with action words are easier to read than lengthy prose.
- Limit the use of punctuation. You should still use commas to separate items in lists, but do not use end-line punctuation. You can eliminate punctuation in most common abbreviations, for example, GPA instead of G.P.A.
- Use lists to make information easy to scan. Break up lengthy sections of dense, packed prose into bulleted lists of parallel grammatical structure.
- Arrange more important information first. **Prioritize** items in lists for example.
- Use technical or industry jargon to help establish your professional ethos. Employers look for keywords to help them decide what type of position you can fill in their organization. Use as many nouns as you can that pertain to your desired field of work.
- **Quantify** your experience wherever possible (e.g., "supervised 15-member staff," "increased profits from first to fourth quarter by 70%")
- Stress your strengths. Don't mention weaknesses.
- Be completely honest. Do not misrepresent or fabricate skills. Your resume will be used to generate interview questions, so don't put yourself in a position where you'll have to deny anything stated in the resume. You won't get the job. You can even be banned from campus recruiting if caught.

Resume Visual Design

Many of the techniques for making your document easier to scan fall into the realm of page design. Because the average HR manager spends 1 to 3 minutes making initial decisions about the resume you need to ensure that your resume will not frustrate the reader and, indeed, that it will invite them to actually spend more time on it. The following page design principles used by technical writers, designers, and desktop publishers will help you do this: **create visual balance**, **apply visual hierarchies**, **use highlighting**, **format consistently**, and **evoke simplicity**.

Create Visual Balance

A page can be divided into four quadrants, as if folding the resume in half twice (see Figure 9.10). Make sure that text appears in each section relatively equally, which is known as symmetry. If you want to create a purposeful asymmetry, fill the right top and right bottom quadrants with more text.

quadrant test
Assesses the visual balance of a resume by dividing the page into 4 equal parts.

Quadrant Test: The quadrant test is a way to evaluate the visual balance of your resume. Fold a printed draft of your resume draft into quarters (as indicated in the picture above). You can also assess an electronic copy of your resume by switching the view to a full-page view, or "One Page" view in MS Word.

- **Bottom quadrants look empty:** If there is too much empty white space in the bottom two quadrants, then you need to add more content to your resume. Resumes should take up the whole page, even if you feel you don't have enough experience to fill the page. You can add lists of courses, computer skills, or personal interests/hobbies, for instance.
- **Right quadrants look empty:** If there is too much empty white space in the top right and bottom right quadrants, then it usually means you need to expand the detail in your bulleted lists (do not have short, three- to four-word bulleted lists of skills). Also notice if you have a consistent 1" border of white space around all four quadrants. You must balance your resume if any border is larger or smaller than 1".

Apply Visual Hierarchies

Designers use headings, indentation, and highlighting (i.e., changing the size or format of text) to create noticeable levels of information (see Figure 9.11). You can signal major section headings by bolding text, for example. You can then signal different parts within sections by indenting and italicizing the next level of information. This creates contrast by creating hanging *indentation*.

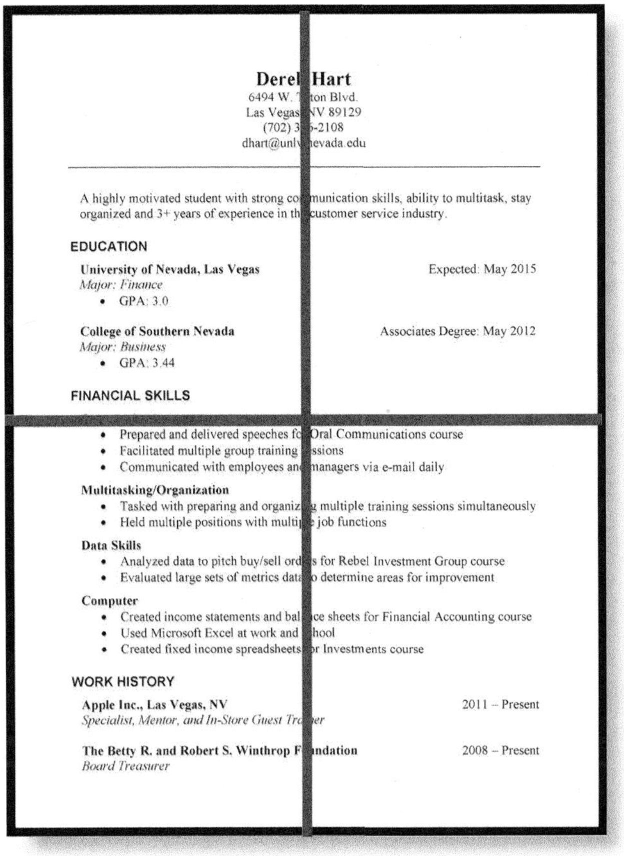

Derel Hart
6494 W. ton Blvd.
Las Vegas NV 89129
(702) 3 -2108
dhart@unl evada.edu

A highly motivated student with strong co munication skills, ability to multitask, stay organized and 3+ years of experience in th customer service industry.

EDUCATION

University of Nevada, Las Vegas — Expected: May 2015
Major: Finance
- GPA: 3.0

College of Southern Nevada — Associates Degree: May 2012
Major: Business
- GPA: 3.44

FINANCIAL SKILLS

- Prepared and delivered speeches fo Oral Communications course
- Facilitated multiple group training ssions
- Communicated with employees an managers via e-mail daily

Multitasking/Organization
- Tasked with preparing and organiz g multiple training sessions simultaneously
- Held multiple positions with multi job functions

Data Skills
- Analyzed data to pitch buy/sell ord s for Rebel Investment Group course
- Evaluated large sets of metrics dat o determine areas for improvement

Computer
- Created income statements and bal ce sheets for Financial Accounting course
- Used Microsoft Excel at work and hool
- Created fixed income spreadsheets r Investments course

WORK HISTORY

Apple Inc., Las Vegas, NV — 2011 – Present
Specialist, Mentor, and In-Store Guest Tr er

The Betty R. and Robert S. Winthrop F ndation — 2008 – Present
Board Treasurer

FIGURE 9.10. Quadrant test

LEVEL OF INFORMATION	FORMATTING	RESULT
Level 1: Section (White space) Level 2: Company Level 3: Job title Level 4: Relevant skills	13 pt. Arial, bold 12 pt. Times, bold, indent 12 pt. Times, italics, indent 12 pt. Times, indent, bullet	**Work Experience** **Acme Finance,** San Diego, CA *Market Analyst* • Analyzed... • Prepared... • Managed...

FIGURE 9.11. Visual hierarchies, i.e., clear levels of information, are created by font style, highlighting, and indentation

You can also use tabs or borderless tables to put lists of information into more readable columns.

Instead of

Related avionics courses

Aircraft Electrical Systems, Aircraft Avionics Systems, Digital Electronics, and Optical Physics.

Other related courses

Computer Programming, Multivariate Calculus, Advanced Composites, Fiber Optics

Use cleaner columns:

COURSEWORK

Avionics Courses	***Other Courses***
• Aircraft Electrical Systems	• Computer Programming
• Aircraft Avionics Systems	• Multivariate Calculus
• Digital Electronics	• Advanced Composites
• Optical Physics	• Fiber Optics

Use Highlighting

Highlighting refers to any manipulation of plain text for **emphasis**, including bold, italics, underlining, all capital letters, or altered size. Choosing a different font style or using a larger font size to create contrast for headings is an example of highlighting.

Avoid overemphasizing text; you don't need to simultaneously bold, italicize, and underline your major section headings. Use highlighting mostly to distinguish levels of text (i.e., for headings and subheadings). Do not use more than two font styles in your resume.

Underlining and italics should be avoided, since it can confuse resume screening software and is generally associated with website links.

Format Consistently

Be sure to follow the principle of consistency when creating visual hierarchies and levels of emphasis. Repeat the same amounts of indentation and spacing for the same levels of information. Use the same font style for the same levels of information and so on. One way to test for consistent design is to use the vertical columns test. Use imaginary vertical lines to check that information of the same-level aligns visually in the same vertical column.

- **Vertical columns test.** You should be able to align a ruler (or the ruler line of your word processing program) to the same level of information. As shown in Figure 9.12 on next page, main section headings, company names and school names, bullet lists, and lists of computer skills all align along the same vertical column indicated by the red lines.

vertical columns test
Assesses how consistently the levels of information in a resume are formatted by using vertical lines to check that all information of equal level (e.g., section headings, company names or schools, job titles or degree type, dates) is aligned in the proper column.

Evoke Simplicity

Create a clean, professional look to your resume. Do not use more than two font families, choose conservative fonts, and do not overemphasize text. Be sure to balance text and white space on the page. Do not pack the page with too much text, but don't have too much white space either.

Remember that design principles are rules of thumb, not absolute laws. You'll have to use your judgment about what looks best based on the above principles, careful study of models, feedback from experts (Career Services Center staff, business professionals, etc.), and feedback from family and friends.

Applicant Tracking Systems

As mentioned previously, ATSs, sometimes called talent management software, are used by many major and midsize companies, the government, and recruiting firms. Employers use ATSs to not only make the job of screening candidates easier, but they also use them to manage the hiring process, including keeping records and scheduling interviews. Depending on the software, it can be unreliable and finicky and reject seemingly qualified candidates due to technical glitches rather than content (see Figure 9.13).

applicant tracking systems
Software applications used by employers to sort and rate resumes prior to screening by a human reader.

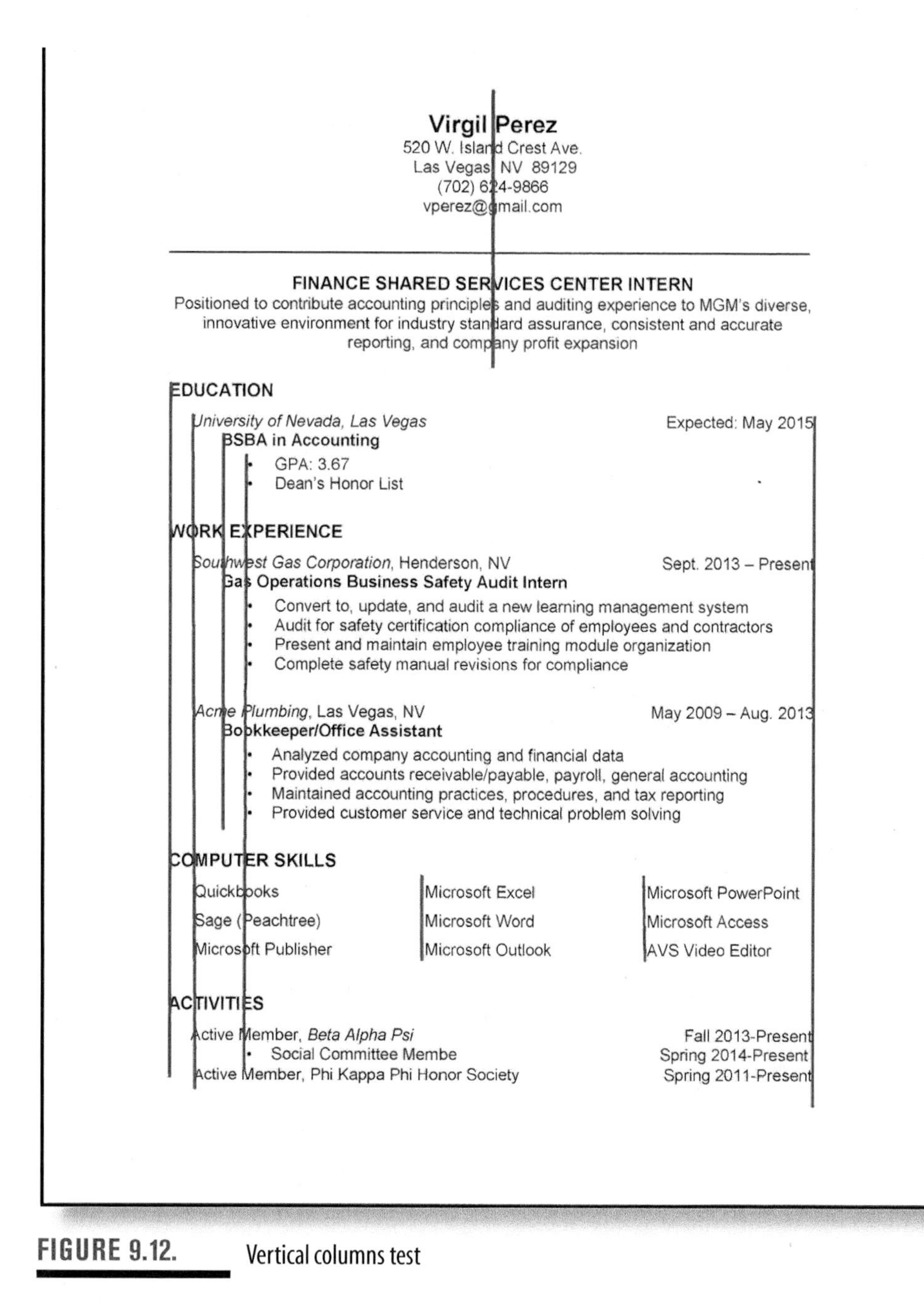
Virgil Perez
520 W. Island Crest Ave.
Las Vegas, NV 89129
(702) 624-9866
vperez@gmail.com

FINANCE SHARED SERVICES CENTER INTERN
Positioned to contribute accounting principles and auditing experience to MGM's diverse, innovative environment for industry standard assurance, consistent and accurate reporting, and company profit expansion

EDUCATION
University of Nevada, Las Vegas Expected: May 2015
BSBA in Accounting
- GPA: 3.67
- Dean's Honor List

WORK EXPERIENCE
Southwest Gas Corporation, Henderson, NV Sept. 2013 – Present
Gas Operations Business Safety Audit Intern
- Convert to, update, and audit a new learning management system
- Audit for safety certification compliance of employees and contractors
- Present and maintain employee training module organization
- Complete safety manual revisions for compliance

Acme Plumbing, Las Vegas, NV May 2009 – Aug. 2013
Bookkeeper/Office Assistant
- Analyzed company accounting and financial data
- Provided accounts receivable/payable, payroll, general accounting
- Maintained accounting practices, procedures, and tax reporting
- Provided customer service and technical problem solving

COMPUTER SKILLS
Quickbooks | Microsoft Excel | Microsoft PowerPoint
Sage (Peachtree) | Microsoft Word | Microsoft Access
Microsoft Publisher | Microsoft Outlook | AVS Video Editor

ACTIVITIES
Active Member, *Beta Alpha Psi* Fall 2013-Present
- Social Committee Membe Spring 2014-Present

Active Member, Phi Kappa Phi Honor Society Spring 2011-Present

FIGURE 9.12. Vertical columns test

However, 50 to 75 percent of employers do use them, so there are strategies that you can use to ensure that your resume is not discarded by an ATS:

- **Use keywords from the job ad.** As shown in Figure 9.13, an ATS will search your resume for language from the job ad and other relevant terms predetermined by the employer. Carefully choose your language to match the most important keywords. You can repeat certain words, but do so in a way a human reader will not think you are padding.

- **Keep the design simple.** Not all ATSs are able to read underlined text or italics. Avoid using fancy typefaces and colors. Use standard fonts such as Times New Roman, Georgia, Arial, Impact, Lucinda, Palatino, Tahoma, Verdana, and Sans Serif.
- **Do not submit a PDF file format.** Many older ATSs are not able to read PDF documents, so there is a chance your PDF resume could be misread.
- **Use a professional summary.** A list of your main qualifications at the beginning of your resume is a valid way to increase your use of keywords that won't be read as a gimmick by the ATS or human readers.

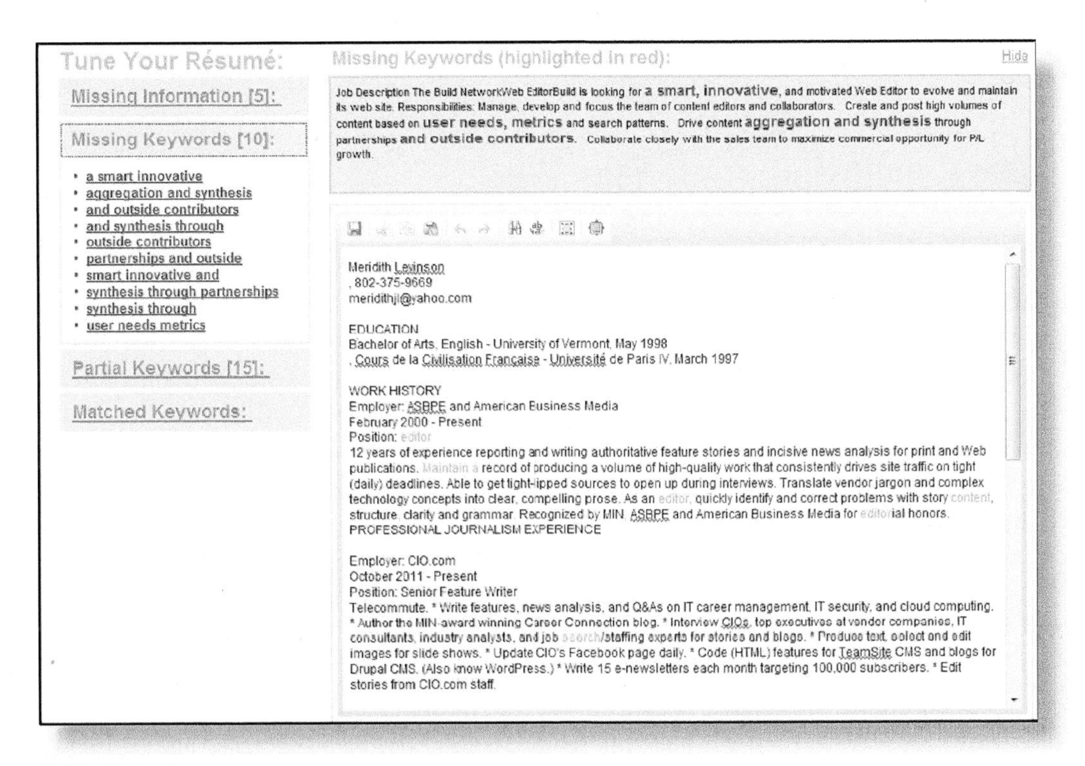

FIGURE 9.12. Applicant Tracking System (ATS) Display. This ATS made many mistakes in displaying this person's resume, including moving the opening profile summary to the work experience section.[14]

- **Do not use tables, graphics, and fancy bullets.** Most ATSs can't read graphics, and tables can confuse them. Boxes, lines, shapes, shading, and colors can also confuse ATS. Use standard bullets as well.
- **Use "Work Experience."** ATSs are programmed to search for this section, and if you use a different title such as Career Achievements or Professional Experience, the ATS can skip over it.
- **Write out dates.** Use a Month-Year format (April, 2014) writing out the full year (2014, not '14). Do not abbreviate the date (04/14) or use a season (Summer 2014).
- **Don't start work experience with dates.** ATSs look for company names first, so always start with the company name, followed by your job title, then the dates of employment.

- **Quantify whenever possible:** Instead of vague statements such as "managed assets," use numbers (#, $, and %) to tell a strong story: "Managed $80K in assets; supervised 20 staff and decreased production costs 50%."

In addition to a statistical view, most new ATSs enable the employer to view a copy of the resume as you originally designed it. You should assume that if your resume makes it through the initial screening process that it will eventually be read by a hiring manager or committee. You should still follow the principles of design discussed in this chapter to ensure your resume is visually appealing for those times when it is being scrutinized by a real reader.

Conclusion: Do it Your Way!

Remember that conventions for resumes vary from field to field. One strategy for learning and adapting to context-specific conventions is by studying models. It is perfectly okay to collect and study samples of resumes and cover letters. But you must imitate models with a critical eye, keeping in mind *your* skills and *your* particular audience. You should never make random or uncritical choices, like plugging your skills into someone else's template. What if the sample you're imitating is in functional format when the combination format could better represent your skills? How can you accomplish your goal of creating a distinctive-looking resume that sets you apart from others if you rely on a Microsoft Word resume template? Like writing any document, you need be able to justify your content and design choices based on document conventions and document design principles.

When it comes to writing resumes, "doing it your way" really means making strategic choices within the boundaries of convention. You need to know enough about the rules to be able to bend them without breaking them.

End Notes

[1] Joel Passen, "Tips for Job Applicants in 2014: How to Submit Your Resume through an Applicant Tracking System Successfully," Newtonsoftware.com, last modified December 11, 2013, http://newtonsoftware.com/blog/2013/12/11/tips-job-applicants-2014-submit-resume-applicant-tracking-system-successfully/.

[2] Susan Ireland. "Four Reasons Not to Use a Functional Format," last modified July 23, 2012, http://susanireland.com/job-lounge/4-reasons-not-to-use-a-functional-resumeformat/.

[3] Trish Thomas, "Why You Should NOT Use a Functional Resume Format," Patch.com, last modified May 28, 2013, http://patch.com/connecticut/simsbury/why-youshould-not-use-a-functional-resume.

[4] Katharine Hansen, "The Demise of the Functional Format," Quintessential Careers, last modified September 16, 2014, http://www.quintcareers.com/functional_resume. html.

[5] Alan Katzman, "Why Social Media Profile Links Always Belong on Your Resume," *Social-Hire*, last modified March 3, 2014, http://www.social-hire.com/career--interview-advice/4224/why-social-media-profile-links-always-belong-on-your-resume.

[6] Dawn Rasmussn, "Resume Objectives Must DIE...Here's Why," last modified June 18, 2013, http://www.pathfindercareers.com/blog/2013/resume_objectives_must_die.

[7] "The Little Things that Will Kill Your Job Search Momentum," Examiner.com, last updated June 13, 2013, http://www.examiner.com/list/the-little-things-that-will-killyour-job-search-momentum.

[8] Louise Fletcher, "The Resume Objective: Why it Needs to Go the Way of the DoDo," Blue Sky Resumes, last modified, August 24, 2014, http://blueskyresumes.com/blog/the-resume-objective-why-it-needs-to-go-the-way-of-the-dodo/.

[9] Katharine Hansen, "Your Job-Search Resume Needs a Focal Point," Quintessential Carriers, accessed September 11, 2014, http://www.quintcareers.com/resume_objectives.html.

[10] Alison Doyle. "How to Write a Resume Headline," About.com, accessed September 5, 2014, http://jobsearch.about.com/od/includeinresume/a/resume-headlines. htm.

[11] Katharine Hansen, "Branding Your Resume: Tips for Job Seekers," Quintessential Careers, accessed August 24, 2014, http://www.quintcareers.com/resume_branding.html.

[12] Hansen, "Branding Your Resume"

[13] Alison Doyle, "Resume Profile Examples," About.com, accessed September 5, 2014, http://jobsearch.about.com/od/profiles/a/resume-profile-examples.htm.

[14] Meridith Levinson, "5 Insider Secrets for Beating Applicant Tracking Systems," CIO.com, last modified March 1, 2012, http://www.cio.com/article/2398753/careersstaffing/5-insider-secrets-for-beating-applicant-tracking-systems.html.

COVER LETTERS AND PERSONAL STATEMENTS

CHAPTER 10

In an age of digital communication and 140-character tweets, many job seekers wonder if a cover letter is still necessary. The answer is yes, though the letter will more likely be in an electronic form such as an attachment or email message. According to Amy Gallo, not sending a cover letter is a sign of laziness to some recruiters. Furthermore, while not all recruiters may have time to read your cover letter in depth, if half of them do take the time, that's a 50 percent better chance that including a cover letter can help you.[1] As one recruiter commented on a recent LinkedIn discussion on cover letters,

> *I work in the career counseling/coaching industry and am astounded by the number of people who erroneously believe that cover letters are outdated. I am constantly coaching my clients on the skill of cover letter writing. Cover letters are never the first thing I read, but if a candidate matches required experience, I will next look to see if a cover letter was included. Extra points awarded to those candidates who took the time to go the extra mile.*[2]

Cover letters, and their close counterpart, personal statements, also serve other functions, such as answering the reader's questions about why you are interested in the particular position or degree program, what your short- or long-terms goals are, and how your qualifications match those listed in the advertisement. Cover letters and personal statements also provide evidence of your written communication skills.

Your cover letter and resume thus should work hand in hand to present your qualifications to an employer as completely as possible. Your resume is a list or outline of your most relevant qualifications, while the cover letter highlights your interest and knowledge of the company and specific skills or attributes that qualify you for the position. Your cover letter should not be a restatement of your resume, and it should not read like a form letter. Instead, it should demonstrate that you took the time to familiarize yourself with the company and show how you meet the qualifications identified in the job advertisement.

Aims of the Cover Letter

Employers want to get as much information as possible before making decisions about whom to interview. They don't want to interview unqualified candidates or candidates with padded resumes. The cover letter gives them one more opportunity to evaluate the quality of their candidates. However, this should be good news to you, for the cover letter gives you one more opportunity to connect your attributes with the needs of that particular employer, one more chance to be persuasive. Of course, to write a truly effective cover letter, you need to research the company in some detail and show readers that you have a working knowledge of their organization. The goals of your cover letter thus are as follows:

- Create a professional ethos
- Persuade through examples
- Show your interest
- Express your personality

Create a Professional Ethos. Researching a job demands a commitment of your time, but it allows you to craft a cover letter that provides a unique introduction of yourself. The cover letter may be your first contact with this employer so you want to be sure that you make a favorable first impression. Your cover letter needs to look like a formal letter and follow appropriate conventions. Pay attention to form and spacing so that your letter is aesthetically appealing.

Moreover, you need to avoid spelling, punctuation, and grammatical mistakes. A poorly written letter can reflect badly on you as a person and as a potential worker, since it implies that you are not interested in taking the time necessary to produce a document of quality. Avoiding errors is hard work, so take your time and check the entire letter closely. Read and reread your letter, and get others to read it. Also, don't be dependent on spell checkers and grammar checkers. Spell checkers, especially, can allow typos to inadvertently slip through.

Your cover letter must create the impression that you are knowledgeable about the company, the job, and your profession. You should be able to write authoritatively about all aspects of the company, especially those that are most important to you. You should be able to discuss in detail how your past experiences enable you to perform the duties described in the job advertisement. To establish your credibility, use keywords mentioned in the job description and drawn from your major/field/profession.

Persuade Through Examples. The main goal of the cover letter is to convince the employer that you can do the job, that you have the qualifications desired by the employer. The cover letter should not repeat the resume, but instead emphasize and elaborate those elements of the resume that are the most important. But you can't just **claim** to possess a certain qualification; you must **demonstrate** your claims with evidence, specific examples that convince the reader you indeed have the desired skill. Instead of writing, "I have tournament planning experience," write

> *I have tournament planning experience. In the summer of 2013 and 2014, I was a little league coach. For those two summers I prepared the round-robin brackets for a 30-team regional tournament. I learned how to set up the brackets, make sure they were fair, and declare an undisputed winner.*

The cover letter allows you to provide specific examples that show the reader your skills and discuss relevant qualifications in more detail than what is merely listed in the resume.

Show Your Interest. A cover letter should never read like a generic form letter. You can write a general resume for use at job fairs, but a cover letter cannot be used in the same fashion. You need to show employers that you are aware of their needs and that you took the time to write a letter directly to them. As Susan Adams, a recruiting expert on Forbes.com writes, "spend at least an hour on the company site reading and thinking, including clicking through every link. If the firm has a blog, read at least a dozen entries. Check out the firm's presence on social media and do a wide-ranging Google search."[3] There should be clear references to the employer and how your skills match the qualifications specified in the job advertisement.

If you have done the necessary research, you should be excited about the possibility of working for this company. Show them your knowledge of the company and that you want to be a part of the company. Point out specific features of the company that attract you. A prospective employer can't help but feel pride when reading about the positive attributes of her or his company. Your goal is to make an impression on the reader that you are sincerely interested in working for their company, which appeals to the reader's interest in hiring an enthusiastic employee (an application of the "you approach" discussed in the Style chapter).

Express Your Personality. Another reason for writing a cover letter is to express your personality more fully than you can in a resume. Since the cover letter is your primary introduction you need to work hard on your writing style. You should write your letter in your own words, avoiding inflated language or words that you are unfamiliar with. Try to be natural and use the language of your field to show your expertise. You don't need to sound more intelligent by using impressive-sounding words. In fact, using words that you are unfamiliar with will often leave readers unimpressed, especially if you use words inappropriately. You should, however, be able to use jargon appropriate to your field of expertise, but don't go overboard.

While you want to express a confidence in your abilities, avoid self-absorption. Unqualified, grandiose statements using words like *ideal, perfect,* or *most qualified candidate* tend to leave the reader unimpressed. Also, don't let humor invade the letter. You have no idea if the reader shares your sense of humor.

Cover Letter Format

Cover letters include the traditional features of a formal letter discussed in Chapter 5, Business Correspondence (heading, salutation, opening, body, closing, signature block, and end notation). Most letters for entry-level positions should be relatively short, **no longer than three-quarters of a page (see Figure 10.1)**. Similar to your resume, the cover letter needs to be easy to read in a short amount of time. Use smaller paragraphs and consider using lists to convey key details (but remember you are not just cutting and pasting from your resume). Writing a cover letter is a good exercise in writing concisely. You simply don't have the room to be wordy.

The Heading

In addition to the date, your contact information (address optional, see Chapter 9 on resumes) and the recipient address of the employer include a bold subject line with the title of the position you are applying to, including a position number if given.

Always try to address the letter to a specific person within the company. Employers don't like to see generic subject lines such as "Dear Sir or Madame:" or "To Whom It May Concern:" If you can't find a contact's name by researching the company's website and social media, consider contacting the company directly. Taking the time to identify the hiring manager's name shows that you are detail-oriented and motivated. If you can't find a specific name, direct the letter to a generic but specific person or department, such as "Dear Human Resources Manager" or "Dear Marketing Department." Be sure to spell the name, title, company, and address correctly.

The Opening

The introduction of the cover letter (1) identifies a specific job and how you learned of the position, (2) persuasively explains your motivation for applying to the position, and

23 North Hampton Road
Henderson, NV 89052

October 7, 2014

Michele Kantor, Internship Coordinator
Women's Monthly
233 S. Beverly Drive
Beverly Hills, CA 90212

RE: Women's Monthly Editorial Internship, LA Office

Dear Ms. Kantor:

I am applying to the editorial internship in the LA office that is posted on the *Women's Monthly* website. I know that your publication is an award-winning magazine that has been a leader in feminist activism in the media since 1972. I am extremely interested in utilizing my writing and organizational skills to assist in providing media coverage to a variety of women's issues. I would love to be a part of this company and learn how the editorial and publication process works from a successful nonprofit publication. I am currently pursuing my bachelor's degree in English and have acquired volunteer experience with both women's centers and student publications. I feel that my volunteer experience coupled with my education has prepared me for this exciting opportunity.

My volunteer positions have given me experience in activism concerning women's issues. As you will read in my enclosed résumé, I volunteered at my university's Women's Center where I helped to plan events, fact-check information for various campaigns, and performed specific office tasks. I learned a lot about organizing, and I developed strong interpersonal skills while interacting with my superiors, other volunteers, and the women that our center helped. Additionally, I have received organizational activism training from Planned Parenthood's Generation Action Power Tour this past summer. During this training, I learned how to use social media and resources such as canvassing to help further a movement. Lastly, I already possess the knowledge of all computer programs that are required for the position such as Microsoft Office, as well as knowledge of fact-checking and proofreading.

I am a senior at the University of Nevada Las Vegas with a 3.5 GPA. I have taken several upper-division writing, women's studies, and sociology courses that will allow me to excel at your magazine. My skills in both writing and organizational activism can be beneficial to your magazine, as this knowledge would provide me with a more inclusive perspective through which to connect with your readership. Additionally, I can apply my experience as editor of my high school literary magazine to help create the best possible content for this magazine.

I hope that you find my education and volunteer experience worth considering as a possible candidate for this internship. As I am graduating this December, I would be available for winter, spring, summer, and fall internship periods. If there is any additional information you require or if you would like to schedule an interview, please contact me at (702) 493-0210 or annewcomb@yahoo.com. Thank you for your consideration, and I look forward to hearing from you.

Sincerely,

Sarah Stewart

Encl.: Resume

FIGURE 10.1. Sample cover letter

(3) summarizes the skills you will develop in the body of the letter. The reader should know immediately why you are writing this letter, and don't make the reader guess what it is that you want. You then need a persuasive statement that explains your motivation for applying, followed by a brief list of your key skills or a preview of the attributes that

the letter will discuss. In this way you should be able to clearly connect your skills with the needs of the company and establish how you will address the qualifications outlined in the job advertisement.

Your opening should start with a strong statement about who you are and why you are interested in the company. Mention where you learned of the position. If you are still in school or graduating, mention where you go to school. If you know someone who works at the company, mention it in the opening as well. If you are applying for a job in another city, state where you are located and that you are planning to relocate or be in the new city during the time of the internship (see Figure 10.2).

> I'm a sophomore at UNLV, planning to major in communication. My communications club forwarded a flyer about a summer internship at MWW Public Relations in New York, NY. I'm a New York native and plan to be in the city from June 1 to late August. It would be exciting to intern at your "Public Relations Agency of The Year" Grand Stevie Award-winning firm. As you can see from my resume, my strong communications skills and prior marketing experience make me a strong candidate for this position.

FIGURE 10.2. Example opening

The Body

The body of your cover letter should be two or three relatively short paragraphs that demonstrate how you meet the most important qualifications in the job advertisement. Imagine your reader/employer sitting across from you in a face-to-face interview asking, "What are your qualifications for this position?" To answer this question, look at the job ad closely. For example, if it states that the company is looking for a person with a strong accounting background, prior accounting experience, and excellent communication skills, then these three qualifications become an organizing structure for arranging the order of topics and paragraphs in the body of the cover letter (see Figure 10.3).

Notice that in Figure 10.3, outline #2 only plans for two body paragraphs, but by combining some topics together into one paragraph, the writer can cover additional qualifications. Writing a topic sentence outline based on important qualifications listed in the job ad prior to drafting can help you write a clearer, more persuasive letter. Assuming your readers are busy and may not read your entire letter, the topics should also be arranged in order of importance.

Outline #1	Outline #2
• Opening • Body paragraph #1: accounting background/education • Body paragraph #2: prior accounting experience • Body paragraph #3: communication skills • Closing	• Opening • Body paragraph #1: accounting background/ education and experience • Body paragraph #2: communication, leadership, computer skills • Closing

FIGURE 10.3. Cover letter outline examples

By writing your cover letter so that it addresses the most important qualifications listed in the job ad, your letter becomes a written argument answering the reader's question about your qualifications. But it is not enough to merely *claim* you possess the most relevant qualifications; you must *support* your claims with evidence. You may have been taught this basic argument structure in your first-year composition course: a basic argument consists of a claim plus support. Without evidence to support a claim, it is considered weak or empty, and thus an unconvincing claim. For example, in response to an internship advertisement at a national woman's magazine seeking "bright, energetic, resourceful people with excellent research and writing skills" and good social media skills, a student wrote the paragraph shown in Figure 10.4.

The paragraph in Figure 10.4 contains five sentences (about the maximum recommended length for any cover letter paragraph). The first sentence, called the topic sentence, makes the claim that the writer possesses relevant volunteering experience. If this writer left it at that, it would be an *empty claim* and an unconvincing argument, but the student backs up her claim by adding four additional sentences of evidence. The paragraph elaborates on her claim about her volunteer experience by discussing specific examples from experience listed on her resume, such as her Woman's Center and Planned Parenthood Generation Action Power Tour experience.

Furthermore, the paragraph in Figure 10.4 includes a number of keywords, or specific terms listed in the job ad that employers are looking for applicants to possess: "volunteering experience," "fact-check information," "strong interpersonal skills," "activism," "social media." The employer (or applicant tracking software system, see Chapter 9) sees these keywords and makes the connection to the criteria listed in the job advertisement.

I possess 2 years volunteer experience in organizational activism concerning women's issues [CLAIM]. As you will read in my enclosed résumé, I volunteered at my university's Women's Center where I helped to plan events, fact-check information for various campaigns, and performed specific office tasks [EVIDENCE #1]. During my time volunteering there, I learned a lot about organizing, and I developed strong interpersonal skills while interacting with my superiors, other volunteers, and the women whom our center helped [EVIDENCE #2]. Additionally, I have received organizational activism training from Planned Parenthood's Generation Action Power Tour this past summer. During this training, I learned how to use social media and resources such as canvassing to help further a movement [EVIDENCES #3–4].

FIGURE 10.4. Claim + evidence structure of a cover letter body paragraph

In the following example, the student supports his claim of communication skills with a specific, concrete example drawn from his extracurricular leadership activity:

> *I see you are looking for someone with strong oral and written communication skills. Throughout my years at UNLV my communication skills have improved. During my presidency in the Asian Film and Drama Club I was required to obtain film rights to screen any movie I planned to use. I contacted distribution companies and film studios located in the United States, East Asia, Southeast Asia, and India by phone and email to obtain those rights.*

In the next example, a student addresses a computer skills requirement by not just claiming "I have knowledge of Microsoft Office" but by backing up the claim with details of specific Office skills she learned in her college courses:

> *My Introduction to Computers and accounting courses helped me gain knowledge of Microsoft Office Suite. I created essays, fictional internal documents, tables, and graphics in Word. I prepared budget spreadsheets, line and bar graphs, and financial statements using Excel. I used PowerPoint to create slideshows with text, audio, and graphics for my class presentations.*

The key to writing a successful cover letter is providing convincing and concrete details to support your claims about your relevant qualifications. Unlike the resume, the cover letter does not just list qualifications. It elaborates them by adding details in the form of examples and stories of past experiences that are not included in the resume.

Another strategy you should use in the body is to customize the letter and demonstrate knowledge of the company by connecting your skills to the needs and interests of the reader/employer. In the sample paragraph in Figure 10.5, the writer explains how her relevant skills relate to the specific needs outlined in the job description (indicated by the underlined phrases).

> I am a senior at the University of Nevada Las Vegas with a 3.5 GPA. During my time at UNLV I have taken several upper-division writing, women's studies, and sociology courses that <u>will allow me to excel at your magazine</u>. My skills in both writing and organizational activism can be greatly beneficial to your magazine, as <u>this knowledge would provide me with a more inclusive perspective through which to connect with your readership</u>. Additionally, I will use the knowledge I have gained from my time as editor of my high school literary magazine <u>to help create the best possible content for this magazine</u>.

FIGURE 10.5. Paragraph that connects skills to employer's needs and interests (underline added for emphasis)

Lastly, it is a good idea to refer to your resume at least once in the course of your cover letter. You want to convince the reader to read your resume and look at specific areas of your resume that they will find particularly interesting.

The Closing

Your conclusion should either request an interview or state a specific time when you will contact the employer about your application. The conclusion should also provide contact information (phone and email) so that the employer can reach you quickly and easily. You can refer to the address at the top of the page, but it is advisable to provide readers easy access to contact information in the closing, as well. Maintain a courteous tone, thank readers for their time, and offer to answer any questions. Your complimentary close can be formal "Sincerely" or slightly more personal "All the best" or similar.

Submission Format

If the job advertisement asks for a resume and cover letter, then that is a good indication to send both documents as attachments in an email message. Follow the employer's instructions for the preferred format documents (e.g., MS Word or PDF) and use similar file names starting with your first and last name spelled the same.

Write a short email message to transmit the attachments. Make sure you use a subject line that summarizes the content of your email (e.g., "Resume: Accounting Internship"). Follow guidelines for writing electronic correspondence, including using smaller paragraphs and lists. Mention in the opening what position you are applying to and what documents you are attaching (put the file names in parentheses). In the body of the message, summarize some of the highlights of your resume, using a bullet list of three to five items (possibly based on a professional summary if you included one in your resume opening). If you referenced any social media, include the link in your bullet list. Include a polite closing, and since the email cover letter doesn't contain a heading with your contact information, you should include your address, phone number, and email address after your name in the signature block.

Save the message as a draft, print a copy, and carefully proofread it. Susan Ireland, a recruiting expert, recommends running an email test by sending the message to yourself to see if the formatting holds up before sending it to the employer. Ideally, you should send your draft to a friend on another email system (e.g., Gmail to Yahoo), since systems process messages differently.[4] When you are ready to send the final version, don't forget to attach the more detailed cover letter and a copy of your resume.

If the job ad states "no attachments," you can copy and paste your cover letter and resume into the body of the email. If your cover is not long, you can copy and paste it into the email and attach it to the message.[5]

Larger companies may ask you to use an online form to submit your application. Some will provide a space to attach or copy and paste a cover letter. Some companies may not request a cover letter directly but will ask you to answer one or more questions similar to a cover letter, such as why you want to work for the company or what your long-term goals are. Never compose your cover letter or short answers about yourself in the form field of the website itself. Take the time to compose your message in a word processor that can help you proofread your answers, and don't forget to get some feedback from trusted friends or mentors.

The Personal Statement

Personal statements are typically written for occasions such as applications to graduate and professional school, for scholarships, and some internships. Personal statements tend to be closer to academic essays in format and appearance but have similar aims as cover letters in terms of trying to persuade readers you possess the relevant qualifications. Graduate programs, professional schools, and nonprofits also often want to know what your goals are and why you chose the particular program. Just as with cover letters, a well-researched and detailed personal statement will be more convincing than a vague, generic, or cliché one.

Personal Statement Structure

A personal statement should be organized similar to a cover letter with an opening, body, and closing. Just as with cover letters, you should determine the most important criteria stated in the application information and arrange your topics in order of importance. However, one common structure to avoid because it is so overdone is the narrative structure. In other words, your personal statement **should not be organized chronologically** beginning with such clichés as "when I was a child...." moving through your school years and ending with your most recent qualifications. This all-too-common structure risks quickly turning off readers who have typically read hundreds of personal statements over several years. Just as with cover letters, carefully studying the application instructions for qualifications and keywords and creating a topic-sentence outline of your key qualifications arranged in order of importance will enable you to draft a stronger statement.

The opening of the personal statement tends to more indirect than the opening of a cover letter. Similar to advice you may have learned about writing an introduction to an academic essay, readers of personal statements still expect to have their interest aroused with what is typically called a "hook" or lead-in. This can be an interesting story, anecdote, or description that provides a compelling reason for the reader to continue reading. A good hook will also be related to the content or theme of the statement. Following the hook, the opening should summarize the qualifications that will be addressed in the body (see Figure 10.6).

The body of the personal statement should develop in specific, concrete details the relevant qualifications outlined in the introduction. Similar to a cover letter, these qualifications should be arranged in order of importance. Readers of personal statements for academic programs or scholarships often want to see evidence of your academic aptitude. They want to know if you have the skills to succeed in their academic program or to maximize the scholarship funds given to you. This usually means including details of your past academic achievements, including grades, awards and honors, scholarships, and participation in academic and professional organizations. Readers of statements of professional programs such as law, business, or medical school want similar evidence of your aptitude and experience in the given area, such as work and internship experience, and other qualifications explained in the application instructions.

Personal statements often include an answer to the reader's question, why do you want to go to this particular program? This is a goals statement or statement of purpose, and is sometimes the sole focus of a personal statement. Here, again, you should avoid vague generalities and clichés by carefully researching the particular program you are applying to so that you can articulate a concrete reason why you want to attend that particular program. Most academic programs have particular specializations or areas of strength. Research the graduate school or professional program so that you can

First line in opening is somewhat direct, but the second line contains an interesting hook statement that also shows knowledge, interest, and some personality.

Last part of opening overviews key qualifications that will be discussed in the body of the statement.

The writer includes several concrete examples that demonstrate relevant qualifications, in this case, using technology and designing curriculum.

Usually a statement for an academic program would provide evidence of academic potential in terms of grades, honors and achievements, and test scores, but the writer may have chosen to leave out such information for strategic reasons.

The conclusion provides a bookend to the opening, reiterating points made in opening.

Personal Statement

My goal is to work in the field of online education. I want to create online environments that are not only educational, but also easy to navigate and fun to use. My academic career includes degrees in both education and graphic design, and I have taken numerous online courses. I believe that this combination has prepared me well for a career as an e-learning professional.

My drive to educate is what led me to obtaining an associate degree in special education. One of the biggest lessons that I took from that program is the importance of alternative teaching methods. While attending the College of Southern Nevada, I was privileged to be part of a NASA project for pre-teachers in which I was responsible for creating a lesson plan and then teaching that lesson. The paint-by-math sets that I created using Adobe Photoshop were a big hit with the children. These sets consisted of outline illustrations of our solar system and math problems to match the correct colors. Seeing children eager to do math is an experience that I will never forget. I want to make education fun for everyone.

The visual appearance of educational programs can have a large impact on the enjoyment of the learners. The skills that I have acquired through my undergraduate degree in graphic design are essential for producing visually engaging materials. I specifically remember an info-graphic assignment where I used the circular logic of the chicken and the egg to illustrate the creative process. Not only is the graphic informative, but also visually appealing.

My college career has also given me numerous opportunities to enroll online courses. One example of this is my independent studies class, through which I am learning to write and illustrate a children's counting book. My instructor and I are communicating primarily through email with written messages and PDF files. In fact, the instructor and I recently discussed the options of publishing the book in an electronic format. I like the idea that my book can be made available to so many readers through technology.

Technology has been a large part of my own education. Not only have I used technology to complete the majority of my assignments, but I have also received much of my instruction by the same means. In fact, my favorite courses are those taught completely online. The ability to attend graduate school entirely online was a major factor in my decision to apply to the Virtual College at FHSU. I also feel that learning how to design electronic courses through electronic courses, will give me a distinct advantage in the field of eLearning.

This will be a culmination of my dreams. I want to become an e-learning professional in order to fulfill my desire to educate through technology, alternative teaching methods, and design. And I believe that FHSU is a necessary part of my success as a future e-learning professional.

FIGURE 10.6. Sample personal statement

determine (1) if the program is right for you and your long-term goals, and (2) how you can best articulate that goal in your statement.

The closing paragraph summarizes qualifications and appeals to the application reader (e.g.., "For these reasons I believe I will be a strong candidate for admission to your program"). Some experts recommend using a "bookend" strategy that brings back your hook in some creative and summative way. If you opened your statement with an anecdote about your goals, end your conclusion with that same point rephrased.

Personal Statement Style

A few other principles you should keep in mind when writing your personal statement (but that also apply to cover letters) are as follows:

- Avoid clichés
- Don't call attention to weaknesses
- Don't damage your image

Always imagine a reader who has read hundreds of personal statements. Phrases like "all my life I have had a passion for…." or "I just want to help people…" may indeed be true for you, but readers are turned off because these expressions are so overused they are cliché. Instead, *demonstrate* your passion by describing your achievements and active participation in relevant organizations, or *show* how you help people through your actions and concrete examples.

Don't call attention to weaknesses. Do not make statements such as "I know my grades are not the best…." Don't undermine your own credibility; instead, focus on the positive traits that you do possess and leave it to the reader to judge where your application may fall short.

Lastly, though it is called a personal statement, you should avoid revealing details that are too personal or that may damage your image with your reader. A common request in personal statements is to discuss how you overcame adversity, but discussing how you overcame substance abuse or dropped out of high school may turn off some readers, so be careful.

Conclusion

Most students have written a resume at some point, but few have much experience writing cover letters since service jobs held in college rarely require cover letters. Yet, cover letters are expected for most professional positions, and even if they are not required, they represent an additional opportunity to inform readers about your relevant qualifications. Just as with resumes, cover letters should never read like form letters or templates. The best approach is to create an outline based on careful analysis of the job advertisement or application instructions and after researching the company, organization, or school. Writing cover letters and personal statements are also good exercises in writing a clear and concise argument responding to the reader's question about why you should be interviewed for the position or admitted to a program. A well-written cover letter usually stands out among the pile of poorly written ones (or absent ones) and can make a strong impression on an employer or application reader that you are an articulate and motivated potential employee, graduate school candidate, or scholarship recipient.

End Notes

[1] Amy Gallo, "How to Write a Cover Letter," HBR.org, last modified February 4, 2014, https://hbr.org/2014/02/how-to-write-a-cover-letter.

[2] Brian De Haaff, "Use These Cover Letters That CEOs Read," Linkedin.com, last modified December 17, 2014, https://www.linkedin.com/pulse/use-cover-letters-ceos-readbrian-de-haaff.

[3] Susan Adams, "How to Write a Cover Letter When You Have No Experience," Forbes.com, last modified January 21, 2015, http://www.forbes.com/sites/susanadams/2015/01/21/howto-write-a-cover-letter-when-you-have-no-experience/.

[4] Susan Ireland, "Should I Send My Cover Letter as an Email Attachment?" SusanIreland.com, last modified August 26, 2014, http://susanireland.com/job-lounge/send-cover-letter-as-email-attachment/.

[5] Ireland, "Should I Send My Cover Letter as an Email Attachment?"

CHAPTER 11

PRESENTATIONS

Most oral presentations in business settings, except for formal presentations such as a CEO's keynote address, are extemporaneous. That is, you need to plan and practice your presentation, but the actual delivery needs to appear as if you are talking off the cuff. *Talking* is the keyword. Business presentations, whether formal or informal, are not read verbatim from 3 x 5 cards. They are not even memorized and repeated word for word. In business settings, where you never want to lose your audience because of the monotone recital of some canned speech, presentations need to be lively and interactive.

In formal presentations you'll likely be selling some idea or product to a small group in a confined conference room, so establishing a positive, personal relationship is critical. In informal presentations where you know your audience more closely, you might be informing supervisors and/or coworkers of the status of a project. In either case, reading to your audience won't do. You need to establish a knowledgeable, confident, and prepared ethos during business presentations. Relying too much on notes suggests that you don't know what you're talking about or didn't care enough to prepare beforehand.

Instead of reading verbatim from notes, you need to talk to your audience, make eye-contact with them, and draw them into your presentation. To help guide you and your audience, you should use *visual aids*, including slides, print literature (handouts), computer demonstrations (projected onto a screen), whiteboards, flip-charts, etc. You should also make your audience active participants rather than passive recipients of your message by asking questions, elaborating with easy-to-understand visual examples, and perhaps adding hands-on activities.

Purpose

Be sure to understand up-front what the desired outcomes of your presentation should be, or what the audience should know about or know how to do at the end of your presentation. Decide what the main point of your presentation should be and define what action you want them to take afterward. Do you want your audience to understand why your product is best for their needs? Do you want them to abandon the planned merger? To buy your product? To know how to use the company's new voice-mail system? Build your whole presentation around getting your point across persuasively.

Audience

Be sure to analyze and define your audience—who they are, what their relationship to you is, what their primary occupations, activities, values, etc. are. Define the context of your presentation as well. Where is your audience working/learning? How will they benefit from your presentation? What do they need to know to use your information effectively?

Be sure to also define your audience's existing level of knowledge about your presentation topic. Consider how well they know the topic you're going to discuss. What can you assume they know? What concepts and terminology will you need to explain, define, spend more time on, etc.? A presentation on how to use Google Mail will have to be tailored depending on whether the audience uses Gmail regularly and wants to learn more about advanced features, has never used Gmail or has never used a computer.

Organization

The structure of effective presentations dates back to Classical times. Aristotle wrote about it in the fifth century, B.C. To get your main point across you need to build into your presentation *planned redundancy*. That is, first tell your audience what you're going to tell them, then tell it to them, and when you've finished, tell them what you just told them. Sound familiar? Of course, it's the standard opening-body-closing structure. It applies doubly to presentations, where your message will be competing with your audience's short, wandering attention span.

The standard organizing pattern of presentations is as follows:

- **Introduction:** Introduce yourself to your audience and establish your presentation's purpose and/or outcomes. Mention your name, some background (e.g., organizational affiliation, if you have one, or your major), and your reason for talking. Establish your credibility with a crisp, practiced opening that gets your audience's attention and gives a clear sense of direction. Include some type of attention-grabbing hook that arouses interest and introduces the subject of your presentation. The hook can be a thought-provoking question, interesting or shocking statistic or fact, an anecdote or story, or possibly a carefully measured joke. If you do decide to use humor, make sure it is audience appropriate and try to test it on a representative audience member beforehand.

- **Overview:** You may decide to start out with some attention-getting opening before or after your overview, but at some point after introducing yourself and your topic and before you start the main part of your presentation, you absolutely must "signpost," or outline, the main points of your presentation.
- **Discussion/Body:** Cover the main points of your presentation in a logical, easy to follow order. You can lend coherence to your discussion by including clear transitions (e.g., *next, secondly, another, Now Jane will...*, etc.).
- **Conclusion:** Tell them what you told them. Give your audience a clear signal that that the presentation is complete by including a concluding or closing discussion that reiterates the main points.
- **Q&A:** You should build time into your presentation for questions and answers. Anticipate and invite questions.

Research suggests that a typical person's attention span lasts anywhere from 10 to 20 minutes. You know this from sitting through boring lecture classes. To secure your audience's attention and help them remember whatever idea or product you're selling, be sure to include *interactive elements* in your presentation. These include such techniques for keeping your audience's attention as the following: questions for them to ponder and maybe even answer; examples, anecdotes, and stories for them to visualize; references to visual aids like charts and figures or handouts; actual activities or demonstrations for them to participate in; and even (appropriate) jokes.

Visual Aids

A visual aid is anything used to enhance the oral delivery of your presentation including a slide deck (a collection of bullet-point slides), overhead transparency, flip-chart poster, handout, or physical object. Presentations are increasingly being delivered over the Internet and through real-time web, video, and telephone conferencing applications. Using presentation software applications like PowerPoint, Google Slides, or Prezi.com to aid the delivery of presentations has become standard, though you need to carefully decide what visual aids will be most appropriate, and feasible, for your particular purpose, audience, and context.

When using presentation software to create visual support for your presentation, as a rule of thumb keep the design of your slideshow *simple* and *consistent*, and *appropriate* and *appealing* to the audience. In "Power Pointless," Rebecca Ganzel discusses how many workers overzealously try to integrate all the bells and whistles of PowerPoint.[1]

Workers get wrapped up in working on the presentation, wasting hours at a time tinkering in PowerPoint when they should be researching their content and honing their script. The result is weak presentations with messages lost in the tacky blur of busy, multicolored slides that make funny noises and contain distracting animated images. Remember, your ultimate goal for visual aids is to support getting your main (oral) message across to your audience.

Slides

When writing your slides, apply the design principles used for writing resumes discussed in Chapter 9. Like the resume your slides are an **outline**. Don't use complete sentences, and avoid text-heavy content. Use concise lists in sentence fragment form. When choosing fonts, sans serif fonts like Helvetica or Arial are best. Choose no more than two font families. Use 24 point font and larger, nothing smaller than 18 point. Avoid the use of italics and all capital letters. Be sure to use **contrasting colors** between the text and background for readability. Do a "back of the room" test to see if there are any problems with your design choices.

Include the following elements in your slide deck:

- **Title slide:** The first slide captures the main point of your presentation and contributes to the initial impression you make with your audience. Try to avoid generic titles. Instead of "Google Slides," how about "see note on top of next page. Google Slides: Revealing the Secrets of Designing with Google Apps"? Include other details like your name, the date, and possibly the setting of your talk. Add specific details to your slide to let your audience know you tailored it to them (e.g., the correct date or the company's logo pulled from its website).

The title slide contains an interesting, descriptive title, names, and date.

It could also mention location and include a company logo.

The title slide layout is typically different from the body slides, but it contains the same design element, e.g., color scheme.

FIGURE 11.1. Example title slide

- **Overview Slide:** The second or third slide should be a bulleted list of the main points of your discussion.

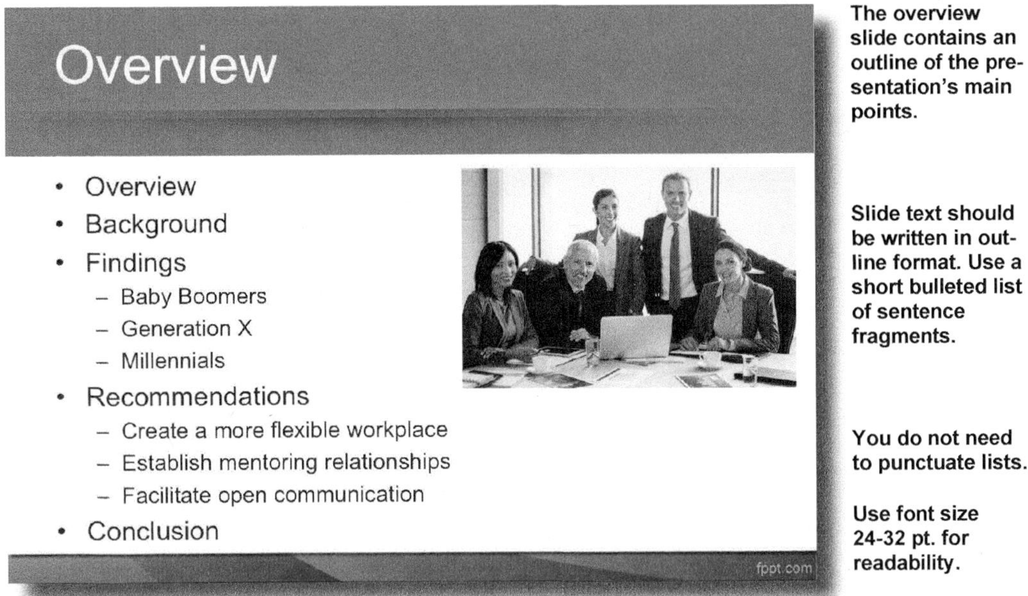

The overview slide contains an outline of the presentation's main points.

Slide text should be written in outline format. Use a short bulleted list of sentence fragments.

You do not need to punctuate lists.

Use font size 24-32 pt. for readability.

FIGURE 11.2. Example overview slide

- **Body Slides:** Format your body slides with consistent use of design elements like font choice, font size, color scheme, location of text, bullets, etc. Use the principle of "talking headings" for each body slide. In other words, give each slide a heading that captures in a short phrase the gist of that particular slide. If you have too many points to fit onto one slide, create a second slide with "title (con't.)."

Generation X (1965-1980)

- Occupy middle management positions
- First generation raised with technology
- Cynical and pessimistic
- Loyal to individuals instead of organizations
- Prefer electronic forms of communication

fppt.com

The body slides elaborate main points outlined in overview slide.

Use capitalization consistently in slide headings (either first word only for all major words).

Keep layout of heading and lists consistent.

Edit slides carefully for mechanical errors. Typos on slides are embarrassing.

FIGURE 11.3. Example body slide

- **Data slide:** If you are including any of the figures or tables discussed in Chapter 8, Reports, make sure the text in the visual is readable to the audience. This generally means using the entire text area of the slide for the visual. Avoid including both a visual and bullet list of text in one slide because this usually makes the visual smaller and harder to read. If you are borrowing the visual, indicate the source in small print below the image.

Personal Values by Generation

Personal/ Lifestyle Characteristics	Veterans (1922-1945)	Baby Boomers (1946-1964)	Generation X (1965-1980)	Generation Y (1981-2000)
Core Values	Respect for Authority Conformers Discipline	Optimism Involvement	Skepticism Fun Informality	Realism Confidence Extreme fun Social
Family	Traditional Nuclear	Disintegrating	Latch-key kids	Merged families
Education	A dream	A birthright	A way to get there	An incredible expense
Communication Media	Rotary phones One-on-one Write a memo	Touch-tone phones Call me anytime	Cell Phones Call me only at work	Internet Picture Phones E-mail
Dealing With Money	Put it away Pay Cash	Buy now, pay later	Cautious Conservative Save, Save, Save	Earn to spend

Source: Danielle Peterson, "Four Generations," Arthur-Maxwell.com, 2011 fppt.com

The visual in the data slide should be large enough so text can be read by the audience.

Avoid combining bullet lists with an information visual, or using more than one information visual.

Indicate the source of borrowed visuals.

FIGURE 11.4. Example slide with data

- **Closing slide:** End by reiterating your main points or outcomes. Afterward, you can consider a clever and/or humorous final slide that leads into the Q&A period, but avoid clichéd endings like a big "The End."

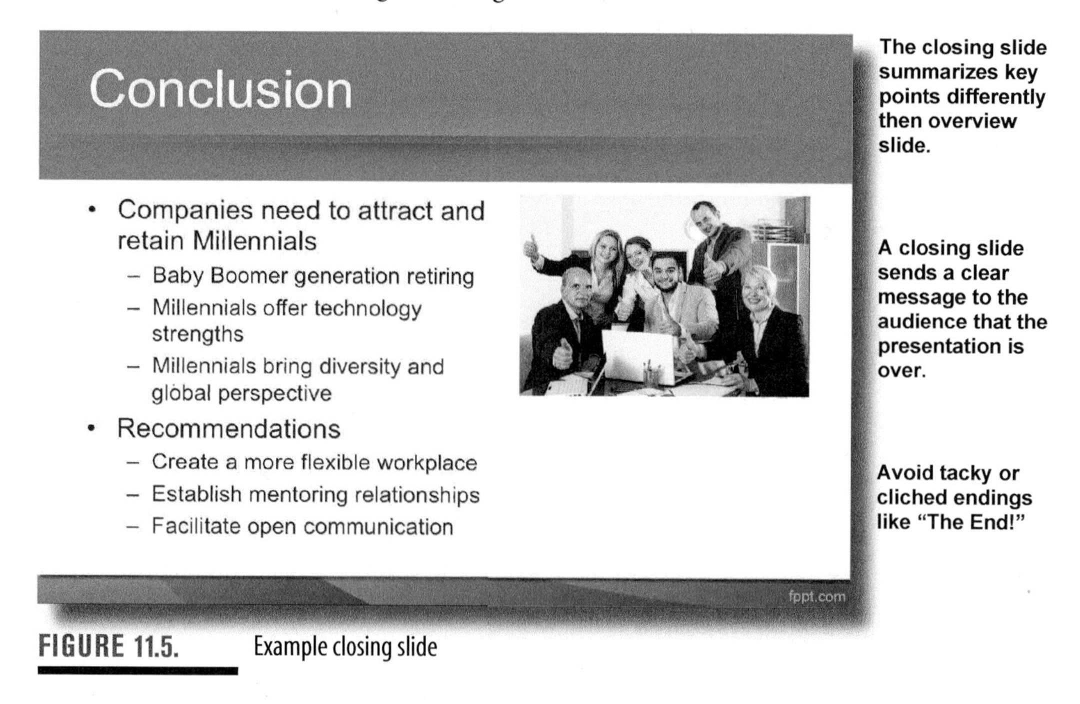

FIGURE 11.5. Example closing slide

The slideshow template used in the example is from Free PowerPoint Templates.com at http://www.free-power-point-templates.com/. This website enables users to search by theme or color. It contains many simple, professional-looking alternatives to the standard set of PowerPoint templates everyone uses.

Other important general guidelines when designing slides include:

- As a rule of thumb, plan to **use no more than one visual aid per minute**; a fifteen-minute presentation will require no more than fifteen slides, and preferably less. Don't overload your audience with specifics they will never remember.
- Templates are helpful for ensuring design consistency. Templates with white backgrounds and contrasting design element in the margins are the most readable and professional. Consider the implications of choosing a template or using a template wizard when producing slides or transparencies. Does the template best represent the image/message you want to convey? Is it appropriate? Is it easy to read? Is it too busy with color and graphics? Very often, novice presenters fail to fully consider the implications of the template they choose. Don't pick the tropical beach-themed template because you want a vacation, but because you're presenting to a group of travel agents.
- Pay attention to the design principle of balance. Your combination of text, graphics, and white space should not appear unbalanced. If items in a bulleted list appear to take up only the top or left portion of your slide, add a graphic (a photo or clipart image) to balance the layout and provide visual appeal. When choosing graphics avoid picking amateurish looking clip art and loading a single slide with too many graphics. Another element of balance is to aim for a consistently sized border of white space around all sides of each slide.
- If you cite statistics, quote someone, or borrow a graphic, be sure to identify your source. In a small font somewhere at the bottom of the slide or near the graphic, write, for example, "Source: U.S. Dept. of Labor Quarterly Report, Oct. 9, 2014." Be prepared to provide the full citation should someone ask, or include it in a handout.
- Move dense, text-heavy content to handouts. Put long quotes, text-heavy tables, or complex figures on paper instead of creating a hard to read slide that will be more distracting than helpful. Use concise slides in conjunction with the printed handouts that contain the dense content.

Handouts

These are among the easiest visual aids to make. They can also be a valuable tool for keeping your listeners on track during the presentation, ensuring that they go away with accurate information, and providing them with permanent reference material. Handouts don't have to be elaborately made, but the more professional looking your handout, the better the impression you make with your audience. As with any document, you should always consider issues of page design as an element of the overall usability and persuasiveness of whatever handouts you distribute.

Should you distribute handouts before, during, or after your presentation? It depends. You should time the distribution of your handouts to coincide with the presentation's purpose. If your purpose is to inform your audience, you can distribute exercises and workshop materials before the presentation. But an article, traditional formal report, or other reference information probably should be distributed at the end (unless you're planning on referencing or explicitly referring to such information during your talk). This keeps your listeners from becoming readers until after the presentation. Sometimes you can coordinate handouts with other visual aids. For example, a procedures manual or a list of features and benefits might be distributed as a model when introduced.

As with every form of visual aid, the most effective handouts are those that are actively incorporated into your presentation. Introduce the purpose of your handouts and tell your audience how to use them. Consider using a fill-in-the-blanks approach: get your listeners actively involved by making them follow along with the handouts as you speak. One literal interpretation of this is to have them fill in information or answer questions on the handouts as you speak.

Always bring enough handouts for everybody plus a few extra. A good rule of thumb is to use the room capacity of your presentation or to find out ahead of time the estimated or real attendance figures.

Delivery

Your goal is to make yourself, your presentation, and your key points as memorable as possible. Preparation is the key. Know your purpose and audience. Prepare your content so the number of key points is limited, so they are presented in a dynamic and convincing way, and so they are repeated strategically (so they are signposted in the introduction and conclusion).

script

A document used to plan and practice a presentation that includes an outline, time line, slide content, verbatim dialogue, and directorial cues.

Develop a **script**, a detailed outline that includes directorial cues such as when to change slides, when to pass out flyers, and who's going to say and do what (see Figure 11.6). Most scripts are written verbatim, but with the knowledge that they will not be read verbatim. The more details you write into your script, the more likely you are to remember them as you review and rehearse. Develop your script based on the total time you have to present, and test and refine your timed outline with actual practice. This will give you greater control later, when you are presenting from your notes or slides.

The other essential ingredient is to prepare yourself by **rehearsing**. The two deadly sins of presenting are being unprepared and speaking for too long. These sins go hand-in-hand. If you are prepared, you will have rehearsed, and if you rehearse, it is easy to time your presentation. You cannot omit this essential element in preparing your presentation.

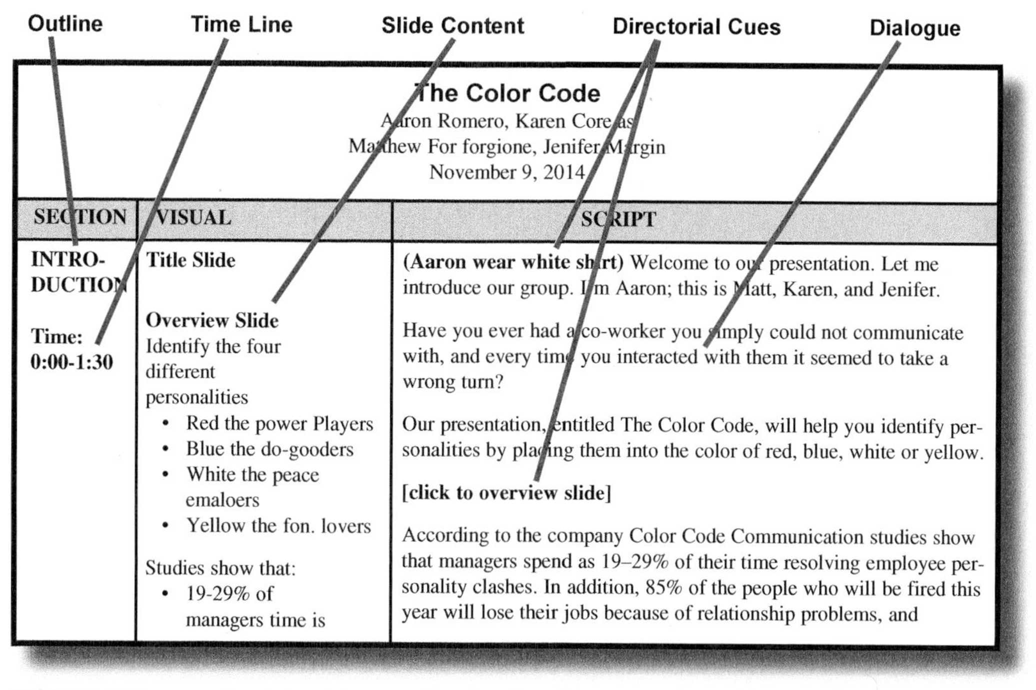

The Color Code
Aaron Romero, Karen Corebas
Matthew For forgione, Jenifer Margin
November 9, 2014

SECTION	VISUAL	SCRIPT
INTRO-DUCTION **Time:** **0:00-1:30**	**Title Slide** **Overview Slide** Identify the four different personalities • Red the power Players • Blue the do-gooders • White the peace emaloers • Yellow the fon. lovers Studies show that: • 19-29% of managers time is	**(Aaron wear white shirt)** Welcome to our presentation. Let me introduce our group. I'm Aaron; this is Matt, Karen, and Jenifer. Have you ever had a co-worker you simply could not communicate with, and every time you interacted with them it seemed to take a wrong turn? Our presentation, entitled The Color Code, will help you identify personalities by placing them into the color of red, blue, white or yellow. **[click to overview slide]** According to the company Color Code Communication studies show that managers spend as 19–29% of their time resolving employee personality clashes. In addition, 85% of the people who will be fired this year will lose their jobs because of relationship problems, and

FIGURE 11.6. A script contains an outline, time line, slide content, directorial cues, and verbatim dialogue.

Once you've written a complete script, practice several times from it. For high-stakes presentations experts recommend a minimum of five trial runs—the last at least two days before the actual scheduled event. You can adjust your practice time accordingly, but you should never walk into an important presentation never having practiced. Stand before a mirror, a video camera, or better yet, an impartial group of observers. Watch and listen to yourself. Play and replay the recordings. Get feedback from the trial-run observers. Ask yourself and your observers whether the main points are clear and receive adequate emphasis throughout your talk. Time the presentation to make certain it isn't too long. Repetition will help you incorporate the specifics of what you want to say and the rhythms with which you want to deliver them into your subconscious so that the actual presentation will be second nature to you.

Other points to keep in mind when preparing and practicing your delivery include:

- **Avoid stuttering** and using too many "ahs" and "ums" and "like, you knows." If you have a tendency to talk like this you need to practice them out of your presentation, and consider taking a speech class to polish your delivery techniques.
- **Don't read from each visual verbatim.** Your audience members can read at their own pace (often faster than yours) when it's convenient to them. Make your role as presenter indispensable: explain, embellish, and add detail. Elaborate on each visual with your discussion. Talk to your audience, don't read to them. Don't expect the slides to do the work for you.

- **Never turn your back to your audience.** Instead, print notes or slide views of your slide deck to keep your slide content in front of you while your actual show is projected onto a screen behind you. If you have to point at something in the slides, keep your toes and shoulders pointed at the audience as you reach back. Avoid using laser-pointer devices. Once a very popular presenter's toy, they're now considered amateurish and distracting.
- **Always try to involve your audience** with eye-contact, questions, and interactive elements within your presentation. While this depends on the context of your presentation, typically audience members are more likely to enjoy and remember your presentation if you involve them as more than passive receptors of your discussion.
- **Include checkpoints** if you're leading an activity or demonstrating how to do something. Stop and check to see if anyone is lost or needs help. Have assistants circulate to assist individuals who fall behind or get stuck.
- **Always get there early** and make sure the equipment/technology is present, working, and suited to the site. Start on time and end on time. Appoint a timekeeper before you begin. If your session is 12 minutes, talk for only 10 and leave time for questions and answers. Dress appropriately and professionally. If you're unsure what's appropriate for a particular context, err on the conservative side.
- **Be yourself when answering questions.** Questions are an opportunity to elaborate on your message around an interested audience member's pointed question. Take advantage by saying more than yes or no. If you don't know the answer, don't be afraid to admit it (but don't get caught not knowing something central to your purpose). If you don't understand the question, ask the person to rephrase it. If you should happen to have someone put an angry or hostile question to you, maintain your composure. Politely and respectfully try to respond; avoid getting into a figural—or literal—boxing match. Maintain goodwill.
- **Prepare a back-up plan for high-stakes presentations.** How will your spectacular Prezi.com presentation fare if there's no Internet connection onsite? What if it dies right in front of you? Always prepare a back-up set of visuals in another medium. Make handout copies of your slides or bring your slides on posters, flip-charts, or easel board. Of course cost is a factor, but when losing a client or professional ethos within your own company is a factor isn't the extra cost worth it?
- **Keep vital presentation elements safely with you if you** are traveling, not in your luggage bag that may wind up in Tahiti. Murphy's law, "anything that can go wrong, will go wrong," applies doubly to presentations and, particularly, computer-assisted presentations.

The Big Day

Delivering high stakes presentations is never easy. It will always require planning and you'll always be worried about making the deal or accomplishing your goal. But it gets easier with each experience presenting. The best way to minimize nervousness is to prepare and practice. If you know what you're going to say and have practiced how you're going to say it, you'll have less to worry about. If you're worried about being the center of attention, know that a well-prepared set of visual aids will help divide the audience's attention between you and the visuals. There's nothing wrong with this, as long as your visuals are professionally designed and contribute to your purpose.

Lastly, don't worry about making mistakes or slip-ups during the presentation. If you slip, just keep going; it's likely your audience won't even notice. If it's a bigger slip, that the audience does seem to notice, make a joke and then go on. Your audience will appreciate your composure.

End Note

[1] Rebecca Ganzel, "PowerPoint? What's Wrong With It? Power Pointless," *Presentations, February 2000, 53–58.*

WRITING FOR SOCIAL MEDIA

CHAPTER 12

If you know what a *hashtag* is, then you're probably aware that a *tweet* is more than just a bird's chirp. Social media has changed the way we communicate, and more and more companies are using it. Marketing, customer service, networking, and employee recruitment increasingly take place through such networks.

The continued popularity of social media seems certain. In 2010, Facebook briefly succeeded Google as the most-viewed website in the United States.[1] It regularly scores among the highest in terms of traffic, along with other social media giants.

With 1.35 billion monthly users as of late 2014, Facebook's value as a communication tool is undeniable.[2] Twitter, LinkedIn, and Instagram are also important platforms that offer strong exposure for professionals and businesses. It's important to know how to navigate the shifting terrain of social media, and to do so with skill.

Writing for social media tends to be more conversational than typical business writing. Humor is often rewarded with a "share" or a "retweet," which passes the content on to new audiences. But the public nature of writing for social media also makes attention-grabbing efforts risky. A joke may backfire if deemed offensive. Piggybacking on "trending topics" like news events for commercial purposes may also be seen as cynical or dishonest. This chapter focuses on the do's and don'ts of social media writing, and includes famous examples of social media communication failures. It also provides overviews of the most popular platforms' features and advice on how to write effectively for them.

Networking

Industry leaders often act as role models on social media, demonstrating the sorts of ideas that are important to succeed in a field. They may link you to valuable news or industry events. In general, social media is a great venue for building a professional community and staying engaged with what's going on in your field.

By connecting with other professionals, you can build relationships that may lead to opportunities down the road. It's important to note that this should happen organically, in the context of a social media relationship or in-person events. You don't want to come off as pushy or be accused of online stalking. Rather, think of sites like Facebook and Twitter as community spaces where people engage in discussions regarding their work and passions.

Sharing your voice should allow you to build a following of your own. Once you've developed a social media community, you will achieve a stronger web presence, especially if readers share your posts or support whatever you promote.

Increased Exposure

According to the Social Media Examiner, 97% of marketers use websites like Twitter and Facebook to share content.[3] One of the reasons for this is search engine optimization, or SEO. Online search engines such as Google and Yahoo consider the amount of social media traffic a website receives when ranking it for searches. A website connected to activity on Twitter and Facebook is generally considered more relevant, or more valuable, than a website with a smaller social media footprint. That high search ranking obviously translates to a higher visibility, which usually means more business.

A study shared by Internet marketing firm Hubspot found that 84% of marketers were able to increase traffic to their websites by spending only six hours a week using social media.[4] So whether you're trying to engage with a community, build brand recognition, or increase online traffic, social media is a highly effective tool.

Major Platforms

In this section we overview four major platforms: Facebook, Twitter, Instagram, and LinkedIn. It's important to keep in mind that social networking sites frequently update their features, leading to changes in use. Be sure to stay current, especially regarding norms and privacy settings.

Facebook

Facebook is regarded as the most versatile social networking site and, as of 2015, is the most popular. According to a 2014 survey by the Pew Research Center, 71% of adults who are online use Facebook compared to 28% for LinkedIn, 28% for Pinterest, 26% for Instagram, and 23% for Twitter.[5]

Although Facebook often functions as an extension of your personal life, it's also used for professional networking and business marketing. That means you should be careful what you post. According to a CareerBuilder survey, two in five companies use social networking sites to research job candidates. Of those employers, 65% primarily use Facebook for their research, and 63% primarily use LinkedIn.[6] According to another survey by CareerBuilder, 51% of companies that use social media to research job candidates find content that causes them not to hire some candidates.[7]

Professionals should tailor posts to ensure their profile on Facebook conveys a positive image that will benefit them in the workplace. This is especially true for college students and those beginning a career. One way to manage your image is to divide your contacts into professional and non-professional lists. Then, in your privacy settings, limit the ability of one or the other list to see certain kinds of posts, such as photos.

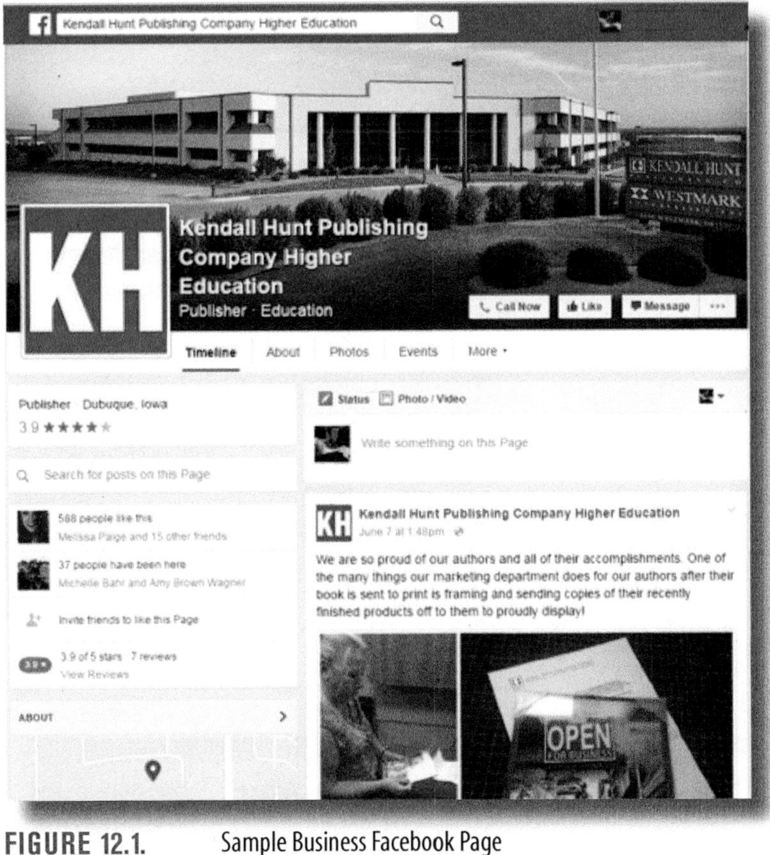

FIGURE 12.1. Sample Business Facebook Page

It's important to be active. Post relevant and compelling content and respond to posts by others. Create or join groups to find people with similar interests that relate to your career.[8] Professionals should also make sure to "friend" others with whom they've had positive interactions, or add them to their network.

Businesses can post content that supports their brand or objectives and offer incentives to interest customers or potential customers. Facebook Insights, a tool to learn about page activity, can help businesses learn how to reach more people.

Twitter

Twitter is similar to Facebook in the way it functions as a platform to share thoughts, photos, and weblinks with friends or followers. One key difference between the two is that each Twitter post, or "tweet," has a 140-character text limit. The site acts as a constant feed of brief comments, including jokes, claims, and news updates.

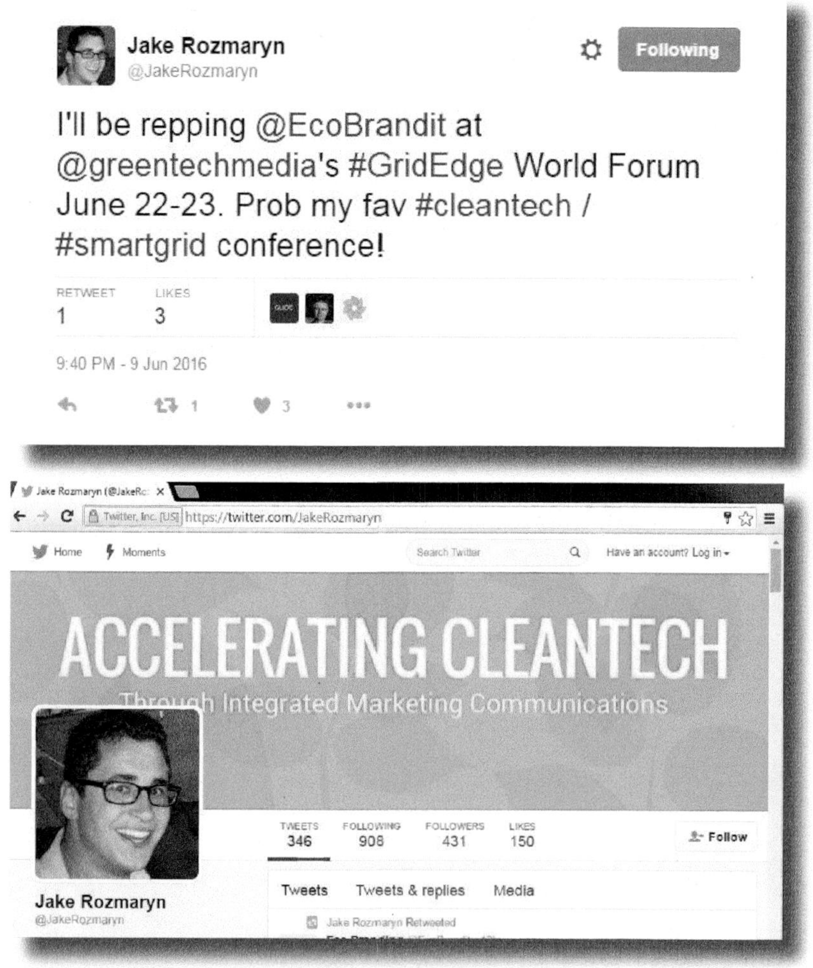

FIGURE 12.2. **Sample Professional Twitter Account.** Public relations and marketing executive Jake Rozmaryn promotes his company via a tweet. Twitter account screen shot contributed by Jake Rozmaryn.

Another feature that makes Twitter unique is that most of its accounts are public. Although a small minority of users may prefer to keep their tweets private, the vast majority embrace Twitter's potential for spontaneous interaction with strangers. Twitter is designed to act as a forum for public discourse. Tweets are searchable, and through the use of "hashtags," indicated by adding the # symbol in front of key words or phrases, users can chime in on trending topics—the most popular issues or events at the time.

Users "follow" each other in order to keep tabs on and converse with specific people; and through the "retweet" feature, a user can syndicate another person's tweet to all of his or her followers, connecting its author with new Twitter users. All of this makes Twitter, which has approximately 236 million monthly users, a powerful messaging tool. Its ability to connect likeminded strangers and broadcast material with the click of a mouse has made it incredibly popular with company spokespersons, journalists, celebrities, activists, and politicians.

Like most social networking sites, Twitter also has its own jargon. Here are a few key terms:

- **Retweet:** Resharing someone else's tweet with all of your users
- **Favorite:** A star-shaped button used to like a tweet
- **Feed:** The stream of posts by users you follow, viewable on your homepage
- **Handle:** Your username, which is always preceded by an @ symbol
- **Mention:** A way to reference another user or communicate with this person. By typing a person's username in a tweet (e.g., @johndoe), this recipient will see your tweet, as will anyone else who follows both of you.
- **Direct Message (DM):** A private message to another user
- **Hashtag:** A way of coding a word so a tweet is more searchable or linked within a larger discussion (e.g., #Superbowl or #2016election). By clicking on a hashtag, you can view every user tweet on that subject.

Instagram

Instagram allows users to share images through tablets or smartphones. Much like Twitter, Instagram users often keep their profiles public, which allows them to connect with other users who may discover their photos through hashtag searches. Beneath photos, Instagram users can write captions and comments, allowing for conversation to develop under each post.

Companies were slow to embrace Instagram at first, when most people used it as an artistic outlet or as a public photo diary. Now, however, many companies take advantage of its popularity by posting photos of items they're selling. Clothing retailers, restaurants, furniture stores, and nail salons are just a few examples of businesses that promote themselves and connect with their customer base on Instagram.

Users with photo-editing skills sometimes overlay pictures with text. It's not uncommon for companies to post coupons or graphics mixed with writing to advertise sales. Instagram also has video capability and built-in editing software. Recently, it surpassed Twitter to reach 300 million active monthly users, and it promises to keep growing.

LinkedIn

LinkedIn is specifically designed for professionals and functions as both a networking tool and an alternate resume. It is especially popular in the following industries: recruiting, marketing, sales, and consulting (ranging from financial to legal). Also, use is high among people who have college educations and incomes higher than $75,000 a year.[9]

FIGURE 12.3. Sample LinkedIn Profile
© Kendall Hunt Publishing Company

Use LinkedIn to research opportunities and potential contacts. Send connection requests when you have a positive interaction to build your network. Once you have a connection, you may find it more effective to contact others through an email address or another avenue given in their LinkedIn profile.

Business representatives and hiring managers can (and likely will) use LinkedIn to find out about you. Make sure your profile supports your professional image (see Figure 12.4). In general, try to be professional and to the point; LinkedIn is not a platform for humor or sarcasm.

Be sure to fill in most of your profile. Tell others who you are, what you do, and what you want to do. As in a resume, include relevant areas of experience, from courses taken to awards to publications. The exact sections will depend on your background and accomplishments. You should include links, photos, and other documents to demonstrate your knowledge and expertise. Learn from how other professionals represent themselves.

Make sure you have an appropriate profile photo. As one expert advises, "Your profile photo should be smart, friendly and professional, and not include things such as bars or beach shorts. Above all make sure that you have one; no one does business with a blue silhouette."[10]

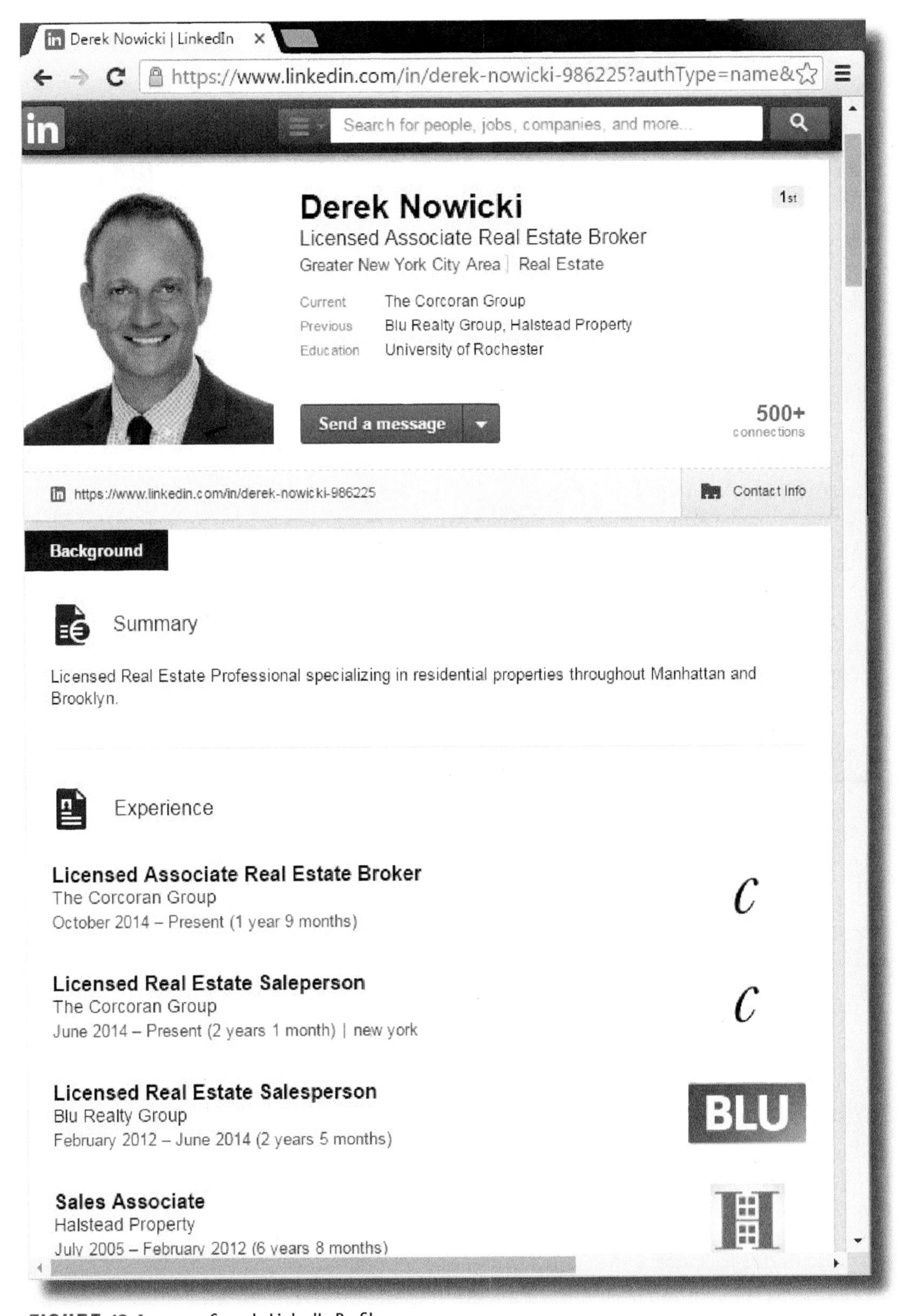

FIGURE 12.4. Sample LinkedIn Profile
LinkedIn account screen shot contributed by Derek Nowicki.

Another expert says you should limit yourself to listing five skills in your profile to focus on your strengths.[11] Utilize keywords that will allow other users to find you via LinkedIn searches.

Principles of Effective Writing for Social Media

As discussed in Chapter 3, to construct your writing to have a desired effect, think about **rhetorical form**. To do this, you need to consider your purpose (what you are trying to achieve), audience (who you are writing to), and context (the external circumstances that affect the reception of your writing).

Many of the general principles for business writing also apply to writing for social media. For example, be concise. Business readers are busy, and attention spans are especially short when it comes to social media. In addition, consider your audience to determine the most effective writing strategy.

Apply the following ten principles to write effectively for social media:

1. Do no harm
2. Write for your purpose
3. Write for your platform and audience
4. Interact
5. Engage
6. Use multimedia
7. Use plain, concise language
8. Find a human tone
9. Be professional
10. Revise before you post

1. Do No Harm

Whether you're on Facebook to keep in touch with professional contacts or on Twitter to promote a company, the same rule always applies: *do no harm.*

Consider the appropriateness of each item you share. Is this weblink or piece of personal information something your followers need to know? Are you comfortable being associated with this material? Before sharing a potentially controversial argument, ask yourself if you've provided sufficient context. Finally, if something you're posting is solely intended to entertain, is there a chance someone might be offended by it?

Be aware of the risks you're taking. For example, on the occasion of the fiftieth anniversary of Dr. Martin Luther King's "I Have a Dream" speech, the Golf Channel asked its followers to tweet their "dream" golf courses. By irreverently using King's phrase on a day when Americans were reflecting on the pursuit of racial equality, the Golf Channel came off looking gimmicky and exploitative. The ensuing mockery was so severe that it deleted the offending tweet and issued an apology, but not before news outlets published stories referencing its "crass attempt at creating social media buzz." Everything published online has the potential to "go viral," just as the Golf Channel's failed attempt at publicity did.[12]

The frequency with which you post suspect material can also hurt you or your company brand. Clogging other people's newsfeeds with dull, rude, or misleading information will at best result in you being ignored, and at worst result in people unfollowing or blocking your account.

However tempting it may be comment on a divisive news topic, remember that the goal is to maintain positive business relations. Don't do anything that may damage your credibility or attract public scorn.

Remember, social media is highly public. Think twice before posting complaints, opinions, or anything that might attract negative feedback.[13] Definitely don't rant. Don't use social media to settle scores, because no one benefits. This includes references to competitors and unfriendly customers.

Be careful of using anniversaries for promotion. During the Arab Spring revolution in Egypt, clothing designer Kenneth Cole used a protest hashtag to promote their sale. The tweet read: "Millions are in an uproar in #Cairo. Rumor is they heard our new spring collection is now available online at http://bit.ly/KCairo-KC." A few hours later, the company was forced to apologize in response to a backlash.[14]

Also, while it may sound obvious, don't use tragedy as an opportunity for promotion. Kmart faced outrage when it included a promotional hashtag in a tweet paying respects to the victims of the Newtown shootings.[15] Similarly, don't use deaths or other tragedies to get "likes."

You may want to postpone a post after related (or unrelated) bad news. Context affects how writing is read. If your writing is read at an inappropriate or inopportune time, it might be criticized or shared for the wrong reasons.

Make sure hashtags, account names, and links are relevant and appropriate. Using hashtags that are irrelevant risks angering or offending readers. In 2009, the UK retailer Habitat came under criticism for promoting its brand by using trending hashtags that were related to the Iranian election.[16] Also, check why a hashtag is trending. You don't want to jump on a keyword for the wrong reason.

2. *Write for Your Purpose*

Remember the four main goals of business communication: develop and maintain positive relations, inform, persuade, and establish a legal record. The first three are especially relevant for social media writing. Ask yourself how using social media can help you accomplish your objectives. Getting likes or followers may not necessarily help your career or your business.[17]

Make sure your social media activity fits into an overall strategy. Be consistent. If you're using a network as an accountant or a financial advisor, then posts about sports or nightlife will probably distract from your professional identity. Readers who follow you for your business knowledge might be turned off by what they deem irrelevant information.

The same rule applies to company representatives who ignore corporate guidelines and write, for example, about unrelated or inappropriate pop culture topics.

3. *Write for Your Platform and Audience*

Identify the audience you are writing to and choose the platform that will best reach this audience. Pinterest, for instance, has a mostly female user base, and so you probably want to choose another platform if you are targeting men.[18] Stay up to date on new platforms and trends in the digital world. However, you don't necessarily need to adopt a new social media platform because it's in the news.

Once you have chosen a platform, tailor length, content, and style based on its features and how it is used.[19] Follow the norms of the platform. Also, don't automatically imitate content posted elsewhere.

Timing is important. Users don't usually scroll far back to read old posts. Do your research and decide on how frequently and when to post depending on how often and when your target audience checks in. Consider the time zone(s) in which your audience lives and other variables. Figuring out the best plan will likely require experimentation. Be willing to adapt or fine-tune your strategy as needed. Also, services are available to help you determine the best time to post. For example, WhenToTweet and TweetStats will analyze your personal following to help you optimize your tweeting schedule.[20]

The results of research vary somewhat on the optimal post times across platforms. According to research cited in one article, Twitter engagement goes up 30% on weekends, and peaks between 1 and 3 p.m. on weekdays. Facebook engagement is highest on weekdays, and peaks as the workday winds down.[21]

Finally, make sure you are posting from the right account if you have access to multiple social media accounts on the same platform (e.g., personal and corporate accounts).

4. Interact

As is evident from its name, social media is about interacting with other people, so make sure you connect with other users and their content. Include hashtags, usernames (such as @Mary), and page links to increase the possibility of views and shares. If your post is especially pertinent, it may be appropriate to ask your followers to retweet or share. In addition, many people are accessing social media from smartphones and tablets, and this number continues to rise. So make sure your content is mobile friendly.

Effective hashtag use increases the opportunity for interaction. Hashtags allow you to join an existing conversation. For example, if you use the hashtag #startups in a post, you will be categorized with similarly tagged content. To use hashtags effectively, search for a hashtag before you use it. That way you can make sure your post isn't associated with undesirable content. Also, use hashtags that people will likely search for. No one is going to think to type in #BestPracticesForStartupsInFirstYear. Finally, don't overuse hashtags—make sure each one is unique. And remember that hashtags are included in your character count.[22]

When sharing links, describe the content with specifics, as in "See this video on the protests in Washington." Also, be careful to link to the actual content and not a home page.

As mentioned, you should like or comment on items posted by other users. But this doesn't mean doing so for every post. Interact when appropriate, such as when someone posts a question. You will have to decide how much activity is enough. When you are sharing a post on Twitter, try not to modify it if you can avoid doing so. If you find it necessary, such as to add context or mention users, use the relevant platform syntax, such as "MT" (modified tweet) for Twitter.[23]

Businesses can think of ways to help people connect with each other or with family or friends. One researcher said, "Despite a brand's best efforts, most consumers are not going to identify with or trust a brand the way they would a friend. To alleviate this problem I recommend that companies focus on facilitating interactions between their customers."[24]

5. Engage

One way to engage users is to make your post useful or beneficial. Another way is to be clever or funny. However, remember to be careful with humor—readers may have a different sense of what is laughable—and to be sensitive to racial and political differences.

Here is some other advice: Don't be afraid of questions. Use detail to draw a reader in. Avoid abstraction and favor specificity. And use the word "you."[25] Other possibilities include contests and giveaways. In addition, on blogs, YouTube, and other platforms or sites, a well-written title or headline can help draw in readers.

Attention spans are short, especially when it comes to social media. Busy users want to know immediately why something is important. Try to state your main point as quickly as possible. Try to also start with the most compelling information to draw in your reader. If you can't fit in all the info, post enough to attract a reader and then include a link.

Context is also important. Provide context through the use of keywords and hashtags, especially on platforms like Twitter. Make sure also to provide all the information a user needs to understand your post. Think of the who, what, when, where, why, and how.

6. Use Multimedia

Including an appropriate and eye-catching photo, video, graphic, or other visual will likely increase exposure. This is borne out by research. According to one study, tweets with images received 18% more clicks and 150% more retweets.[26]

When including images, movies, or other multimedia, make sure to give credit and respect copyright laws.

7. Use Plain, Concise Language

Expressing your ideas with plain language facilitates understanding and makes writing more compelling.[27] This means using simple, conversational words that are easily understood. Avoid long polysyllables, pompous words, and clichés. Keep away from jargon unless you are writing for a specialized audience. In such cases, jargon or industry terms can help you establish credibility.

When writing for social media, less is often more. Getting rid of extra words saves your audience time and helps ensure you will be understood. Ask yourself: Can I cut anything and retain my message? Shorter posts generally produce maximum engagement. According to Kevan Lee, the optimum length of a tweet is 71 to 100 characters, while the optimum length for a Facebook post is only 40 characters.[28] Also, remember that keeping your posts short allows readers to add comments or other content when sharing.

Finally, divide longer posts into small paragraphs. Readers are less willing to wade through large chunks of text.

8. Find a Human Tone

Social media means being human. Find a human tone that will support your professional image and objectives. Make sure this tone is friendly and diplomatic.

Social media users want to read posts that show personality. Content that is more personal and less promotional is more likely to be shared.[29] Try to focus on people rather than abstractions in your writing.

Finally, be honest. If you write out of character or misrepresent, there is a good chance it will have a negative impact on you or the reception of your content. In addition, it's better to admit when you're wrong or if you made a mistake in a post.

9. Be Professional

Even though social media writing is more informal or casual than email or print letters, it should support your professional image. Be conversational, but make sure your writing contributes to your credibility.

Avoid online abbreviations like "lmao" or "brb," which often make writing seem unprofessional. Common abbreviations such as "&" are more acceptable. Emoticon use may be OK, within limits and depending on context.

Do not break grammar rules and make sure to punctuate (but not to overpunctuate). One researcher has found that retweets are more likely to contain punctuation than non-retweets.[30]

Also, be consistent when it comes to spellings, abbreviations, verb tense, etc. Use parallel syntax if you are writing a list.

Try to have the same username across platforms, or to have similar usernames. Make sure the username is relevant and supports your image. Finally, try to make it short and easy to remember, especially for Twitter.[31]

10. Revise Before You Post

All writing should be strengthened through revision, including writing for social media. Revise for larger problems of organization and sentence-level issues. Make sure you have applied the previous nine principles. Also, try to put yourself in the reader's position and see your writing objectively. Is the message clear? Is there enough context?

Proofread to double check word usage, spelling, grammar, punctuation, names, and numbers. If you unsure about something, check a reference guide or ask an expert. You may find it helpful to compose in a software program like Microsoft Word, which includes features like spell check. Then you can copy and paste into your platform. It may also be helpful for you to print out your content. Many famous writers draft before tweeting or posting on other social media platforms.

Exercises

1. Pretend you are working for the corporate owner of a new restaurant. Write three different posts promoting the restaurant's brand, products, or services. Each post should reflect a unique approach and be written for a specific platform discussed in this chapter. Make sure you follow platform guidelines and take advantage of unique features. Include ideas for visual elements.

2. Revise the following tweets and Facebook posts to reflect the guidelines for effective social media writing:

   Twitter

   a. This article is dumb. www.businessnews.com/office-policies-that-limit-productivity

   b. #Sales is #fun but if #clients aren't rich you don't do #good. Recessions $uck :(We do #advertizing & want your big #corporate account!

   c. No Juan ever said "I hate tacos". Hehe. Come by for happy hour Latin food. & pretty Latina girls say hi to bartender Ken. He's single ;)

   d. ATTN @billgates, Plz retweet this job call. I just finished business major now looking for internships. I have skills. Also Plz follow back.

   e. @salesmanBob hi again, nice to meet you at the conf. Sorry I criticized your speech, but I do think you need delivery help. Visit salesguys.com

   Facebook

   f. Inspired Business Strategy based on extensive expertise and knowledge should lead every Corporation. Currently, CEOs are more PR-leaders and manage the corporation to deliver profit per P&L with satisfies the Board, shareholders and stock market daily/monthly monied interests www.insiderceo.com/why-you-should-question-the-cult-of-the-charismatic-CEO

   g. Uber has once again run the red light that is the law and this time the trouble has come in the Golden State – they're HQ! -- but really they're in trouble with regulators around the world, and that's the BIG story. Pump the brakes fellas because: www.CAupdate.com/California-regulators-have-threatened-to-suspend-Uber's-license

   h. Hello Desert Casino fans! Today is election day, so don't forget to vote for Arianna Tran. We at Desert Casino Incorporated don't want our employees to lose their jobs, and if that "other" candidate wins, he'll destroy the economy. Thank you for your loyal patronage. Good luck with all your bets!

End Notes

[1] Julianne Pepitone. "Facebook Traffic Tops Google for the Week." CNN Money. Accessed March 16, 2010, http://money.cnn.com/2010/03/16/technology/facebook_most_visited/

[2] Statista. "Number of Monthly Active Facebook Users Worldwide from 3rd Quarter 2008 to 3rd Quarter 2014 (in millions)." Accessed January 15, 2015, http://www.statista.com/statistics/264810/number-of-monthly-active-facebook-users-worldwide/

[3] Jayson DeMers. "The Top 10 Benefits of Social Media Marketing." Forbes. Accessed August 11, 2014, http://www.forbes.com/sites/jaysondemers/2014/08/11/the-top-10-benefits-of-social-media-marketing/

[4] Sam Kusinitz. "16 Stats That Prove Social Media Isn't Just a Fad." HubSpot Blogs. Accessed June 6, 2014, http://blog.hubspot.com/marketing/social-media-roi-stats

[5] Maeve Duggan, Nicole B. Ellison, Cliffe Lampe, Amanda Lenhart, and Mary Madden. "Social Media Update 2014." Pew Research Center. Accessed July 10, 2015, http://www.pewinternet.org/2015/01/09/social-media-update-2014/

[6] "Thirty-seven percent of companies use social networks to research potential job candidates, according to new CareerBuilder Survey." CareerBuilder. Accessed July 11, 2015, http://www.careerbuilder.com/share/aboutus/pressreleasesdetail.aspx?id=pr691&sd=4%2F18%2F2012&ed=4%2F18%2F2099

[7] "Number of Employers Passing on Applicants Due to Social Media Posts Continues to Rise, According to New CareerBuilder Survey." CareerBuilder. Accessed July 11, 2015, http://www.careerbuilder.com/share/aboutus/pressreleasesdetail.aspx?sd=6%2F26%2F2014&id=pr829&ed=12%2F31%2F2014

[8] Boris Epstein. "How To: Use Facebook for Professional Networking." Mashable. Accessed July 11, 2015, http://mashable.com/2009/08/14/facebook-networking/

[9] Laura Shin. "How To Use LinkedIn: 5 Smart Steps To Career Success." Accessed July 12, 2015, http://www.forbes.com/sites/laurashin/2014/06/26/how-to-use-linkedin-5-smart-steps-to-career-success/

[10] John Hillman. "How to Use LinkedIn to Boost Your Business." Accessed July 10, 2015, http://www.theguardian.com/small-business-network/2014/nov/19/how-use-linkedin-boost-business

[11] Jeff Bullas. "10 Insights: How to Use LinkedIn to Build Business Success Faster than Your Competitors." Accessed July 13, 2015, http://www.jeffbullas.com/2014/05/23/10-insights-how-to-use-linkedin-to-build-business-success/#du8Qux3DwH2dhI3r.99

[12] Linda Lacina. "What You Can Learn from the Golf Channel's MLK Speech Twitter Slip-up." Entrepreneur.com. Accessed February 8, 2016, http://www.entrepreneur.com/article/228133

[13] Charlotte Ward. "Oops! How to Avoid Clangers on Social Media." The Marketing Donut. Accessed January 10, 2015, http://www.marketingdonut.co.uk/blog/2013/10/oops-how-avoid-clangers-social-media

[14] Nathan Olivarez-Giles. "The Business and Culture of Our Digital Lives, From the L.A. Times." Accessed July 12, 2015, http://latimesblogs.latimes.com/technology/2011/02/kenneth-cole-tweet-on-egypt-protests-sparks-controversy-.html

[15] Arielle Calderon. "19 Companies that Made Huge Social Media Falls." BuzzFeed. Accessed January 12, 2015, http://www.buzzfeed.com/ariellecalderon/19-companies-that-made-huge-social-media-fails#.euW5VvvO0

[16] Kevin Anderson. "Habitat Ticks off Twitter with Spam." The Guardian. Accessed January 10, 2015, http://www.theguardian.com/media/pda/2009/jun/22/twitter-advertising

[17] Mikołaj Jan Piskorski (interview). "Social Media and Branding: A One On One with a Harvard Business Professor." Forbes. Accessed January 13, 2005, http://www.forbes.com/sites/steveolenski/2014/03/17/social-media-and-branding-a-one-on-one-with-a-harvard-business-professor/2/

[18] Carol Han. "How Older Brands Can Reach New, Younger Audiences Through Social Media Branding." The Huffington Post. Accessed January 13, 2015, http://www.huffingtonpost.com/carol-han/how-older-brands-can-reac_b_4942409.html

[19] University Communications, University of Colorado-Boulder. "Social Media Guidelines." Accessed January 15, 2015, http://ucommunications.colorado.edu/node/288

[20] Leonhard Widrich. "Top 5 Tools to Better Time Your Tweets." Mashable. Accessed January 16, 2015, http://mashable.com/2011/10/28/best-time-to-tweet/

[21] Shea Bennett. "When is the Best Time to Post on Twitter and Facebook?" Social Times. Accessed January 15, 2015, http://www.adweek.com/socialtimes/best-time-post-twitter-facebook/500286?red=at

[22] Jennifer MacDonald. "How do I use a hashtag?" LinkedIn. Accessed January 16, 2015, https://www.linkedin.com/pulse/20140429124921-23085126-how-do-i-use-a-hashtag

[23] University Communications, University of Colorado-Boulder. "Social Media Guidelines." Accessed January 15, 2015, http://ucommunications.colorado.edu/node/288

[24] Mikołaj Jan Piskorski (interview). "Social Media and Branding: A One On One with a Harvard Business Professor." Forbes. Accessed January 13, 2005, http://www.forbes.com/sites/steveolenski/2014/03/17/social-media-and-branding-a-one-on-one-with-a-harvard-business-professor/2/

[25] Ben Donkor. "How to Write for Social Media." Link Humans. Accessed January 15, 2015, http://linkhumans.com/blog/sweet-retweets-how-write-social-media-sookio-tips-video-smlondon

[26] Shea Bennett. "Tweets with Images Get 18% More Clicks, 89% More Favorites and 150% More Retweets." Social Times. Accessed January 13, 2005, http://www.adweek.com/socialtimes/twitter-images-study/493206?red=at

[27] "CDC Guide to Writing for Social Media." Centers for Disease Control and Prevention. Accessed January 10, 2015, http://www.cdc.gov/socialmedia/tools/guidelines/guideforwriting.html

[28] Kevan Lee. "Infographic: The Optimal Length for Every Social Media Update and More." BufferSocial. Accessed January 16, 2015, https://blog.bufferapp.com/optimal-length-social-media

[29] Carol Han. "How Older Brands Can Reach New, Younger Audiences Through Social Media Branding." The Huffington Post. Accessed January 13, 2015, <add url here>.

[30] Dan Zarella. "The Science of Retweets." Danzarella.com. Accessed January 15, 2015, http://danzarrella.com/science-of-retweets.pdf

[31] Mitt Ray. "6 Tips to Choosing the Perfect Twitter Name." Social Marketing Writing. Accessed January 17, 2015, http://socialmarketingwriting.com/6-tips-to-choosing-the-perfect-twitter-name/

CHAPTER 13

DEFINITIONS

Technical and business communicators are often asked to define concepts, terms, and processes for audiences who have little or no understanding of a particular field's jargon or a company's product, system, or process. Technical and business writers are also frequently asked to provide precise definitions to explain complex legal, social, ethical, and moral issues.

Knowing the level of technicality of an audience (non-technical, semi-technical, or highly technical) helps writers determine which concepts within a document may require definition. It may be helpful to view some document types, such as an instruction manual, legally binding policy, or even a business prospectus, as an instance where definition is one of the underlying purposes of the document. The strategies for writing definitions discussed below can help writers in any situation when the task at hand requires explaining complex ideas clearly.

Types of Definitions

When explaining a complex idea to a reader unfamiliar with the concept, there are three types of definitions at the writer's disposal:

- Parenthetical definitions
- Sentence definitions
- Expanded definitions

Parenthetical Definition

A parenthetical definition is a word or brief clarifying term placed within a sentence. These are not intended to be comprehensive but are mainly meant as a short definition of an unfamiliar term.

Example:

The aircraft accident report concluded that the crash was the result of faulty maintenance on the right aileron (moveable wing flap).

Sentence Definition

At times you will need to write a more detailed definition for a complex or unusual concept. The sentence definition, which may be more than one sentence, is based on a pattern of composition that seeks to include as much information as possible in a minimum amount of space. Sentence definitions are composed of three parts:

- The **term**, either a word or a phrase, to be defined
- The **class** to which the term belongs
- The **distinguishing characteristics** that make it different from all other terms in its class

When developing a sentence definition use the following pattern:

Term	Class	Distinguishing Feature
Angina pectoris	*heart condition*	*severe chest pain caused by insufficient blood supply to the heart.*

These three elements are then combined to form one or more sentences:

Example:

Angina pectoris is a heart condition caused by an insufficient blood supply to the heart resulting in severe chest pain.

Expanded Definitions

There are times when parenthetical or sentence definitions are not satisfactory. This more often than not occurs when defining vague or abstract terms. Terms such as *rental agreement* or *auto lease* can have a number of meanings and interpretation depending on the knowledge and previous experience of the writer and the intended audience. Therefore these terms and others like them need to have a clearly stated definition so that the writer and the reader have an understanding and a consensus on what terms mean and how they are used.

An expanded definition may be a paragraph, a multi-page document, or even an entire manual. Here is an example of an expanded definition of *deductible* from the Glossary of Insurance Terms on the Insurance Information Institutes website:

> *Example:*
>
> *"The amount of loss paid by the policyholder. Either a specified dollar amount, a percentage of the claim amount, or a specified amount of time that must elapse before benefits are paid. The bigger the deductible, the lower the premium charged for the same coverage."* (2005)

When composing an expanded definition, you will typically start with a sentence definition of the term and then move to one or more of the following strategies to develop a useful definition.

- **Etymology:** This method of expansion uses the origin of the word to help the reader understand the meaning.
- **History of the Term:** The meaning of a specialized word or phrase can often be defined by providing the reader with a discussion of how the term developed.
- **Compare/Contrast:** You may compare and contrast the term to information that the reader already understands.
- **Negation:** At times you may expand a definition by telling the reader what it is not.
- **Visuals:** Providing clearly labeled visuals will help to expand a definition.
- **Example:** It is frequently useful to provide examples that are familiar to expand your definition.
- **Process Analysis:** For complex multipart definitions, providing the audience with a list of the parts that contain an accompanying description of how the parts work and interrelate is one method of expansion that many readers find useful.

Writing Effective Technical Definitions

As with any workplace writing, composing and publishing an effective technical definition requires planning. Before composing a technical definition you should be able to answer the following questions in detail:

1. Who is the audience for the document?
2. What level of technicality is appropriate for this audience? (non-technical, semi-technical, highly technical)
3. Given the audience's level of technicality, what concepts within a document require definition?
4. What type of definition is suitable for each concept? (parenthetical, sentence, expanded)
5. For sentence definitions, have you considered the concept's class and distinguishing feature?

6. For expanded definitions, what method(s) will be used to fully develop each concept, and why?
7. Are there any ethical considerations for any of the definitions?
8. Are there any potential legal consequences for any of the definitions?
9. Are there any design and format issues that need consideration?

Potential Problem Areas

When writing definitions there are some potential problem areas watch out for:

- Be careful not to write a **circular definition**. This happens when the writer uses the term as part of the definition. For example, a writer wants to define *sexual harassment* and begins the definition with, "Sexual harassment is a form of harassment..."
- Don't be **overly technical**. To be effective a technical definition should not use terms that are too technical for the intended audience. Understanding the level of technicality that the audience possesses will help you to determine the type of language that is appropriate for the purposes of the definition.
- Avoid the use of **broad, unfamiliar, and abstract terms**. One of the purposes of a technical definition is to be useful, to be understood by the audience. Using terms that are too broad, that are unfamiliar to the reader, or that are too abstract undermines the effectiveness of the definition.
- Do not use ***is when* and *is where***. These adverb phrases do not work well as an introduction to a definition. A useful rule of thumb is to use nouns and noun phrases to define nouns, verbs to define verbs, and adjectives to define adjectives.

Useful links

This list of links will help you to better understand some of the techniques for writing definitions.

- The Writing Definitions handout at the Purdue OWL provides another approach to writing a sentence definition.
  URL: http://owl.english.purdue.edu/owl/resource/622/01/
- How Stuff Works.com. One of the important things that you need to be aware of about this website is the way it "defines" and "illustrates" complex technological concepts using "everyday" language and illustrations. An excellent example of this is the "How Blu-ray Discs Work" by Stephanie Watson: URL: http://electronics.howstuffworks.com/blu-ray.htm
- Duncan Kent & Associates Ltd's Technical Communicators Resource site is "a comprehensive resource for writers of technical publications." URL: http://www.techcommunicators.com/techcomm/index.html

"Definitions" contributed by Homer Simms

CHAPTER 14

INSTRUCTIONS

This chapter provides an overview of one of the most crucial and common types of all technical writing—instructions. It includes a detailed examination of the purpose and function of instructions as well as a discussion of their structure and design, including page layout, styling, and graphics.

Aim and Importance

Although writing instructions seems like a relatively straightforward matter, providing a clear step-by-step explanation of how to do something is far more involved than it first appears. In order to lead a user through the process—usually to construct, operate, or maintain an item—the writer will often need to make use of descriptions, definitions, and process discussion while maintaining a coherent organizational pattern.

To complicate matters further, writers must also thoroughly consider the audience and the situation for which they are writing. Failure to do so has more profound and serious ramifications than for other forms of technical writing, because faulty or ineffective instructions for an item can lead to user injury and property damage for which the company is often held liable.

In order to properly instruct the reader, you must do more than provide clear, simple writing. The writing must also be effective. You must put yourself in the user's position and consider both their needs and their level of familiarity with the subject. For instance, experts and technicians already have an extensive understanding of how to use an item and do not need the extended definitions a non-specialist would need to understand that item. Likewise, non-specialists do not need a detailed discussion of all the esoteric technical aspects of an item because it would only lead to confusion.

Common Forms

While all instructional documents provide some degree of step-by-step explanation, the level of depth and even the structure of the text can take on a variety of forms based on their perceived purpose and audience. There are many types of instructional documents, including:

- **Reference cards:** These small documents, usually no more than a page, provide the user with key information about an item or condensed notes about specific aspects of that item. Used either to provide quick setup, such as pairing a Bluetooth speaker, or to outline common rules or features, such as keyboard shortcuts, reference cards principally serve as a mnemonic and visual aid for the user.
- **Instructional brochures:** Usually between 5 and 15 pages, these documents contain more detailed instructions for installing, assembling, and operating something than a reference card. These guides typically address specific aspects of a product but can contain nearly all the common procedures for a particular item. For instance, an instruction guide for a rice cooker could provide explanation of all the basic settings on the machine as well as on how to correctly cook rice.
- **User manuals:** Often a book-length document of 100 pages or more, user manuals provide comprehensive and highly technical instructions on nearly every aspect of operating an item. In addition to providing step-by-step directions, user manuals typically contain detailed reference information as well as troubleshooting and maintenance guidelines. For instance, a car repair manual provides a comprehensive overview of all the procedures necessary in the process of maintaining that specific car.
- **Online help:** User manuals tend to be written with the intention of being read sequentially, but online help is usually written in smaller chunks of information intended to be called up through an application's help search feature or associated with the application's feature in a context-sensitive way (i.e., via a help button or right-click menu option). Online instructions also allow for the ability of creating hyperlink cross-references to related instructions and supplemental references.

The key feature of instructions is that they are tailored in both content and form to the background knowledge and experience of the intended readers. Often, companies provide two types of instructions with the same perceived purpose, but with modified content based on the type of reader they seek to address. Each aspect of the document must be considered in regard to its contribution to the user's understanding first since the sole aim of the document is to guide the reader in the proper and safe use of the given item, product, or service.

How to Organize Instructions

Organizing steps into a logical and easily readable form for the user is one of the principle aims when writing instructions. However, depending on the audience and the purpose of the instructions, you can organize those steps either by *task* or by *function*. Like all other aspects of the document, your choice depends on the intended user and the purpose of the instructions.

One Procedure, Many Tasks

When you write instructions, you guide a user through a sequence of steps. Each step specifies a particular action the user will need to perform. Cumulatively, those actions make up the *task* that you want the user to complete. A *task* is nothing more than the series of actions needed to finish a particular activity. For instance, changing the oil in a car is one task comprised of a series of steps (e.g., draining the oil, replacing the filter, filling the oil). Each action is dependent upon the previous one, and all steps are needed to complete the task at hand.

Most instructions, particularly for electronic devices and advanced machines, contain a number of tasks that are independent from one another. Even the instructions for a relatively simple appliance like a microwave contain a large collection of unrelated tasks (e.g., setting the clock, power level, or timer). The sum total of these separate tasks is known as a *procedure*. In the case of the microwave, these tasks collectively form the general *procedure* for operating the appliance.

Task-Based Organization

Typically considered the best approach to arranging instructions for non-specialists, task-based organization orients the instructions around the activities the user will *most likely* perform. Generally, the phrasing for task-based organization follows the simple problem/solution format. The instructions for using a camera are often organized in this way (see Figure 14.1).

Problem	Task-Based Solution
Is there an easy way to take snapshots?	"Point-and-Shoot" Photography
Can I use special effects while shooting?	Shooting with Special Effects
How do I copy photos to a computer?	Copying Pictures to the Computer

FIGURE 14.1. Task-based solution

One of the issues with this particular form of organization is it requires studying how users use the product and then determining what activities are most important or necessary to them. Directly observing and interviewing the users is not typically feasible until the final stages of product development, which leaves little time for document adjustments prior to launch. Often, you must resort to educated guesswork when using task-based organization, which always leaves the possibility that you misanalysed the potential uses of the item and/or the needs of users.

Another issue with task-based organization is repetition. Separate tasks might require identical steps to complete or the current task might require the completion of some prior, unrelated task. In order to deal with these situations you have the option of either writing duplicate steps for separate tasks, which leads to increased document length, or redirecting users to prior tasks through the use of notes, which leads to increased document complexity. Effective instructions often make use of both as needed in order to provide balance and mitigate user impatience and fatigue.

You should also consider grouping related steps within a task together into *phases* (e.g., framing the shot, selecting the shooting mode, adjusting the aperture), to make duplication easier and the instructions more uniform. Users will skip over particular phases if they recognize them. Likewise, grouping semi-related tasks together (e.g., all advanced photography modes) will make redirection easier and smoother for the user. The user will not have to fumble through the entire set of instructions to find what is needed.

Identifying and Organizing Tasks

When considering how to group tasks together, your primary goal is to avoid having the user search blindly through your instructions. You need to make sure the user can clearly follow the instructions sequentially as well as quickly locate individual tasks without frustration. Grouping tasks based on two different types of reading can prove difficult at times, but you must strive to make it easily readable no matter how it reads. Therefore, you should choose broad categories based on the user's familiarity and experience with an item as well as its potential uses. For example, consider the following groupings from a phone user manual shown in Figure 14.2.

These categories are *skills based,* with more complicated tasks coming after the user has mastered the basic functions of the phone. In addition, the tasks are organized around particular *uses* for the device. All music-related tasks (e.g., playing music, buying music, organizing music) are conveniently collected together under "Music," so the user can quickly locate individual tasks based on demand.

Contents

Chapter 1: Getting Started

Set up and activate your Phone

Connect you Phone to a computer

Connect to Wi-Fi

Connect to the Internet

Set up other mail, contacts, and calendar accounts

Date and time

International settings

Usage Tips

Chapter 2: Basics

Use apps

Customize your phone

Type text

Dictate text

Voice Control

Search

Sharing

Transfer files

Chapter 3: Phone

Phone calls

Voicemail

Contacts

Call forwarding, call waiting, and caller ID

Ringtones and vibrations

International calls

Phone settings

Chapter 4: Messages

SMS and MMS messaging

Send and receive messages

Manage conversations

Share photos, videos, your location, and more

Messages settings

FIGURE 14.2. Phone user manual table of contents

Source: Apple, *iPhone User Guide*. (San Francisco: Apple Inc., 2014), 2.

Function-Based Organization

Function-based organization, also sometimes referred to as tool-based organization, systematically and sequentially deals with every function, feature, and element of a particular product. Product specifications for electronic devices are an example of this type of organization and are particularly useful for experts and specialists since they provide quick reference to all pertinent information at a glance (see Figure 14.3).

Product Specifications	
Operating System:	Google Android 4.4.2 "KitKat"
CPU	1.9 GHz Quad-core Cortex A53
GPU:	600MHz Quad-core Krait 450
System Memory:	3GB Low-Power DDR3

FIGURE 14.3. Cell phone product specifications

Function-based organization can be more thorough than task-based instructions, because it typically covers all functions no matter how inconsequential to the user. In addition, it is typically more concise, because it does not focus on any tasks the user might need to complete and thus avoids unnecessary repetition. These strengths come at a high cost, however; specifically, complete lack of consideration for the user's needs.

While developers attempt to make their products intuitive, some functions are not readily apparent or require the user to complete additional tasks beforehand. The Crop tool in Adobe Photoshop is a good example of this issue. The tool will not function unless the user has first selected the item to crop with the "Rectangular Marquee" tool. Function-based instructions often require the user to have extensive background knowledge in order to use them effectively. Even those with the necessary experience might grow irritated or fatigued by this organization because of the energy needed to complete certain tasks.

Nevertheless, a function-based organizational approach can be quite useful or preferable in certain situations due to its thoroughness. Some companies include function-based instructions as a supplement to their general task-based instructions in order to cover more complicated features or to allow users to quickly access information once they have mastered using an item.

Common Components

Instructions vary in length and format. Most instruction sets are the result of a considerable investment of time and resources, because they are designed for commercial publication. That is, instructions contain the elements typically found in published books, including a cover, table of contents, and a uniform page design or theme. Most instruction sets contain the following sections and elements:

Introduction	Introduction Precautions Information Technical Background Equipment and Supplies
Body	Discussion of Steps Supplementary Explanation
Back Matter	Technical Notes Support

Although you commonly find these sections in instruction sets, do not assume that every element *must* be present or that these are the *only* sections possible. Reference cards would contain very few of these sections, whereas user manuals may contain many more sections than the ones provided here, such as a title page, table of contents, and an index. Remember, all instructions are tailored to suit the intended readers.

Introduction

The introduction includes all information and guidelines that are necessary *before* the user begins to follow the instructions. These sections include the introduction, important precautions, necessary technical background, as well as equipment and supplies.

- **Introduction:** All instructions contain a carefully crafted introduction to help users determine, often even before opening the document, if this document will help them accomplish their task at hand. You should state the purpose of the document, who should read the document, and under what circumstances it should be read. Make sure you explain the scope of the document's coverage, including what will *not* be covered.
- **Precautionary information:** Instructions typically highlight any possibility where the users might harm themselves, ruin equipment, or inadvertently disrupt the procedure. To highlight these issues you use special notices—danger, warning, and caution notices—to alert users before they begin following the step-by-step directions.
- **Technical background:** Not all people can complete all tasks. So, at the beginning of highly technical instructions, or even before critical procedures, you need to discuss the background and skill level necessary to do what is discussed. For instance, to fully operate a professional camera, users must have some prior experience in photographic theory. The terms *aperture, exposure,* and *ISO* have limited meaning to novice photographers. By clearly indicating what level of expertise is necessary you help ensure that users know what is required of them.

- **Equipment and supplies:** All building and construction instructions provide a list of items the user will need to gather before starting the procedure. This typically includes both the tools needed for the procedure (e.g., cordless drill, saw, screwdriver, hammer) and supplies the user will need during the procedure (e.g., wood, glue, paint). Generally, for prefabricated construction or assembly kits, you would also specify the amounts, types, and sizes of all parts included with the model (e.g., 2 x 1½" M8 bolts) so the reader can check if anything is missing. Although the structure and information provided vary, the equipment and supplies are typically formatted in a two-column list for easy reading.

Body

The body, or main part of the instructions, contains the actual discussion of the steps as well as any supplementary explanation of the order or structure of the procedure.

- **Discussion of steps:** This section contains the actual procedure for constructing, operating, or maintaining an item. As the principle part of the document itself, the format, point-of-view, and writing style must remain uniform and clear to the user. If you choose a task-based orientation, the tasks should be organized in categories according to the necessity and skill level of the user and all steps with each task organized into phases. This form of *nesting* tasks and steps allows you to break down the rather complex procedure into manageable chunks for the user. You should format these nested lists as you would style lists in other documents with appropriate indenting and font styling.
- **Supplementary explanation:** Part of guiding users through a task is providing necessary feedback. Explaining how an item should look before and after a particular step allows users to avoid making easy mistakes. You should also provide additional explanation on the purpose or necessity of a particular step if it is not readily apparent. For example, if you direct the user to use Paste Special in Microsoft Word, you need to explain why it must be used instead of the regular Paste function. Typically, this additional explanation comes in line with the step, but excessive or unnecessary supplementary discussion might hide the specific actions the user needs to take. To avoid this situation, clearly separate the instruction from the supplement through paragraphing and font styling.

Back Matter

All reference material as well as support documents should appear at the end of the report. While user manuals make use of appendices and indexes, instructional brochures typically only contain technical notes and support.

- **Technical notes:** This section contains all material that does not fit within the main body of the text. Any alternative steps or additional explanatory information not directly relevant to the construction, operation, and maintenance of an item should be placed in this section. Alternate steps include both additional ways of accomplishing an action and how-to's for carrying out a task under atypical conditions. Additional explanatory information includes safety, handling, and regulatory information for a particular item. Typically, you would provide any warranty and liability information in this section as well as a list of any revisions made to the document or the item after initial publication.

- **Support:** Despite the considerable investment of time and resources that goes into product development, issues will occur. This final section of the instructions aids the user with those difficulties. Regularly organized in the form of a Frequently Asked Questions (FAQ) or a Question and Answer (Q&A) guide, this section contains a flowchart of common user problems and the specific answers or actions to take. Additional support resources as well as customer service and contact information for the company are also listed if the user needs support beyond what is provided.

Readability

In addition to the design and layout of the instructions, you must pay particular attention to how you write each step. Unlike academic essays, and even other forms of business or technical writing, you must write instructions in the *imperative mood,* which uses the second-person subject "you" to directly address and command the user. Typically, you write each step phrased as a response to the following question: "What should I do next?" Consider this with the example in Figure 14.4 from a camera user manual.

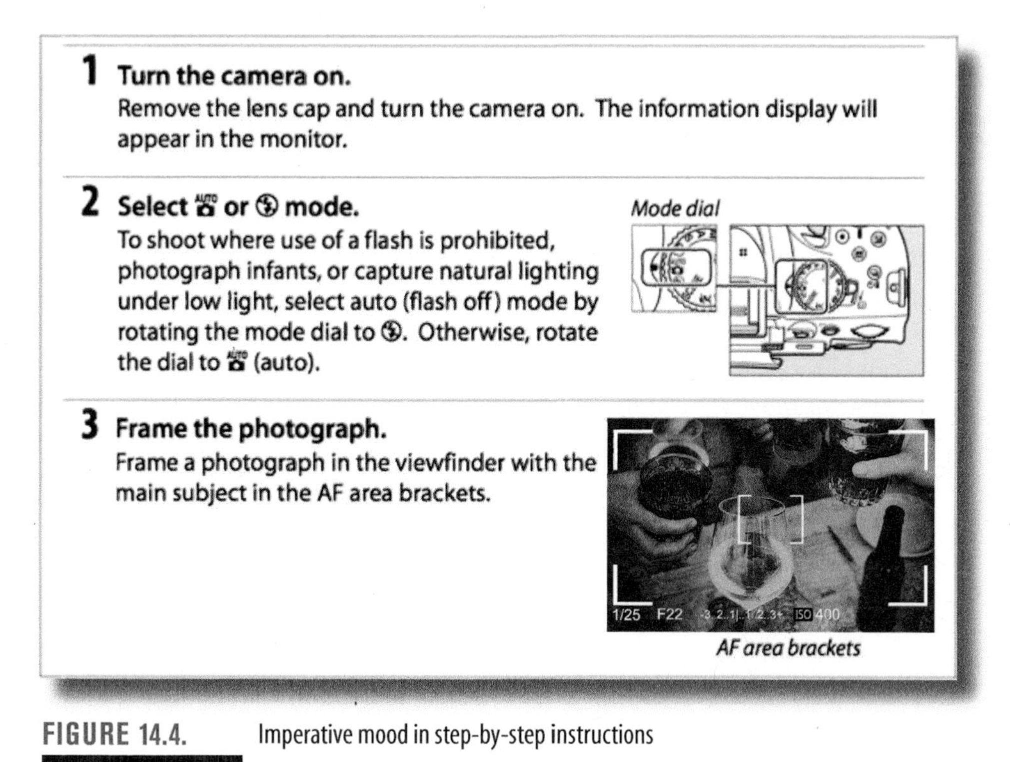

FIGURE 14.4. Imperative mood in step-by-step instructions

Source: Nikon, *D5200 Digital Camera Reference Manual.* (Tokyo: Nikon Corp., 2012), 21. Photo: Rawpixel.com/Shutterstock.com

The purpose is to grab the user's attention and keep it. For this reason, imperative mood always uses active voice. Not only does it clearly identify the subject and the action in a sentence, but it also eliminates excessive wordiness. You do not want the user to grow lost or disinterested in the instructions, so write steps as concisely as possible.

Effective Design

Instructions pay careful attention to visual presentation and uniformity, because they are designed for commercial publication. Aside from aiding readability, careful layout demonstrates the level of investment the company put into the document in consideration for the user.

Headings

Your instructions need to make proper use of headings. You need headings for all the major sections of the document such as the technical background, equipment and supplies, and the support section. The instruction section should have a general heading followed by subheadings for each of the individual tasks as well as the phases within each task. Follow these guidelines for headings:

- Use headings to mark off all major sections and subsections of the document.
- Use the same design for headings throughout the document (e.g., font size, color, and style).
- Use parallel construction in all headings and make sure the headings are self-explanatory.
- Avoid "lone" headings—leaving a heading alone in a section without another heading of the same level (e.g., having only one second-level heading in an entire section or only one third-level heading).
- Avoid "stacked" headings—leaving two or more consecutive headings together without providing intervening text or explanation.
- Avoid "orphaned" headings—leaving a heading isolated from its associated section because of a page break. Always keep at least two to three lines of body text with a heading or move the heading to a new page.

Lists

Instructions contain a variety of lists, particularly extended, vertical lists for the step-by-step discussion of tasks. Not only do lists help emphasize particular information to the reader, but they also allow readers to easily jump from reading a step to doing the associated action. Unfortunately, as easy as lists are to follow, many writers have difficulty writing them. Follow these general guidelines for lists:

- Use the same layout for lists throughout the document (e.g., spacing, punctuation, indentation, and font style).
- Use a lead-in to indicate the purpose and contents of the list to the user.
- Use each type of list consistently across the document.
- Avoid lists of more than six or eight items (e.g., subdivide or consolidate long lists).

There are three types of list—ordered, unordered, and simple—commonly found in instructions. Each type serves a specific purpose and function. One of the most widespread problems with instructions is the incorrect use of these list types. Keep in mind the following rules while writing instructions:

- **Ordered lists:** Use ordered, or numbered, lists for actions and tasks that require strict sequencing to be successfully completed (i.e., step-by-step directions). For instance, you cannot take a picture with a camera without first turning the camera on. *Fixed-order* steps like these need to be written using a hierarchical, numbered list. Sub-steps or sub-lists within the ordered list often use lowercase letters to differentiation list levels.
- **Unordered lists:** Use unordered lists for information that has no order or need not be completed sequentially (e.g., function-based instructions cataloged information based on convenience). When discussing the various tools under the Edit menu in Microsoft Word—Undo, Redo, Cut, Copy, and Paste—you organize the discussion based on their position in the dropdown menu. FAQs and Q&As also present information in this manner. These guides contain *variable-order steps*, which can be performed in practically any order, to troubleshoot problems. For these types of applications, using a bulleted list is more appropriate than a numbered list.
- **Simple lists:** Use simple lists for items that do not need any additional emphasis or to aid the user in reading. Equipment and supplies are often simply listed, because the format provides a compact and quick reference for users to check that they have all the materials needed to complete the procedure. Often these lists are written in two-column form so users can see the amount and the materials side-by-side.

Abbreviations, Acronyms, and Symbols

Instructions commonly make use of abbreviations, acronyms, and symbols to make the document more efficient for the writer and reader. Although abbreviations and symbols can distract or confuse the user, since they interrupt the flow of the sentence, excessive repetition can be equally burdensome. For instance, replacing measurements with symbols (e.g., replacing 4 feet and 6 inches with 4' 6") or using acronyms for long phrases (e.g., using SDHC Card instead of Secure Digital High Capacity Card) would be preferable if the term appears multiple times throughout the document. In all cases, use standard or accepted abbreviations, acronyms, and symbols. Do not make up your own jargon. In addition, always write out the full term the first time it occurs in the text followed by the appropriate abbreviation, acronym, or symbol in parentheses just after it so the user clearly understands the meaning.

Graphics

Unlike other forms of business and technical writing, which ordinarily rely on graphics to emphasize a point or catch the reader's attention, graphics are a crucial and fundamental part of all instructions. Nearly all instructions, and nearly all steps within them, provide some form of visual guide or feedback to the user. Rather than graphics underscoring the importance of the words, the words underscore the importance of the graphic direction. Without illustrations, most users will have difficulty visualizing what they are being asked to do. This does not mean that you should discard written instructions, but rather keep in mind that graphics and texts within instructions form a cohesive, coherent, and deeply integrated whole.

Illustrations, Diagrams, and Photographs

Instructions need to provide exact or near exact representations of user actions, or the results of those actions, to guide the readers. Users need to visually understand the relationship between various parts and actions of a task in order to complete that task. To provide this type of accuracy in depicting objects, and the relationships between them, most instructions rely on illustrations, diagrams, or photographs. Each type of graphic has its benefits and its disadvantages. Knowing them allows you to choose the most appropriate graphic for a particular step or task.

- **Illustrations:** Depending on the purpose and audience of a particular set of instructions, illustrations can vary in complexity from simple line drawings to highly detailed, photorealistic images. One of the primary benefits of illustrations, as opposed to photographs, is they allow you to maintain a consistent perspective across all steps. Another benefit is they simplify objects and allow the user to focus on key details you have purposefully highlighted with increased detail. The primary drawback to illustrations, however, is the cost and time involved in producing them.
- **Diagrams:** Similar to illustrations, diagrams are used to clarify the relationship between abstract concepts and provide a more schematic overview of all elements of a system. Electrical wiring schematics are a good example of the usefulness of these graphics. The typical automotive interior lighting system has dozens of relays, switches, and circuits, which can prove far too complicated to describe in words or visualize in any other way. The main obstacle to using diagrams, however, is user understanding. Most diagrams do not resemble the actual physical item at all and require prior experience, or at least a general familiarity, with the elements of the system to be useful.
- **Photographs:** Providing the most detail of all, photographs promote a more accurate sense of scale and perception compared with other types of graphics. Because a photograph captures an image of the real item, users will not need to decipher what they see in the instructions with what they see in reality. Additionally, the perspective of a photograph often corresponds directly with the users' own perspective as they carry out the particular task. The user spends less time interpreting steps and more on carrying them out. However, there are several issues with photographs as well. First, background clutter and the inclusion

of inessential details in photographs can lead to user distraction and confusion. Second, photographs may obscure important components or features due to bad lighting, angle, or focus. Finally, line drawings can display more intricate and miniaturized components than a photograph, because they can illustrate what is not visible to the unaided eye.

Integrating Graphics

Instructions maintain a high level of integration between graphics and text. Visual graphics both support and elaborate on what is provided in the text and the text explains what is depicted in the graphic. For this reason, the text must reference the graphic and connect it to the overall task being completed. Typically, each step in a task contains at least one corresponding graphic. Although there are no strict placement requirements for graphics and text on page, graphics and text are dependent upon one another. You should place them within close proximity of each other. The graphics and text for a particular step should also never be on separate pages. In addition, because graphics and text are interspersed throughout the document, the size of photographs and illustrations rarely exceeds half a page. Larger images overwhelm the written steps. Consider the example in Figure 14.5 from a camera user manual.

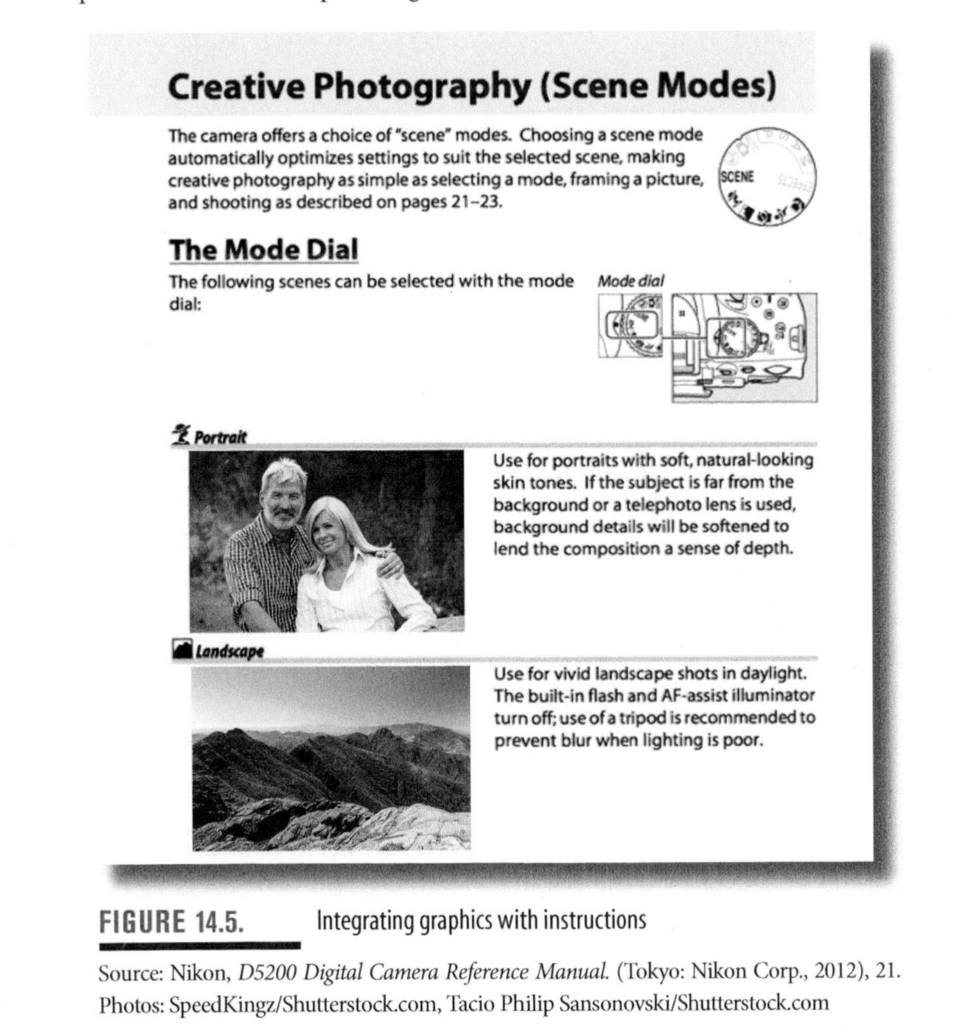

FIGURE 14.5. Integrating graphics with instructions

Source: Nikon, *D5200 Digital Camera Reference Manual.* (Tokyo: Nikon Corp., 2012), 21.
Photos: SpeedKingz/Shutterstock.com, Tacio Philip Sansonovski/Shutterstock.com

The graphics and text in Figure 14.5 occupy equal space on the page and the design remains simple and uncluttered. If the reader cannot determine how a graphic corresponds to the text in five to ten seconds, the relationship is not clear enough. Moreover, these instructions also demonstrate the use of multiple graphics, and multiple types of graphics, within a single step. The mixture of illustrations and photographs, in addition to the corresponding mode symbols, allow the instructions to guide multiple readers through the various "scenes" available with the camera.

Special Notices

When guiding the reader in the proper and safe use of the given item, product, or service, instructions need to highlight special situations where personal injury or equipment damage may occur as well as important supplementary information necessary to understand the procedure. These special notices typically fall into one of four categories:

1. Dangers
2. Warnings
3. Cautions
4. Notes

Each type of notice serves a specific purpose and they are hierarchical in nature. In order to provide visual differentiation, and add extra emphasis over the procedural tasks, each type also has specific formatting requirements. You must carefully choose which type of special notice to use and when to use it. Imprecise usage can cause readers confusion and disregard and lead them into potentially life-threatening danger.

Dangers

Danger notices alert users to eminent hazards that will result in serious or fatal injury to themselves or others if not avoided. Therefore, the use of this type of notice is limited to only the most extreme situations. Dangers are the highest level of special notices and thus should have the most emphasis in the text. Typically, you would separate this type of notice from the text using bold font, capitalization, and bright font color. In addition, they often provide a corresponding symbol depicting the danger, as shown in Figure 14.6.

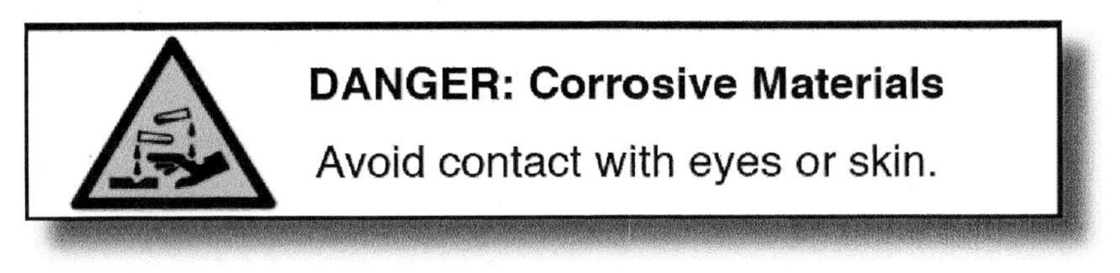

FIGURE 14.6. Example danger sign

Source: German Federal Institute for Occupational Safety and Health (BAuA). *Technical Regulations for Workplaces: Safety and Health Signs.* ASR A1.3. Dortmund, Germany, 2013. http://www.baua.de/cae/servlet/contentblob/669216/publicationFile/62400/ASR-A1-3.pdf (accessed June 1, 2016).

Notices use universal or standardized symbols when possible. For instance, the previous example uses the German Institute for Standardization (DIN) symbol for corrosive materials. However, symbols are not always required. Both signs in Figure 14.7 conform to the U.S. Occupational and Safety and Health Administration (OSHA) regulations, which require specific formatting only.

FIGURE 14.7. Example OSHA danger signs

Source: American National Standards Institute. *American Standard for Product Safety Signs and Labels.* ANSI Z535.4-2007. Rosslyn, VA: National Electrical Manufacturers Association, 2007.

Within instructions, the word Danger should always align with the text in reference unless an entire procedure, and all the tasks within it, could cause serious harm or death. Like the headings Warnings and Cautions, Danger must come before the point at which it is relevant. You need to indicate the hazard before the user performs an action. Do not place Danger at the end of a task or at the end of the guide. Unlike other types of notices, danger notices always specifically identify the danger itself and indicate what actions to avoid, to provide additional clarification.

Warnings

Far more common than dangers, warning notices alert users to potential hazards that may result in injury to themselves or others if the user does not take proper precautions.

Typically, this type of notice deals with injuries that are not fatal but could still cause permanent damage or bodily harm.

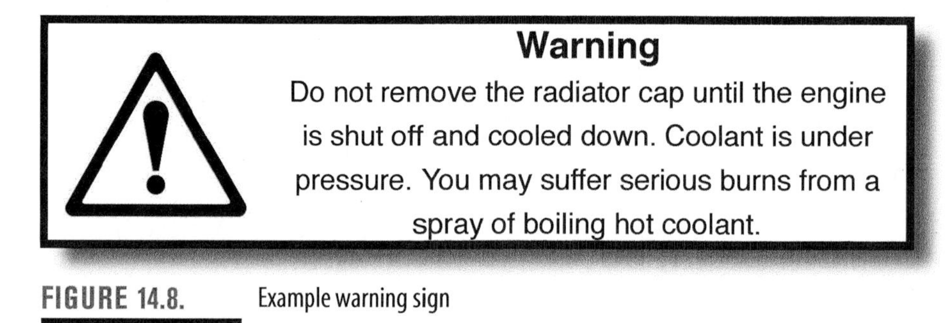

FIGURE 14.8. Example warning sign

As Figure 14.8 shows, warnings are more explanatory than danger notices. Each warning states what action should not be taken and indicates the consequences for ignoring the warning itself. If you have multiple hazards, you need multiple warnings. Each notice should state only one action and its associated consequences. The formatting of both warning and danger notices should be consistent throughout the document, but warning notices often use smaller fonts and avoid capitalization to indicate their lesser importance.

Cautions

Caution notices alert users to potential hazards that may damage machinery, equipment, or problems associated with an action. This type of notice also warns readers about actions that may jeopardize the outcome of the task at hand. Like warnings, caution notices indicate both the action that should not be taken and the consequences associated with it. For instance, consider the caution in Figure 14.9.

Caution

Do not splash the engine coolant on painted parts. The alcohol contained in the coolant may damage the paint surface.

FIGURE 14.9. Example caution sign

As with other types of notices, you must maintain an active voice and imperative mood. In addition, align the heading Caution with the text it references, but place it after any danger or warning notices that might also be present. Caution notices should also be less visually distinct than the previous two notices. They stand out from the step-by-step actions that must be performed in the instructions, but they do not have the immediacy of the previous two types of notices. You can think of caution notices as advisory statements against costly mistakes.

Notes

As the last and lowest type of notice, a Notes heading calls attention to important supplemental information and emphasizes important points in the procedure and its outcome. Notes provide feedback to enhance the user's understanding and performance following a procedure. You can also use them to provide alternative steps or additional explanation when the purpose of a step is not readily apparent. For this reason, notes typically appear in line with the step-by-step instructions or directly after the associated step. For instance, consider the example from an automotive maintenance manual in Figure 14.10.

Checking the Coolant Level

Check the coolant level at each fuel stop.

1. Check the coolant level on the outside of the reservoir while the engine is cool.
2. If the level is close to or lower than the "LOW" level mark, add coolant up to the "FULL" level mark. If the reserve tank is empty, remove the radiator cap and refill coolant up to just below the filler neck as shown in the following illustration.
3. After refilling the reserve tank and the radiator, reinstall the cap and check that the rubber gaskets inside the radiator cap are in the proper position.

Note: Vehicles are filled at the factory with Super Coolant that does not require the first change for 11 years/137,500 miles (11 years/220,000 km). Should it be necessary to top up the coolant for any reason, use only dealer provided Super Coolant. Mixing with a different coolant will reduce the life of the coolant.

FIGURE 14.10. Example note for checking coolant level

Source: Subaru, "Maintenance and Service," 2015 *Forester Owner's Manual.* (Cherry Hill, NJ: Subaru of America, Inc., 2013), 11.13.

In order to avoid distracting the user, you should keep the visual differentiation between notes and the steps limited. Notes should naturally flow with the instructions. For this reason, notes can come before or after the steps in reference. Unlike other types of notices, placement for a note depends on the relative importance of the information and whether the outcome might be adversely affected. If you have multiple notes associated with a particular action, collect the notes into a hierarchical list based on their importance and place them at the end of a task so they do not disrupt the structure of the task.

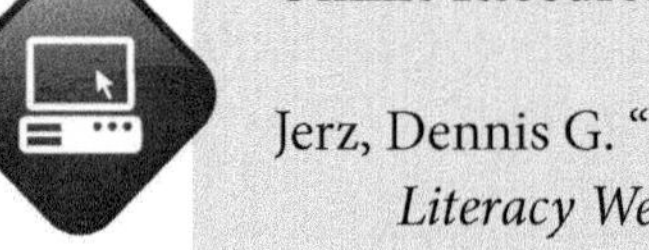

Online Resources for Writing Instructions

Jerz, Dennis G. "Instructions to Write Guides for Busy, Grouchy People." *Jerz's Literacy Weblog* (blog). http://jerz.setonhill.edu/writing/technical-writing/instructions-how-to-write-for-busy-grouchy-people/

Price, Jonathan, and Lisa Price. "Help." *Hot Text: Web Writing That Works*, 305-22. San Francisco: New Riders, 2004. PDF e-book. http://www.webwritingthatworks.com/HTprocedures.pdf

Ross, Derek G. "Writing Instructions: Guiding a Reader Through a Process." *Auburn University Office of Writing*. https://www.auburn.edu/~dgr0003/Classroom%20PPT/instructions_process.ppt

Resources for Special Notices

Environmental Protection Agency. "Pesticide Labels." *EPA.gov*. http://www.epa.gov/pesticide-labels

New OSHA/ANSI Safety Sign Systems for Today's World. Milford, PA: Clarion Safety Systems, 2013. PDF White Paper. http://www.ishn.com/articles/99405-new-oshaansi-safety-sign-systems-for-todays-workplaces

McMurrey, David. "Very Special Notices." *Online Technical Writing*. https://www.prismnet.com/~hcexres/textbook/special_notices.html

Sattler, Barabara, Bruce Lippy, and Tyrone Jordan. "Hazard Communication: A Review of the Science Underpinning the Art of Communication for Health and Safety." *United States Department of Labor*. https://www.osha.gov/dsg/hazcom/hc2inf2.html#3.1.3.1

CHAPTER 15

USABILITY TESTING

Usability generally refers to how well a tool or technology helps someone perform a task. Website usability deals with effective user interface design. The Usability.gov site defines website user testing as:

> *Usability testing encompasses a range of methods for identifying how users actually interact with a prototype or a complete site. In a typical approach, users—one at a time or two working together—use the website to perform tasks, while one or more people watch, listen, and take notes.*
> (http://usability.gov/basics/index.html)

How are Usability Tests Conducted?

Usability experts generally agree on a six step process for usability testing:

1. Determine the goals of the study
2. Develop a profile of typical users
3. Write the user tasks
4. Conduct user tests
5. Evaluate the data
6. Recommend or implement changes

Step 1: Determine the Goals of the Study

Your evaluation should target common tasks that users are expected to perform on the website. Identify the purposes or aims of the site. What are the intended uses of the site? Why do people come to the site? What tasks might they attempt to perform? Break these uses down into primary and secondary tasks. For example, a travel website's primary function is airline booking, a secondary task is finding information about travel destinations.

Step 2: Determine the User Profile

The goal is to match the people you user test with the target user profile. Identify typical users' attributes: Who will come to the website? What is their age, gender, education level, etc.? What will be their knowledge level about the website and its purposes? What will be their knowledge level about the Internet in general? What other websites are related to this and will the users be familiar with them? If you're using classmates as your user group, justify how they fit with the user profile.

Step 3: Write the User Tests

User testing involves observing actual members of the target group using the website and interviewing or surveying them before, during, and after they use the site to gather their feedback. Most experts agree that a well designed test of only a small group, 5 to 10 people, reveals any major usability issues with a website (see Jakob Nielsen's "Why You only Need to Test With 5 Users" http://www.useit.com/alertbox/20000319.html)

Once you've brainstormed a list of the site's primary and secondary functions, translate them into a list of tasks or scenarios you will ask your users to perform during the tests. There are two basic types of user tasks:

1. **Directed tasks:** These are specific, close-ended tasks related to primary or secondary purposes. For example, to study the usability of an airline's website, you might ask the user: "Book a flight from Cleveland to Phoenix for your Christmas vacation. You want to leave the morning of December 20 and return anytime on

Online Example:

- Jakob Nielsen's 1994 Usability Report, which used all directed tasks. http://www.useit.com/papers/1994_web_ usability_report.html

December 30."

2. **Undirected tasks:** These are good for general evaluations/assessments (a good choice if you notice lots of problems with a website). Undirected tasks are more open-ended. They are scenarios designed to mirror how intended users might typically use the site such as: "You're at home browsing the Internet, interested in finding the latest news about Iraq. You decide to use the latest search engine being promoted by your ISP." Once the user is finished with the task you can collect their opinion about general and specific issues.

Online Example:

- System Usability Scale (usability net) http://usabilitynet.org/trump/documents/Suschapt.doc

After you have developed your user tasks/scenarios, decide on a method for collecting user feedback. Choose a written protocol (distributed to a user who writes his or her own feedback) or an interview protocol (with answers recorded by researcher), or a combination of both.

Make sure you plan on asking enough of the right questions before, during, and after your user tests:

- **Pre-Test Questions:** Used to collect basic demographic information and possibly information related to purposes of your website (e.g., if it is an auto manufacturer's website, ask users if they have ever visited such a site before). Other demographic questions used to verify the validity of the user group include age, gender, and computer experience. You could also ask questions such as: How often do you use the WWW? For what length of time do you typically use the WWW? For what purposes do you use the WWW? Have you ever used to the WWW to investigate products you wish to buy? Have you ever purchased anything over the WWW? Have you ever purchased anything you saw advertised on the WWW?
- **Test Questions:** Used to prompt users' reactions to specific tasks: What are you doing now? Why? What is the first thing you're attracted to? What is a particular feature, for instance, background color, like? NOTE: some of the post-test questions below might work better as in-test questions given during or immediately after specific tasks.
- **Post-Test Questions:** Prompt users' overall reactions to the test and website: What was your general impression of each page? What problems did you have using the page and why? What do you believe caused this problem? What, if anything, helped you find the information you were looking for? What, if anything, made finding the information more difficult? What information would you like to see on the page that was missing? Was the page easy to use? Rate the pages you saw—which was best? Give three reasons.

Lastly, consider using other methods for collecting data:

- **Observation.** You should record key user actions, e.g., how long it takes to complete a task, where the user clicks, how many clicks it takes the user to complete a task, and what the user's page-to-page navigation is.
- **Think-Aloud.** You can record what users say while performing tasks and verbalizing decision making process involved in performing a certain directed or undirected task.

Step 4: Conduct the Tests

This is where you get to actually see how users perform your tasks. Keep Keith Instone's advice, watch and learn, in mind as you conduct your tests. That is, don't interfere with the users as they attempt to perform the tasks. Avoid giving verbal or visual cues that might influence the users' interaction with the website. Do not give them any background information or help them if they get stuck. That's one thing you're looking for, if they get stuck. Take notes—write down what they say, what they get stuck on, what links they use, how much time it takes to perform tasks, etc. Ask follow up questions about what they had trouble with, their overall opinion of the site, etc. Use the questions from the protocol you developed beforehand. At the end of the test you can demonstrate for users what they had trouble figuring out.

Step 5: Evaluate the Data

The most important skill of the usability consultant is the ability to recognize patterns. As researchers and usability analysts, it is your team's task to make sense of the results of your user testing. You should first compile your findings (one of the sections of your final report will be a findings section). You can group your findings by tasks and by answers to protocol questions.

For more on what it takes to be a good usability tester, see Jakob Nielsen's "Becoming a Usability Professional." https://www.nngroup.com/articles/becoming-a-usability-professional/

Once you compile the data, you'll want to look for significant patterns: what did users have the most difficulty with? How severe were their difficulties? How satisfied were they with aspects you were testing? Once you start to ask questions like this, then you can begin to brainstorm solutions or fixes (your final recommendations). As you start to discover patterns in the results of your user tests, consult website design principles for expert opinions on how to improve these elements.

Keith Instone's "First User Test" article is an excellent discussion of the findings and recommendations of his evaluation of the Internet Travel Network website http://web.archive.org/web/20030219012940/http://www.webreview.com/1997/05_30/strategists/05_30_97_8.shtml

Step 6: Recommend Changes

Most usability consultants can implement recommendations firsthand. Your team will have to instead write your recommendations in a credible and persuasive report. Again, Keith Instone's "First User Test" article is an excellent example of how detailed your recommendations should be. Notice that Instone makes three recommendations, all of which are linked to more detailed discussion.

Sample User Tests

- Melissa Cheung and Anuja Dharkar's protocol draft for testing their "Policy Maker" software http://ldt.stanford.edu/~mcheung/policymaker/appendix/protocol.html

PROJECTS

INTRODUCTORY MEMO PROJECT

CHAPTER 16

Project Objectives

- Practice the conventions of writing business correspondence
- Learn the business memo format
- Inform your instructor of your background
- Practice using computer technology to compose a document from a template

For your first assignment you must write a memo that informs your instructor of your current academic status and other features of your background. This will allow your instructor to (a) find out about you, (b) assess your ability to write and follow directions, and (c) help your instructor place you in collaborative groups and situations throughout this course. Additional aims of this assignment are to introduce you to the basics of business memos and get you used to working in the course's computer environment.

Steps for Completing this Assignment

First, read the Style and Correspondence chapters in this textbook. Your introductory memo should demonstrate knowledge of the principles reviewed in these chapters.

Then, using the Microsoft Word memo template and your knowledge of the required reading for this assignment, write a formal business memorandum that provides your instructor with the following information:

- **Personal information:** What is your phone number and email? Who is your advisor and how can your business writing instructor contact him or her?
- **Career objective:** What is your major, and what are your career goals?
- **Computer experience:** What is your knowledge of and level of with computers?
- **Computer access:** What is your access to computers? What kind of computer(s) will you be using for this course? Be as specific as possible about the type, processing speed, and version of Internet browser. Where are those computers located (home, school lab, work)? If you have any concerns about your access to a computer during this semester, please inform your instructor so he or she can help resolve any problems you might have.
- **Work/internship experience and workplace writing experience:** What work experience do you have and have you done any writing for/at work?
- **Writing courses:** What other writing courses have you taken (e.g., composition) and where did you take them? How did you do in your writing courses? How would rate yourself as a writer?

You can also include in your memo whatever you think helps achieve the two aims identified in this memo's opening. For instance, if you have any concerns or anxieties about this course or if you have certain goals/expectations, it would be worthwhile to express them.

Follow the Required Memo Format

Unless your instructor specifies a different format, your memo should be no more than 1.5 pages and should follow the format, or document specifications, exemplified in the memo template provided by your instructor. Include the memo heading from the template and use only 11 point Arial font (for headings) and Times New Roman font (for body text), single spacing within paragraphs, double spacing between paragraphs. Include bolded headings to help guide the reader through the memo, and use bulleted lists where appropriate. Unless otherwise instructed, you should follow this format whenever your instructor asks for a writing assignment from you.

Write Using a Business Style

This is your instructor's first impression of you as a person, a student, a writer. Treat it like you would other high-stakes, first-impression contexts such as a job interview or sales meeting.

Write using complete sentences and use a business style: write clearly and concisely, avoid overly pretentious diction, and edit the memo for neatness and correctness. Try to use headings and bulleted lists to present information clearly and quickly (as this memo does).

If there are too many mistakes your instructor will simply ask you to revise the memo until it is appropriately and satisfactorily written, just as a supervisor would in the workplace. But the damage to your identity as a careful and conscientious student/writer will have already been done (just like any other blown first impressions).

Submission Format

Your instructor will tell you how he or she wants you to submit your completed assignment. Most instructors will ask you to submit only an electronic copy.

Templates

Introductory Memo Template

Evaluation Criteria

Purpose: **Does the memo provide all of the requested information and thus meet the reader's needs:**

- Does it provide contact information?
- Does it provide career information?
- Discuss computer experience?
- Discuss computer access?
- Discuss work and workplace writing experience?
- Discuss past writing classes?

Conventions of the memo: **Has the writer used a template and does the memo include all the basic parts of a memo:**

- Was the memo template file used so that the logo appears at the top of the memo and the same original font styles are used? (Times New Roman for text and Arial for headings, both 11 pt.)
- Is there a heading with information about sender, receiver, date, and subject?
- Is there an opening that briefly states purpose? (no heading necessary)
- Is there a clear body divided into subtopics, arranged by headings?
- Is there a closing that restates main points and politely requests action of the reader?

Readability devices: Are page design techniques used to improve the readability of the document?

- Does the memo use single-spaced paragraphs?
- Does the memo use short paragraphs?
- Are heading used for each main topic section?
- Is at least one list used to present information? (e.g., bullet list of courses taken, list of contact information, list of computer specifications)

Style: Is the memo written clearly and concisely and in a professional tone?

- Has wordiness been reduced or eliminated in a way that makes the memo easier to read?
- Has the writer used complete sentences and not written in a "telegraphic" style that uses only fragments, omits articles such as "a" and "the," and overuses bullets? (A memo should be written in complete sentences and should not read like an outline or resume.)
- Has the memo been carefully proofread for obvious grammatical and mechanical mistakes?

CHAPTER

17

SITUATION ANALYSIS

Project Objectives

- Practice analyzing a business writing situation
- Apply concepts of effective business communication to the analysis of a specific writing situation
- Demonstrate awareness of context, audience, purpose, ethos, and document design

Business and technical writers need to be aware of the following situational factors when composing documents:

- Context
- Audience
- Purpose
- Writer
- Document Design

Use the following prompts to brainstorm important factors that influence how documents are written. Writers who reflect on their writing situation before writing are more successful and efficient than those who don't. As you become more familiar with the situation analysis exercise, it will take less time and energy to complete.

Consider *situation analysis* as an informal prewriting exercise. Focus on getting your thoughts recorded via the prompts. Worry more about content than grammar or mechanics. The list of questions is not prescriptive. You may use as many or as few of the prompts as seems appropriate.

Preliminary Considerations

List the following:

- Writer's name, title:
- Department, project number or name:
- Subject/Assignment:

Audience

1. **Who is the primary audience for this document?**
   Be specific. If you are writing for an individual within an organization provide a name and title. If the document is for a broader audience, who are they? What assumptions can you make about the audience?
2. **What level of technical knowledge or expertise about the subject will the primary reader(s) have?**
   Is the reader an unfamiliar novice or an informed expert in the subject that you are writing about? Will you need to write a highly technical document, a semi-technical document, or a non-technical document?
3. **What preconceptions will the primary audience(s) have?**
   Are there any preconceptions that you can infer from your knowledge of the primary audience? How will the preconceptions influence the reader(s)? Will they be resistant to the information contained in the text? Will they agree/disagree with the information provided?
4. **Are there any cultural considerations that the writer needs to consider?**
   The WWW and advances in communication technologies have made the business environment in the new millennium global. How likely is it that someone from another country will read the document? Do you need to tailor the content to accommodate a worldwide audience? Do you need to at least acknowledge that this document will be read by an international audience?
5. **Who else will read the document?**
   You must assume that others, besides the primary reader(s) will have access to the document. Consider your supplementary or secondary audience. Who might that be, and why would they be reading the document as a secondary reader? For example, the company's lawyer may need to review your response to a client.

6. **What is the supplementary reader's level of knowledge or expertise?**
   Is the supplementary reader an unfamiliar novice or an informed expert in the subject that you are writing about? Will you need to write a highly technical document, a semi-technical document, or a non-technical document?
7. **What assumptions or preconceptions will supplementary readers have?**
   Are there any preconceptions that you can infer from your knowledge of the supplementary audience? How will the preconceptions influence the reader(s)? Will they be resistant to the information contained in the text? Will they agree/disagree with the information provided?

Purpose and Intended Use(s)

1. **What is the main purpose of this document?**
   Do you wish to convey information? Are you seeking to inform the reader(s)? Instruct them in a policy, method of operation, and/or a procedure? Are you offering a solution to a problem or concern?
2. **What are the secondary, tertiary, etc. purpose(s) of this document?**
   You may identify and explain any other purpose(s) that you have or that you wish to accomplish with this document. What else do you hope to achieve with this text? Don't forget about the important purposes of maintaining positive relations and creating a legally binding record.
3. **What information and/or content does the audience expect to find in the document?**
   Another way of approaching this question is to consider what the primary reader(s) are expecting to find in the document. Have you provided the necessary information and/or content required by the audience?
4. **How will the audience use the document?**
   What are the identifiable uses of the contents of the document? Having an understanding of how the audience will use the document will help you to compose an effective (i.e., useful to the reader) document. Will users rely on the information to build something, buy something, act differently, avoid harm or death, etc.?
5. **What is the best way to organize the information in the document?**
   Given the purpose and audience's intended use, how should you organize the document? Should your organization be direct or indirect? You could sketch a preliminary outline here.

The Writer

1. **What is your relationship to the primary reader(s)?**
   Defining and expressing the relationship between writer and reader(s) is important since this will influence much of the information in a document. Is the primary audience your teacher, manager, employee, client, etc.? What are you to your audience: superior, subordinate, friend, enemy, etc.?
2. **Given your audience and purpose, what tone should you adopt to convey your message?**
   In general, your tone should always be professional and sincere, but other moods may be appropriate such as humility to admit a mistake or courtesy to welcome a new employee, etc. You should generally avoid hostile or arrogant tones.
3. **Given your audience and purpose, what level of vocabulary should you employ?**
   Unless you are on a first-name speaking basis with your reader, assume a more formal style of language use. Should you avoid jargon, slang, or colloquialisms (conversational style)? Should you use first-person "I" and second-person "you" or should your voice be in the more traditional third-person "they" and "it"? Are there any concepts or ideas in your message that may have to be defined or clarified depending on your audience's technical background knowledge? For each term that needs clarification, what level of definition should be used?
4. **Are there any political or ethical considerations?**
   Do you risk stepping on anyone's toes with your message? For instance, how will you report a costly error to your company during your first assignment at work? Will your message impact others negatively or positively? Are there any financial, legal, or ethical factors related to your message that you should consider? How should you adjust your message?

Document Design

1. **What are the specific design components that are required to effectively complete the document?**
   This focuses on the issues of the use of a required template or format, type style, font size, inclusion of charts, graphs, illustrations or other visuals. Are you required to follow a specified format? Are there any complex ideas that could be clarified through the use of tables or charts, e.g., numbers, instructions, systems, or specifications?
2. **Are there additional design considerations that will influence the production/composition of the document?**
3. **When is this document due? Are there any other deadlines (drafts, etc.) that you need to consider?**
   List the exact days, dates and times.

Notes

Record any additional thoughts relevant to the project at the end.

Optional: Situation Analysis Memo

Your instructor may ask you to write a more formal **Situational Analysis Memo (SAM)** as a way to help you plan your documents and also as another instance of simulated professional writing. In other words, the Situation Analysis Worksheet is intended as an informal prewriting exercise. The SAM, on the other hand, is a formal representation to your instructor of the results of informal prewriting you did using the worksheet.

1. Using the **Memo Template**, translate the information on the Worksheet into a narrative memo.
   - You will notice that there are boldfaced headings on the SAM Worksheet. Use these headings in your narrative translation of the Situation Analysis Memo.
   - Plan on spending enough time on the Worksheet to compose an effective SAM. The more time and effort that you invest in completing the worksheet, the more you will benefit. You will make more economical use of your available time, and your writing will be concise and focused and therefore more effective.
2. Make decisions about the type and the amount of information that you include in the SAM. You will not need to include everything from the Worksheet in the SAM. Some of the information on the Worksheet will only be useful to you as you draft, revise, proofread, edit, and publish the project document(s).

Be sure to revise and edit the SAM. As a formal presentation of your work it is important to pay attention to both content and form.

PROJECT ASSESSMENT MEMO

The Project Assessment Memo (PAM) should be a short memo (under 2 pages) that provides your instructor with an explanation of your approach to writing a response to a specific case. The goal is to persuade your instructor that your composing decisions were based on your critical analysis of the rhetorical situation of the case and were not just arbitrary or pulled out of a hat. You response should be based on your careful analysis of the case and understanding of the course principles introduced during the case.

> Use the PAM template, do not exceed two (2) pages, and include the following sections with these same section headings

Overview

Answer the reader's question: *What is this memo about?* Include a statement of purpose, a summary of main points, and a list of the items submitted along with this document.

Context

Answer the question: *What was your understanding of the project?* This section should include a summary of your understanding of the case including what your role was, what the situation was, and what your tasks were.

This section is also your chance to include information about how you approached the project that your instructor might not otherwise be able to get by looking at the documents you are submitting. It may also include details about the following:

- Assumptions you have made about the background or scenario
- Connections with the material or the project you may have that I don't know about
- Problems, conflicts, or contradictions you faced with gaps in scenario's background
- Information about the audiences your documents are addressed to
- An overall statement about your approach to a project or case

Documents

Answer the question: *What documents did you produce and why?* Subdivide this section by document, explaining the decisions behind each. This is the place where you do an analysis of the documents you are turning in and explain the decisions you made about them as well as how these documents relate to your overall communication strategy. Your decisions about purposes, argument, style, and format/layout are all possible topics. The idea is for you to explain why you have written the document as you have. Don't simply list the decisions you have made, you want to explain why you made them. For example, write: "The purpose of the Philips report is to X and Y. I wanted to X because....To accomplish X, I included details about...."

Production

Answer the question: What steps did you go through, *what strategies or techniques did you use to produce/make/write your final documents?* This section is where you give me details about how you produced the documents. Production generally leads the reader through the stages of development of the project. It includes such things as research conducted, planning or invention/analysis work done, accounts of group meetings, descriptions of group member roles and contributions, problems you had during the process, and ideas about how the production process may have influenced the work you are turning in. What new production strategies did you find particularly effective that you might use in the future?

Summary

Answer the question: *How did you accomplish/meet the project and course objectives?* This is the section where you review your analysis of the project. It is good to end with an evaluative statement regarding the success of the project. Remember to keep in mind the specific project objectives and the general criteria for professional business communication printed in your syllabus. Your assessment of the project should refer to these criteria and show how, specifically, your work meets or does not meet those criteria. Also remember to keep arguments/evaluations realistic; weak or overstated claims can actually do more harm than good.

DEFINITIONS PROJECT

CHAPTER 19

Project Objectives

- Learn to write clearly by using definitions and other rhetorical strategies
- Learn and apply principles of writing technical memos
- Learn to integrate research into technical documents
- Learn to apply ethical principles in the planning of technical documents
- Gain additional practice using electronic writing technologies

In this project individual students research, plan, and write a policy memo for a fictional company. Beyond simply summarizing policies discovered through research, students must develop a *persuasive definition* that satisfies an organizational need for clear policy and that minimizes any resistance to such policy. In developing a usable policy for all parties you must carefully consider and apply ethical principles as well. In this way, you'll learn how clear and accurate definitions help readers understand and evaluate complex technical and social issues.

This project is a good exercise for business students who hope to be involved in the development and management of corporate policy. For students preparing for technical careers this project is a good exercise in clarifying complex information through definition, expansion, and example. Technical writers often convey complex, highly specialized information through the same strategies of definition (see Definitions chapter). Any business or technical document must be conveyed in language and concepts that readers understand, and learning how to write definitions helps writers translate complex ideas in reader-friendly ways.

Background: Bad Culture – New Management – New Policies

For this project assume that you work for a medium-sized software development company with 300 employees. The company has recently undergone several changes in upper management due to a company buyout and some rather nasty internal problems related to sexual harassment, corporate malfeasance, and media relations.

Several former employees threatened to sue the company amid allegations of sexual harassment. Women workers claimed that the previous management had fostered a hostile work environment. Among the allegations is that workstations were sabotaged, causing equipment malfunction and loss of productivity. A small few complained of physical assaults and verbal abuse. The complaints were settled quickly, primarily because the company was not in compliance with federal and state regulations regarding sexual harassment. Specifically, the company did not have a sexual harassment policy in place. Beginning several weeks ago, the new management has tried to bring the company into compliance by sponsoring several workshops on sexual harassment and announcing plans to implement a zero tolerance policy. However, some employees have claimed the anti-sexual harassment training was treated like a joke, with some male coworkers making inappropriate and harassing comments during the training. Others took the comments as tension relieving humor. Many employees—including both men and women—are fearful, distrustful, or resentful that the company is becoming "PC" (politically correct) and overreacting to a few isolated incidents within an otherwise good workplace environment. Questions about the meaning of zero tolerance have fueled all kinds of speculation and innuendo about permissible behavior within the company.

Many employees feel the company has been poisoned by the old management, which seemed to purposefully mislead the rank-and-file about the financial state of the company. Most employees believe the buyout—and subsequent restructuring and layoffs—could have been avoided if the management had cared more about the well-being of its employees than the profits gained by selling off a deliberately mismanaged company. This profit-at-all costs mentality has seemingly permeated the culture of the company. There exists more an attitude of competition rather than cooperation among employees. Distrust is the norm rather than the exception. The employees are worried about how the new management will treat them and about their futures with the company. In turn, the new management feels strongly that an ethics policy that covers issues of corporate malfeasance and employee conduct would help quell cynicism, repair morale, and foster a more positive workplace environment.

Another problem plaguing the company is bad public relations. The company earned its reputation in the IT industry for developing innovative spam and pop-up blocking programs that have been incorporated into many lesser-known email and browser applications. It was only a matter of time before the company's products would be sought by the big players in the email and browser markets. The company's future prospects roused the interest of many Wall Street analysts, business journalists, and IT industry executives. Employees found themselves inundated with phone, email, and face-to-face requests for information about how its products work and what the future holds for the company. Some details of internal unrest related to sexual harassment and corporate greed also found their way into the mainstream media. The temptation to share insider information came to a head during the buyout, when some employees quoted in trade publications and national newspapers said too much. The company's new management wants a policy setting guidelines for employees dealing with the media and public.

At this stage the employees are confused, resentful, and paranoid about these issues. Your charge as a member of the communications group is to develop the company's new policies about sexual harassment, corporate ethics, or media relations (you choose one). Your specific task is to develop policy as a form of expanded definition. You have full backing from the new management, which wants to create a model workplace. Beyond compliance with legal requirements your organization seeks to improve relations between coworkers, management, and the public in hope of boosting productivity. Thus, you face an informative, persuasive, and ethical challenge: to move beyond the usual matters of clarity so that your definition promotes real understanding and reconciliation. You need to expand upon any legalistic definitions so that employees are able to understand clearly what does and what does not constitute inappropriate behavior related to any of the aforementioned issues. Thoughtless readers are likely to change their behavior only if they feel coerced. But appeals to fear almost always have limited success. You want your definition to convince people to accept the values that underlie the issues. This requires an expanded definition of each issue.

Online Resources: Writing Company Policy

"User Guide to Writing Policies" by University of Colorado Office of Policy and Efficiency https://www.cu.edu/sites/default/files/APSwritingguide.pdf

Exercises

1. Research and find examples of company policies online. In small groups or as a whole class, discuss their common features. Which policies seem effective, and what elements contribute to their effectiveness? Which policies seem ineffective, and what elements contribute to their ineffectiveness? Develop a checklist of criteria for writing effective policy that covers content, organization, style, and format.

2. Write a policy memo that does not exceed 5 pages. Make sure the technical document is user friendly, both in content and form. Include a list of sources and/or copies of any memos you used as a guide or reference during your planning process.

3. If your class is doing exercise 2, have a discussion about how much can be borrowed from existing policies when you as a company writer are charged with writing a company policy. What might constitute plagiarism in this context? How should outside sources be cited in this context? What sources should be incorporated into policies and what sources shouldn't?

4. Conduct a situation analysis and/or write a situation analysis memo for this project. This can be done as a standalone exercise to reinforce situation analysis or in preparation for completing exercise 2.

CASES: OVERVIEW OF GOALS AND STRATEGIES

CHAPTER 20

Case projects ask you to assume a role in a fictitious situation (often based on actual events) and respond as if you were in that situation. Case assignments help engage students in real world communication problems by simulating the ambiguity and political complexities of writing in workplace organizations.

Cases help develop the necessary critical thinking, problem analysis, and close reading strategies that are part of the writing process. Background readings will introduce you to principles of business writing, including ethical awareness and the conventions of formal business memos, letters, and reports. The cases give you the opportunity to apply these principles in simulated situations that are as close as a classroom setting can get to real world writing.

Students often find cases very challenging because they are being asked to think through ambiguous situations that do not have clear-cut right or wrong solutions. Background readings, class discussion, and instructor feedback should help you develop the critical thinking skills that lead to writing more effective solutions to such problems. The good news: if you mess up you can't be fired or sued as you might be in the real world. You can learn from your mistakes and be more prepared for future writing tasks.

Your instructor may also ask you to write a Project Assessment Memo, or PAM, that describes and justifies your response to the case. This is also a useful exercise, as it forces you to articulate your approach to the case. Good writers make strategic choices that are appropriate to a given situation. Writing a PAM helps make you a more deliberate and strategic writer.

Case Project Objectives

- **Learn to see multiple purposes and audiences in particular contexts.** A writer's choices have real-world consequences that go beyond the page. Cases are purposefully complicated and ambiguous: you need to draft documents that please your supervisor as well as maintain relations between your organization and the client company. These audiences may have different and conflicting values that, as a writer, you must discover and make decisions about. The act of writing begins at this stage of problem analysis and definition.
- **Learn to see the organizational context. You're asked to consider who you are in each case.** In other words, as a writer, you work in a complex web of relations involving overlapping and sometimes contradictory audiences. This organizational context puts the writer in multiple positions—employee, company representative, ethical agent to name a few. Your position changes as your audience changes: if you write to your manager, your position is quite different than it would be if you were writing to an external client.
- **Learn to develop a communication strategy.** Cases try to get you to see business writing as more than just writing a single text or document. Cases encourage you to begin to see how multiple documents—memos, letters, notes for interviews, research reports, presentations—come together in response to some need or exigency.
- **Learn to establish credibility, trust, and goodwill through business writing.** Maintaining goodwill is the most important strategy for all business communication. Coercion and hostility generally make business problems worse. You should certainly keep practical consideration in mind (i.e., money) as you think through case problems, but you should do so in a way that treats people fairly and avoids harm to others. Looking for win-win solutions is a cliché, but it will certainly help you navigate complex business problems where multiple interests are at stake.
- **Learn to see generic features/format concerns of letters, memos and reports as rhetorical.** Letters are primarily external communications, whereas memos are usually internal. Surface level features of the letter—headings, margins, salutation/closing—as well as clarity and tone all contribute to the persuasiveness of the document.
- **Learn to use collaboration.** Cases place you in collaborative situations, helping you see how seeking feedback from others is a valuable strategy when solving complex problems.
- **Learn to use computer technology to design and revise documents.** Many of the cases include MS Word templates that are consistent with the organization setting presented in the scenarios. You have access to these electronic files, which facilitate business writing in new but not necessarily easier ways. Remember to reflect on how the technology affects your project.

Questions for Analyzing Cases

These questions are designed to help you better understand the writing situations of each case. These questions can help you better understand the problem, anticipate possible solutions, and reveal any questions you or fellow students may have about the limited background information presented in each case.

Organizational Context

- Who are the **players** and what are their relationships? (Try mapping/diagramming)
- What are the lines of power within your organization and within the client organization?
- Where in the case do you see problems arising based on power?
- What information don't you know about the case that you have to assume? What information would help the situation? What are the sources of this information?

Rhetorical Context

- Who are you? What is your job? According to your job title, who do you think you typically write to?
- Why are you being asked to do this?
- What are your **purposes** here? How do you **prioritize** them? How can you accomplish them?
- Who should you write to (audience)? What are his/her/their interests and values?
- How many and what kinds of documents will you produce? What level of formality is suitable?

Ethical Context

- Do you agree with your boss? Is simply following orders the best possible response here?
- If you don't agree, what are your options?
- Who is likely to be harmed and how will they be harmed? Whose interests are being privileged? How okay are you with this situation?
- What is your socioeconomic background? How does gender, race, or class potentially factor into this case?

Focus on Managing Ethical Dilemmas

The complex, impersonal, hierarchical nature of contemporary organizations present some difficulties for communication *ethics*, the guidelines or values people should follow based on ideas of right and wrong. How does an individual mesh his or her personal ethic with the ethical norms and expectations of the organization and of society? Most ethical systems consist of the following two principles:

- **Respect others:** "Treat others as you wish others to treat you" is a maxim central to virtually every religious and secular ethical system. Some insist on a stronger version: *care for others*—work proactively to improve others' well being, including the people you work for and with, the people you write to, and the people you write about.
- **Do good, do no harm:** Think about purposes of writing in terms of a larger good or an overall life goal. What good are you trying to accomplish? Why are you working? What are your goals—personal, social, organizational—and how does your work for an organization fit in with those life goals? What are the effects of your work?

These principles become very complicated in cases where unethical actions pose a threat or actually harm others. What is your position—what *should* your position be—regarding the following questions:

- Within organizations does responsibility and accountability reside at the top with the president, chairman, or corporate officer?
- Does responsibility reside with the immediate actors within the company?
- What ethical responsibility should be borne by managers and employees?
- Is the responsibility to be shared equally, or to varying degrees, by members of an organization?

When, for example, is it ethically responsible to "blow the whistle" on your own organization, on competitors, on clients, coworkers, etc? Is it acceptable to whistleblow when you become aware of problematic actions or policies? How serious is the problem? Can it be resolved without legal or police action? Who should you "report" the problem to? How can you make superiors aware of your dilemma and not risk harm to yourself (your credibility, job security, etc.)?

The cases in this textbook will present ethical challenge to you, the writer. You will have to weigh whose interests should be served when considering possible solutions to the problems posed by each case. Keep the above ethical principles in mind as you deliberate on each case: (1) tread others with respect, and (2) do good, do no harm. Also, remember the first maxim of business communication: maintain positive relations, or maintain goodwill.

Focus on Following the Chain of Command

Another communication issue to pay attention to when analyzing business and technical writing cases is the chain of command, or the hierarchy of responsibility within an organization. Business owners or CEOs are at the top of the chain of command. Vice presidents and upper managers report directly to the owners or CEOs. Department managers and supervisors report to the upper managers, and front-line employees report to their supervisors. Each company has its own structure or chain of command, and employees are made aware of the structure through various means, such as job titles, organizational charts, and communication infrastructure (e.g., employees might not be able to instant message upper management within a company intranet).

totallypic/Shutterstock.com

It is conventional to follow the chain of command when communicating within organizations for several reasons:

- **Authority:** When an employee doesn't follow the chain of command, the authority of the immediate supervisor is undermined.
- **Responsibility:** Each person in the chain of command is responsible for a particular area of the organization and may not be able to solve problems as readily as someone at a different level more familiar with the specifics of the problem.
- **Efficiency:** It is generally faster to report problems to the manager at the lowest level before escalating the issue higher up the chain of command because it is less disruptive to the lines of responsibility within an organization (i.e., reporting the problem to an upper manager will require that manager to spend extra time investigating the issue at the lower levels).
- **Morale:** When employees ignore the chain of command it can create uncertainty and resentment among all levels of the organization.[1]

There can be drawbacks to traditional hierarchical reporting, such as lower departments may be hesitant to pass "bad news" up the chain of command, or new procedures may not get communicated clearly as they pass down through the chain of command. Decision making can also be slowed as multiple departments with various interests must be consulted in the process. In such cases, techniques for improving organizational communication can be used, such as seeking feedback from employees and supervisors on the quality of communication within the company and using multiple channels to transmit messages up and down the chain of command (e.g., email, blog, intranet).

When is it okay to go over the boss's head? Following the chain of command has important consequences for the daily operation of an organization as well as the maintenance of a professional image for the employee communicating within the organization. In some cases, however, it is the chain of command that causes problems and in those rare cases an employee may be justified in "going over the head" of the direct supervisor, such as an employee who perceives certain employee rights, ethical standards, or organizational policies are being violated by the behavior or inaction of the direct supervisor. Even in such cases, the employee should try to resolve the problem at the lowest level, while being conscious of the record being created by such communication in the event the issue needs to be reported to higher levels.

Notes:

[1] Luanne Kelchner, "The Importance of Following the Chain of Command in Business," *Houston Chronicle.* Accessed May 1, 2016, http://smallbusiness.chron.com/importance-following-chain-command-business-23560.html

BIG-1 RENTAL AGENCY CASE

CHAPTER 21

Case Overview

The Big-1 Rental Agency case puts you in the difficult position of having to help your hot-headed boss solve a problem between two partner companies that could, depending on your actions, involve people at the highest levels of both companies. How will you choose to solve the problem in a way that not only pleases your boss but also maintains the business relationship of both companies? Whose interests should be addressed and how will you prioritize them? Do not forget to pay careful attention to the information in **Big-1 Case Notes** section as you think through your response to this case.

As the assistant manager at the Pleasanton assembly plant of the Continental Car Corporation (CCC), you are responsible for helping the plant manager, Frank Page, in a variety of tasks. Today he calls you in to take care of a problem that has him visibly annoyed—probably more annoyed than you have ever seen him.*

"The Big-1 Car Rental Agency at the airport has really done it this time," Page says. "Yesterday we had a couple of VPs from a Japanese firm in here, along with an American who represents them out of San Francisco. After lunch, McConkey, the American fellow, and I were walking together out of the restaurant when suddenly he starts telling me about the terrible problems they'd had with the CCC rental car we'd reserved for them at the airport. Seems the heating system didn't work after the first five minutes, so they were freezing during the whole of the 45-minute ride from the Minneapolis-St. Paul

Source: From *Cases for Technical and Professional Writing* by Barbara Couture and J. Rymer Goldstein (Boston: Little, Brown Co. 1985). Reprinted by permission of the author.

airport. But that wasn't all. The car made some funny noises and just wasn't running well. Naturally, the polite Japanese didn't say a word, but McConkey didn't hesitate to say how unhappy and embarrassed he was about the heater not working. Well, if he was embarrassed, can you imagine how embarrassed I was? I didn't know what to say to the Japanese. What a fine way to advertise our products. Yes, sir. Let them see for themselves how wonderful CCC automobiles are!"

You shake your head. "They must have been miserable riding without a heater all the way back from the airport in the 10-degree weather...really impressed with CCC's quality!"

Frank Page nods. "And it isn't as if this is the first time that the Big-1 Agency at the Twin Cities airport has given our visitors lousy service. Several vendors have had problems lately, and have let people in the plant know about it. Why, Manuel Lopez was just complaining to me last week about the poor maintenance on the cars from the airport outlet. Some incident with the reps from the Zorelco Company who got a car that had ten things wrong with it. Lopez finally called in and had the agency come out with another car."

"Actually, the complaints have been coming in for several months at least," you reply. "I remember Joe Bomarito telling me some horror story about a high-mileage car from the Big-1 Agency about the time the addition was being put on the front reception area."

"That's a full seven months ago," Page says tersely. "Enough. We have a sufficient history of complaints on the airport franchise. Now we're going to do something about it."

"So what do you want me to do?" you ask. "Go see the manager at the airport agency?"

"No. There's no point in dealing with him. I believe he's proved his incompetence beyond a doubt. Any outfit that consistently rents poorly maintained cars is running a sloppy operation. I want something to happen. The best-made car in the world won't run well if it's not serviced properly. This Big-1 Agency is causing us an image problem."

"And of course we can't take our airport business anywhere else," you note.

"We have no choice," replies Page. "We can only rent CCC products at the Twin Cities airport by dealing with this outfit. That's why I want to blow the whistle on them. Make sure that they get their act cleaned up, NOW."

"So we go to the top, to corporate headquarters?"

"Right," Frank Page says firmly. "I want you to write a letter to the general manager of the Rental Division of Big-1 in New York City."

"You don't want to start giving him chapter and verse on our problems, do you?" you wonder.

"No, I sure don't. We haven't kept any records anyway. Look, Continental Car Corporation and the Big-1 Rental Company do a lot of business together. I just want to let the corporate people at Big-1 know that this is an intolerable situation and they'd better do something about it. Get their guys out here to look at this airport franchise and see how the place is being run."

"Got it," you reply as you head back to your desk to draft the letter for Page's signature.

Background Information

When formulating your response to the case, please keep in mind the following assumptions you must make:

1. **CCC Uses Big-1 Rental Exclusively.** Standard procedure at Continental Car Corporation is to rent CCC model cars for visitors from the Big-1 Rental Corporation. Big-1 is the major rental firm for CCC products across the United States. Like most automobile manufactures, CCC views renting its own brand of cars for visitors as a public relations effort, and so naturally expects the vehicles will be in top operating condition.
2. **What Frank Page is asking you to do is reasonable and within the bounds of your job** responsibilities. The first paragraph, in fact, states this: you "help" Page. Keep in mind that as an assistant manager you are part of the CCC managerial team at the local Pleasanton plant. You answer to Page, but at the same time, Page seems to think you capable of representing him and CCC to external business partners.
3. **You have all the names and contact information** of all the players in this case at your disposal. You do not have to consider gathering contact information as part of your response to this case. When writing your response documents, just add reasonable names and contact information to your documents. Use this information consistently and be sure whatever information you include conforms to the conventions of business writing (e.g., be sure to follow correct heading formats for letters and memos).
4. **You only know what is available to you in the case.** You cannot make up facts to suit you or to fill in the blanks with information absent from the case scenario. You cannot, for example, assume that you have received records from some other department within CCC or from Big-1 that proves someone's culpability in this case. You can, however, make up specific dates for the rental incidents mentioned in the case.
5. **Your response represents the next step in this case.** If there is some gap in the information, assume it is a legitimate gap that exists at the time of the case. Finding out information not available in the case might be one purpose of some part of your response. In line with Note #3, you cannot, for example, imagine you've received a reply to one of your documents, then formulate another document based on this invented reply. The documents in your response portfolio should represent a communication strategy, a set of documents that simultaneously attempt to address the situation.

Exercises

1. Write a formal complaint letter addressed to someone at Big-1 informing them of the problem. Frank Page asks you to write to the corporate headquarters. Decide if this is the best solution. Discuss other possible solutions and choose the most appropriate before writing to someone at Big-1.

2. Write a memo to Frank Page explaining a communication strategy that would best address the problem. This memo could include asking permission to gather more information and/or asking Page to reconsider his hasty plan, if necessary. You'll have to be tactful if you outline a communication strategy different than Page's.

3. Write a Project Assessment Memo (PAM) written to your instructor reflecting on decisions made in writing the case documents

Templates

Big-1 Planning Worksheet
Continental Car Corporation Letterhead

Big-1 Case Evaluation Checklist

Purpose: How effectively do the documents accomplish their intended task as defined by the writer and the multiple purposes of business writing?

- How likely will your response, the letter to Big-1 and memo to Page, address this problem while maintaining goodwill between
  - CCC and Big-1?
  - Page and CCC?
  - Page and Big-1 (local agency)?
  - Page and you?

Product: How well constructed are the documents?

- Do they follow the generic conventions of business letters and memos?
- Do they maintain a professional tone?
- Are they clearly and concisely written?
- Do they follow conventions of standard written English?

Production: How effectively was the document produced?

- Have all the required documents been produced?
- Have they been produced using the required templates?
- Do the documents reflect careful planning, drafting, revision, and editing? Does the writer's PAM reflect careful analysis and conscious effort to integrate course principles into problem analysis and document production?

* **Source:** From *Cases for Technical and Professional Writing* by Barbara Couture and J. Rymer Goldstein (Boston: Little, Brown Co. 1985). Reprinted by permission of the author.

A BUSINESS FAUX PAS CASE

CHAPTER 22

Case Overview

The Business Faux Pas case is ultimately concerned with cross-cultural communication issues between a French and American company. But it also deals with ethical issues about what should be done in this situation, given the context and the tasks at hand. Obviously, tension between companies causes stress within companies. Bill Nestor, your immediate supervisor and family friend, has offended Madame Marie Jeaneaux, the president of an important French company (see Jeaneaux Letter). She sends a letter to your company's president and the problem finds its way to you, working in BellCom's Accounts Management department. To what extent is this case about cross-cultural communication, basic business etiquette, or poor judgment? You must decide.

You work as an account representative for BellCom Corporation, a growing telecommunications company that specializes in mobile phone payment applications. BellCom is the nation's sixth largest mobile phone service provider and among the fastest growing in its industry. Its annual revenue last year was 1,100,000,000, a 63.5% increase from the previous year. Its net income grew 47.3%, to $74,800,000. BellCom has been looking to partner with overseas companies that specialize in mobile phone payment applications. You and your senior account manager, Bill Nestor, have recently been sent to Marseilles, France to tour several facilities operated by Téléphone de Nice (NT), a major European 5G end-to-end systems provider (i.e., infrastructure, terminals, applications and expertise). This trip was the latest in a series of negotiations over BellCom acquiring the rights to market NT mobile applications in the U.S. as a BellCom product.

This trip was significant to you for many reasons. Not only was this your first business trip abroad, having been on the job only a few weeks, but it was also your first trip with Nestor, an experienced manager for BellCom who is also a good friend of your family. Your parents and Bill grew up in the same neighborhood together. Nestor's been close to your family for as long as you can remember. He was instrumental in securing your BellCom employment, under the condition that he would be your mentor. For you, it was a chance of a lifetime to secure this job and work under such an experienced businessman.

Nestor knows telecommunications and has had a lengthy tenure at BellCom. After working closely as a business associate of Nestor's, however, you've concluded that he's a brilliant buyer and manager, but seems a bit old fashioned in his values and ways of doing business. You've wondered on several occasions if Nestor was suited to overseas acquisitions. (Recently promoted to senior account executive, Nestor's responsibilities have expanded from regional U.S. sales to all overseas accounts.)

During your trip to France you have actually become a bit disillusioned about Nestor as a mentor. It began at the French airport when Nestor said in a voice loud enough for others to hear, "Geez, do the French stink or what!" Even though several more such incidents left you quite embarrassed for Nestor, you decided that saying anything would be awkward.

After visiting the NT facilities as planned, you and Nestor were invited to join the brand new President of NT, Marie Jeaneaux, at a luncheon with several other French business representatives and managers. Expecting to arrive at a boardroom filled with platters of veggies, finger sandwiches, and cans of soda, all surrounding the imminent PowerPoint presentation, you and Nestor were shocked when your NT limousine took you to the world renown Palm d'Or restaurant. This would be a lunch unlike any other.

At the Palm d'Or, Marie Jeaneaux stood to greet her American clients and offered an accented, "Hello, how do you do," as she shook each of their hands. Nestor looked at you and winked then turned to Madame Jeaneaux with a wide, toothy smile and said, "What a lovely scarf you have on." A look of confusion fell on Madame Jeaneaux's face. An older man to her right said, "Bonjour Monsieur Nestor," and introduced himself, "Pontelle, Emmanuel," as they shook hands. "Nice to meet you, Pontelle," replied Nestor. After several awkward handshakes and "hellos" and "bon jours" around the table, lunch began. You recall striking up a conversation with a French woman sitting to your right and asking about the French Riviera. During this conversation you overheard Nestor ask Madame Jeaneaux, "So, what does your husband do?"

The waiter circled the table pouring red wine into each glass with care and grace. It took a few minutes to finish the pouring, and by the time he was done Nestor had finished his glass. No one else had yet taken a sip. You looked around and were sure there was tension amongst the French clients, despite the fact that their whispers and mumbles were beyond him. Not much of a drinker to begin with, you joined in Madame Jeaneaux's toast and took a sip.

Though the two-hour lunch was, as Nestor had told you, to talk about BellCom concerns, most of the time was spent eating and drinking. When Madame Jeaneaux asked you and Nestor if the food was enjoyable you replied, "Yes, it was delicious. Thank you." Nestor, with his mouth full, gave Madame Jeaneaux the OK sign with his fingers.

As the group stood from the table to say their goodbyes several French businessmen kissed Madame Jeaneaux on each cheek before departing. Nestor, following suit, faced Marie Jeaneaux and slurred, "Let me thank you for lunch by doing it your way." Then he put his hands on Jeaneaux's shoulders and leaned in to kiss her on the cheeks. Marie Jeaneaux pulled away in shock but didn't dare broach the subject with her inadequate English. Again, you watched the faces of Jeaneaux's employees wrench and gape in disgust. Not surprisingly, Nestor failed to pick up on the palpable tension in the room. When Nestor said, "Then we'll talk soon. Pleasure to meet you," Madame Jeaneaux simply nodded and walked out of the restaurant.

Though little business was done at lunch it seemed to have been a successful trip at the production plants. Nestor turned to you and said, "A job well done, kid!"

Two Weeks Later at BellCom's California headquarters...

Sherri Philips, Vice President of Operations, has asked to see you in her office. "Good morning," says Ms. Philips. "I assume you've seen the letter from Ms. Jeaneaux." **(See Jeaneaux's Letter.)** So you know she's threatening to back away from our company because of Mr. Nestor's behavior. I'm sure you know how important this deal is. If our company doesn't acquire rights to NT we'll be bought out for sure. We really thought that you and Nestor's tour would be a harmless way to get both your feet wet on this deal. And though we're certain you had less to do with this, you were there, which makes you involved in this mess."

You dip your head, showing some distress.

"You were with Nestor and your name is associated with the uproar. Tough luck. What we need from you is a brief report about what happened. I want you to figure out what Nestor did wrong and give me your recommendation on what the the company can do to avoid future problems in dealing with NT. Do some research if you have to."

"This document, along with Ms. Jeaneaux's letter, will be discussed in an executives meeting about Bill Nestor," Philips continues. "Look, Bernstein (BellCom CEO) and the other vice presidents are livid about this situation and want us to make amends and ensure this never happens again. I know you just started here, and we're glad to have you, really, but you need to help us clean this up or it could end up dragging you down, too."

"Nestor is to draft a formal letter of apology, and we'd like your name on it, too. I have already discussed this with Nestor and he should have a draft to you by the end of the day. I need both the letter and the report by Friday."

Later that day in your office...

"Can you believe that letter?" Nestor angrily murmurs. "I could be fired for what she said happened, you know that?" He pats you on the shoulder and hands you a draft of his apology letter. (**See Nestor's draft.**) "Look kid, I screwed up but I need you to cover for me. This is the letter I want sent, so just put your name on it. And what's in the letter is what I'm going to tell Philips and Bernstein. So, please, make this what really happened, if you know what I mean." And at that point Nestor turns to leave. "Thanks kid," he says winking, "I'll see you at your folks' house on Sunday."

Jeaneaux Letter

12 Month 2003

Rick Bernstein, CEO
BellCom Corporation1
390 5th Avenue
Santa Monica, CA 90405

Dear Rick,

I am pleased to have had the opportunities to learn about your company and have generally been treated gently. As the new president, I must write to you to explain my concerns.

Weeks ago, you sent two associates of yours to tour our facilities. After their visit, we joined for lunch at the very nice Palm d'Or restaurant. The young one was very quiet and one woman said many nice things of him. The other man is the severe problem. Not solely was he rude in numerous areas of etiquette, like with the wine and food, but he made no effort to know my culture. When I come to the United States, Mr. Bernstein, I do not act badly like he did. I do not:

- Drink too much
- Speak badly of the food
- Invade personal space
- Ask personal questions that are sexist
- Talk about someone's dress

It is true that Americans and French have different ways to communicate, but I believe Mr. Nestor should have shown more respect. I am sure you should feel the same.

I very much respect you and your grand company, but I am not sure if Mr. Nestor is the right person to be representing your company during our negotiations. At the very least, I hope that he would receive some kind of education in conducting business in my country. I do not want to quit your company, but I will not be disrespected as a French person, as a competent businessperson, or as a female.

I strongly believe there are greater things in the future for our two companies. I am impressed with BellCom's growth potential, but I hope that you would take care to see how important understanding our cultural differences is if we are to move forward.

Sincerely,

Marie Jeaneaux
President, Téléphone de Nice

Téléphone de Nice

4 Rue de l'arbre, 6th Arrondissement
Nice, France 78911-3444
tel: (1) 5564 5565; fax: (1) 5565 7848; www.NT.fr

Nester Letter

BellCom Corporation
1390 5th Avenue
Santa Monica, CA 90405
Tel (310) 458-8316
Fax (310) 458-8310
www.bellcom.com

Month 16, 2002

Ms. Marie Jeaneaux, President
Téléphone de Nice
4 Rue de l'arbre, 6th Arrondissement
Nice, France 78911-3444

Dear Mrs. Marie Jeaneaux--

In this letter I hope to make amends as well as address and respond to your accusations. First of all, I did not drink too much. The rest of the company members all shared several bottles of wine, during lunch I might add, and all I did was join in the festivities. That, I think is totally unfounded. Second, the food was fantastic and I gave you the "OK" sign when you inquired and I graciously thanked you for lunch, if you remember properly. When I asked about your husband, I was merely trying to get to know you. I apologize for wanting to establish a friendly business relationship. In America, we try to create a rapport. The only time I might have "invaded personal space" was when I kissed you good-bye and thanked you. I've seen enough of French culture to know that's how you all great each other hello and goodbye. You shook our hands, which was very "American" and so I tried to be "French" by doing the cheek kiss.

We hope your company will become a valued partner for many years, and I wouldn't want an overreaction such as this to cause a chasm between companies. [Your name] and I enjoyed our visit immensely and feel that we have been misunderstood; in fact, your misinterpretation of the events has jeopardized our positions at BellCom. We both agree that our behavior during our visit, and especially at the lunch, was appropriate and quite acceptable.

Please accept my apologies and know that we value your company. I won't be so friendly next time if you promise to be less paranoid. Obviously, you've earned your position at Téléphone de Nice and I respect that.

Sincerely,

Bill Nestor
Senior Account Manager

[Your name]
Account Representative

Exercises

1. Individually or in small groups, research and identify the *faux pas* that Nester committed during his trip to France. Based on your research, discuss strategies for avoiding cross-cultural miscommunication.
2. Write an internal report memo to Philips about the incident. You must write a report of what happened at the business luncheon where the offending incident occurred. How should you construct your account of what happened?
3. What are your responsibilities to Nestor, to BellCom, to Téléphone de Nice (NT)? Where, in the case, do you see problems arising based on authority, ethics, and communication issues?
4. Write an apology letter to Mrs. Jeaneaux. You have also been asked to sign an apology letter drafted by Nestor (see Nestor letter). The letter is written from Nestor's point of view and may not be the most accurate or appropriate, but Nestor asks you to cover for him. You must decide what to do for both the apology letter and the report. Should the letter be sent as is?
5. Write a memo to Nestor that responds to his request to sign his letter draft. What if you believe the letter should be revised? How will you persuade him not to send his version and to accept yours?
6. Write a project assessment memo (PAM) addressed to your instructor reflecting on the decisions you made in completing any of the assigned documents.

Online Resource: Cross-Cultural Communication

Executive Planet - Covers many aspects of cross-cultural communication etiquette for specific countries http://www.executiveplanet.com/

Faux Pas Case Templates

Use the following templates as you draft your response documents

- Nestor Letter/BellCom letterhead
- BellCom memo template

Case contributed by Jenny Bania

THE ERROR AT OTS CASE

CHAPTER 23

The Error at OTS Case puts you in a situation, based on actual events, where you have to repair a mistake made by your company. Thomson, a computer products manufacturing company, has partnered with your company, Online Training Services, Inc. (OTS), to run a one-month web-based training program aimed at the retail sales associates (RSAs) working in retail stores across the country. The program offers a free all-in-one printer to the first 500 individuals to score 100% on the training program's final quiz. However, due to a program error, OTS sends the email winning notification to 1,000 people instead of the first 500.

Although the Error at OTS case arises from a software glitch, it is ultimately concerned with the precarious balancing act in the business world in which a company must decide how to correct its own mistake. This case also involves ethical decisions on the part of you, the project manager, who as the voice of the company is heavily relied upon to smooth errors over through active communication with the clients.

You work as a project manager for Online Training Services Inc. (OTS), a web-based training company connecting computer products manufacturers to retail sales stores through comprehensive training solutions. The company's business rationale is that frontline retail sales associates significantly affect the sales of products, as most customers make their purchase decisions based on the feedback and advice given by point-of-sale professionals. Hence, it's good business practice on the part of computer products manufacturing companies to invest in the training of these sales personnel.

Bob O'Brian and Sheila Gallagher started OTS in Santa Fe, California in 2007 with a small but strong client list, two computers, and a rented studio that became their first office. Thanks to Bob and Sheila's initial connections within the industry, the company achieved a small turnover of $300,000 during its first year. With the reputation of the company spreading through the industry, its annual turnover in 2014 grew to $15 million. Its clients now include all the major names in both the computer products manufacturing industry—including IBM, Dell, Toshiba, Sony, HP, Intel, and Motorola—and the retail outlets—including Best Buy, Dell, Target, and Staples.

The main service of OTS is a website platform created, designed, and hosted by OTS where companies offer training programs for the retail sales associates (RSAs). These programs are offered on a monthly basis and include comprehensive training courses, knowledge assessments (quizzes), and valuable prize incentives. The training focuses not only on product knowledge but the enhancement of selling skills.

OTS's company structure is headed by Bob and Sheila, whose titles are, respectively, director of marketing and director of business development. It also includes a project manager (your role), a legal manager, an it supervisor, sales trainers, and a group of web content developers. As the company's project manager, you report directly to Bob, and you oversee all projects including communications with clients.

OTS is faced with a serious problem—one that could potentially make or break your growing position with the company. Thomson, the manufacturer of a wide range of all-in-one printers has teamed up with OTS to offer an online training program for RSAs at top U.S. retail chains across the country. The participants in the program are required to complete a one-month training program and take a quiz upon completion. The first 500 RSAs to score 100% in the final quiz of the program will win one of Thomson's top-of-the-line all-in-one printer called the ELEGANCE A960.

OTS and Thomson have worked together in the past on smaller promotions and have established the beginnings of a strong relationship. This promotion marks an upward shift in the companies' relationship, in that Thomson is transferring a hefty sum of their marketing budget from their prior public relations agency and placing more trust in OTS. Thomson provides an undisclosed rebate to the retail outlets for each RSA that completes the training program. The online training program requires and average of 4 hours to complete. RSAs can complete the training in intervals or all at once; however, RSAs must earn a perfect score on the final knowledge assessment to complete the training. RSAs can retake the final knowledge assessment as many times as necessary. The average completion rate for such training is 98%. Thomson paid OTS $500,000 for the Elegance A960 printer promotion, which enrolled 19,897 RSAs. The wholesale cost of an Elegance A960 printer is $200.

Now, as the contest is wrapping up, you discover a major error has been made by your IT manager. OTS has sent out 1,000 email winning notifications to RSAs instead of 500, as initially planned. The email notification **(see Email Winning Notification)** was supposed to be an automatic response sent to the first 500 RSAs that achieved a 100% score on the final quiz of the training program. This condition was intended to be built into the computer program. However, something in the program went awry and twice as many winner notifications went out. The error has been fixed from the technical end, and now must be dealt with from a P.R. angle, which is your job.

Upon discovery of the glitch, Bob and Sheila pull you into the conference room.

"What do you think we should do?" Sheila asks you.

You consider your options."Well, we could just keep quiet on the issue, purchase the 500 extra printers ourselves and award them to the winners," you suggest.

"I'd rather not," Sheila says. "I'd prefer to use that as a final option. Perhaps we could write an email to the 500 false winners, explaining the situation. Maybe we can offer them some kind of consolation prize instead."

"That might compromise our relationship with our retail partners," Bob interjects. "What if we wrote a letter to their marketing director admitting our mistake, but persuading her to double the prize total considering the success of the training program?" "Both of these options involve our admitting to the error," you point out to your directors. "It may affect our credibility in the future."

"Do you have a better idea?" Sheila says sharply.

"Why don't you think about it a while," Bob says to you.

Bob and Sheila send you on your way, and you immediately consult with Leo, OTS Legal Manager. You explain the situation to him, and he gives the following reply: "Frankly, we're not under any legal obligation to grant the prizes," he says. "The email notification was simply a public relations ploy, not a legally binding contract. It even says so in the small print. All we'd have to do to prove our case is print a winners list from the database showing exactly when each of the 500 winners completed the quiz." **(See official rules.)**

You walk away from Leo's office, your head spinning. You realize, despite your short time on the job, that this is the type of decision making that will help determine your success in the business world. The company has put their trust in you. You return to your office and ponder your options.

Official Rules: Thomson's Sales Training Program

The participation in the program is subject to the following rules:

Participants may register for training at OTS's website, www.ots.com.

The participant must be a U.S. citizen over 18 years of age and employed as a retail sales associate with an OTS recognized retailer (see the drop down list on the home page).

All assignments and quizzes in the program are mandatory for the award of the prize.

Any employee of OTS, or a family member thereof, is prohibited from participating in the program.

You must score 100% on the final quiz of the program to qualify for the prize.

Only the first 500 winners are eligible to win the prize.

The winners will automatically receive an email notification. However, the actual award of the prize will only be made after documentation and verification on our end. The official documentation and a legally binding affidavit will arrive shortly after the initial email. The decision of OTS remains final in the matter. Any disputes related to the prizes will be subject to the jurisdiction of the federal and state courts in Clark County, NV.

The delivery of the prize will be directly from the sponsor company and OTS is not responsible for any delay or damage in the delivery.

Email Winning Notification

Date:___________________

From: service@otsonlinetraining.com

To:________________

Subject: Your Thomson Training Prize

Dear (Participant's name),

Congratulations! We applaud you on completing the Thomson Sales Training Program, and we are pleased to inform you that you are one of the 500 sales associates all over the U.S. to win Thomson's ELEGANCE A960, the latest in all-in-one printers featuring dazzling clarity, speed, and durability.

By completing this exclusive training program you have further boosted your already strong selling instincts, and we are pleased to have you representing our products at the retail frontlines. We are happy to reward you for your dedication to the program. Please allow six to eight weeks for the delivery of your prize.

Congratulations again. We look forward to working with you in the future and wish you the very best in your selling career.

Best,

Online Training Services Inc.

Exercises

1. Write a recommendation memo to Bob O'Brian, OTS Director of Marketing, that explains the best solution to the award notification error. Be sure your report explains OTS's options and recommends the one you think is best for the continued success of the company. The two main options—writing to Thomson or writing to the RSAs—involve an admission of error by OTS. Should you write a letter to Thomson requesting 500 additional printers? Should you send a corrective email to the false winners? Are there other options? What is the best solution to this problem? And what do you base your recommendation on? Remember, you cannot afford to harm your company's relationship with either Thomson or the RSAs, so you will want to choose your words very carefully.

2. Write a letter to the marketing director at Thomson that explains the error and informs them of OTS's solution. Before you write your letter, consider if you should you offer your audience any incentives to sell them your proposition, as is sometimes the case with adjustment letters. What will you write to establish credibility for the future, to ensure that both retailers and manufacturers continue to rely on OTS for their web-based training needs?

3. Write a letter to the 500 retail sales associates who mistakenly received notification of winning the training promotion. Before you write your letter decide if it is best to withdraw the notification or offer some alternative solution.

4. Write a project assessment memo (PAM) addressed to your instructor explaining any of the documents you wrote.

Use the following templates as you draft your response documents

- OTS, Inc. letter template
- OTS, Inc. memo template

Use the following peer review guidelines to evaluate your drafts

- Peer review guidelines for Error at OTS case

Original case contributed by Kelle Schillaci and Anish Dave

INSURANCE FRAUD AT MEDTECH CASE

CHAPTER 24

This case puts you in the role of a human resources manager at MedTech, a company that manufactures medical supplies. It is reported to you that while an employee is earning compensation for a work-related injury, she is also running a bar, earning income, and performing duties inconsistent with her injury, including tapping kegs, cleaning tables and waiting on customers.

Fellow employees within the company become angered when they learn that the company is investigating this woman's potentially fraudulent worker's compensation claim. Your task in this case is to justify the company's actions while building an atmosphere between employees and employer that is based on trust, mutual respect, and the achievement of common goals.

You work as a human resource manager for MedTech Manufacturing, a small but successful Midwestern medical equipment manufacturing company. MedTech produces 80% of the regional hospitals' waste equipment, including medical gloves, biohazard storage and disposal containers, catheter bags, IV bags, etc. You have just taken over for the previous HR manager, Bill Smith, who worked at the company for over 20 years. Since you are new to the area, you haven't met many people, nor do you socialize with the plant workers, most of who have grown up together, and are loyal, hardworking people. While MedTech is in a small town, it is a big business and a vital mainstay in the regional economy.

A female employee, Susan Seer, developed a work-related repeat trauma injury while working on a company assembly line manufacturing medical catheter bags. She strained the tendons in her wrist and sprained her elbow; since she assembles items on a line, this injury made it impossible to continue with her job until she healed. She went to the doctor, then applied for and received worker's compensation based on her alleged temporary disability, which amounted to a certain percentage of her normal income.

This employee also owns a bar and lives upstairs above the bar. She and her live-in significant other normally run the bar nights and weekends. It is reported to you anonymously that while she is earning compensation for being disabled, she is running the bar, earning income, and performing all normal bar duties, including tapping kegs, cleaning tables and waiting on customers: activities that indicate she is not, in fact, disabled pursuant to her physician's advice.

Under state law, the employee is supposed to report supplemental income, which would then reduce the amount of compensation she receives for the work-related disability. In addition, if it is found that the duties she performed in her tavern job are similar to her previous duties and ones that she legally claims not to be able to perform due to the injury, she may in fact be denied the work-related compensation altogether.

In order to establish evidence and get a first hand account of her activities in the tavern, you dispatch an insurance claims adjuster representing the company to visit the bar one evening to document her activities and review their potential as disqualifying factors in her compensation claim. He views the employee serving beer, cleaning tables, waiting on customers, etc. When the claims adjuster pulls out a video camera and begins filming the employee in the tavern, the adjuster is roundly beaten and thrown into the street by the customers in attendance. In addition, the adjuster's vehicle is vandalized and he has to take a taxicab home.

Despite this unpleasantness, the insurance claims investigator filed his paperwork and the insurance company stopped payment of Seer's injury compensation. Susan Seer, in turn, has written a letter to MedTech manufacturing, suggesting that she will file suit for if her benefits are not reinstated to compensate her for her work-related injury **(see Susan Seer's Letter)**.

News of these events spreads rapidly on the factory floor, causing immediate employee relations problems. Fellow employees and friends of the individual in question are demanding to know why she is being harassed by the company, and what right the company has to send out an "agent" to investigate her. The employees are united in opinion and have little trust in you, the new Manager, to handle the situation. Among the comments muttered by the workers:

"Bill never would've harassed Susan like this!"

"Why is the company trying to cheat us out of worker's comp benefits?"

"It's this new guy causing all the trouble."

McKay, the Vice President of MedTech and your boss, has received Susan's letter and is understandably upset. He calls you to his office and says, "This Seer business is turning into a real mess. Why the hell is she collecting compensation and working? And if she's able to work, why hasn't she come back?"

"I want you to draft a memo to all our workers explaining why we had to take the actions we did. I don't want this to create unnecessary distrust among our workers. Tell them as little as possible but enough to make them understand why we have to follow the law."

"I also want you to draft a letter to Susan making it clear to her that she doesn't have a legal leg to stand on. We take care of our workers, but if follows through with this lawsuit she won't have compensation or a job to return to. I'd like you to write me a summary report once you've drafted the paperwork, and copy me on the memo and letter you send out."

You leave his office with a frown on your face. The sheer delicacy of the problem baffles you. This isn't how you wanted to establish yourself at the company, and your future may depend on whether or not you're able to gain the employees' trust.

Letter from Susan Seer

Susan Seer
821 Main St.
Waupun, WI 53963

January 15, 200X

Mike McKay
Vice President
MedTech
5000 Industrial Lane
Beaver Dam, WI 53916

Mr. McKay:

Your management has taken away my workers compensation benefits. Im not able to return to work for several more weeks as per my doctors orders. I am sure there has been some misunderstanding, but I need these benefits or I will not be able to pay my bills. If you are unable to do something to help me I will take you to court to enforce your payment of my benefits. It is my rights. I expect a timely resolution to this problem am look forward to hearing from you soon.

Sincerely,

Susan J. Seer
Susan Seer

Susan Seer's Physician's Report/ Certification of Disability

PHYSICIAN'S AND CHIROPRACTOR'S PROGRESS REPORT CERTIFICATION OF DISABILITY

Claim Number: 22-0142

Social Security Number: 392-84-9574

Patient's Name: Susan Seer

Date of Injury: 8/20/04

Employer: Acme Manuf.

Name of MCO (if applicable):

Patient's Job Description/Occupation: Assembly Line worker

Previous Injuries/Diseases/Surgeries Contributing to the Condition: n/a

Diagnosis: torn ligaments, sprained elbow

Related to the Industrial Injury? Explain: yes - strain from repeat stress action

Objective Medical Findings: 8 weeks for muscle repair

☐ None - Discharged Stable ☒ Yes ☐ No Ratable ☐ Yes ☒ No

☐ Generally Improved ☐ Condition Worsened ☐ Condition Same

May Have Suffered a Permanent Disability ☐ Yes ☒ No

Treatment Plan: Hunk Quinine prescrip, pill form, splint on left arm, no use of arm for 6-7 weeks related to lifting carrying

☐ No Change in Therapy ☒ PT/OT Prescribed ☐ Medication May be Used While Working

☐ Case Management ☐ PT/OT Discontinued

☐ Consultation

☐ Further Diagnostic Studies:

☒ Prescription(s): Phenol 2x daily for 4 weeks

☒ Released to **FULL DUTY**/No Restrictions on (Date): 10/20/04

☒ Certified **TOTALLY TEMPORARILY DISABLED** (Indicate Dates) **From:** 8/20/04 **To:** 10/13/04

☐ Released to **RESTRICTED**/Modified Duty on (Date): **From:** ______ **To:** ______

Restrictions Are: ☐ Permanent ☐ Temporary

☐ No Sitting	☐ No Standing	☒ No Pulling	☐ Other: ______
☒ No Bending at Waist	☐ No Stooping	☒ No Lifting	
☒ No Carrying	☐ No Walking	☐ Lifting Restricted to (lbs.): ______	
☐ No Pushing	☐ No Climbing	☒ No Reaching Above Shoulders	

Date of Next Visit:	Date of this Exam:	Physician/Chiropractor Name:	Physician/Chiropractor Signature:
10/13/04	8/20/04	Mike Knight	Dr. Mike Knight

D-39 (Rev. 7/99)

Online Resources

To help write responses to the issues in this case, you should research information about federal employee compensation laws, the legalities of company policy, and insurance company policies.

Resources for Worker's Compensation Laws and Issues:

- DisabilityClaims.com, a resource for long term disability claims and litigation, sponsored by the law firm of Riemer & Associates http://www.disabilityclaims.com/
- Wisconsin Department of Workforce Development's information on worker's compensation http://dwd.wisconsin.gov/wc/

Exercises

1. Write a Situation Analysis Memo summarizing your analysis of the rhetorical situation for one or more of the documents you might write in exercises 2–4.
2. Write a letter to Susan Seer, the injured employee.
3. Write a memo to all company workers tactfully addressing the issues.
4. Write a memo to McKay, detailing how you approached the previous two documents. The documents require McKay's approval before they are sent.
5. Write a Project Assessment Memo informing your instructor of your approach to writing one or more of the documents from exercises 2–4.

Original case contributed by Heather Lusty

COMPLAINT AT THE COLONNADE HOTEL CASE

CHAPTER 25

Case Overview

In The Complaint at the Colonnade Hotel case, you work as a general manager at a medium-sized hotel who receives a complaint from an angry guest, upset that a family reunion was ruined because the hotel was overbooked and unable accommodate all family members. In addition to an email complaint, the guest has let his frustrations be known through a series of negative reviews posted on several social media sites. Worried about the damage these reviews will cause to your hotel's image, you must try to persuade the guest to understand the situation and revise the negative social media coverage.

Scenario

You are the general manager of Colonnade Hotel and Resort, an independent property in Fort Lauderdale, Florida. Having worked at the Colonnade for four years, you are no stranger to customer service complaints. As the general manager, you rely on your front desk agents to provide quality service and prevent most issues from being escalated to you; however, of course, not all customers can be satisfied.

When you arrive at the property on Monday morning, you are faced with one of those customer situations that requires your attention. As you open your email inbox, you see a message with the subject line, "Shut out of the Colonnade." You take a deep breath, sip your latte, and open the message (see Figure 25.1).

To: Pat Fielding, General Manager

From: Kevin Ward

Subject: Shut out of the Colonnade

Dear Pat Fielding,

I'd like to bring to your attention a horrible experience I had with my family last week at the Colonnade. For my parents' 50th anniversary celebration, 15 of us booked rooms at the hotel eight months ago. We had heard good things about the Colonnade and were all looking forward to the trip.

However, when my wife, son, and I arrived at about 10 p.m. last Friday, we were told by Kathy, the front desk agent, that the hotel was booked, and no rooms were available for us. How could this be, when we pre-paid for our rooms so far in advance? My brother and his family also were shut out (although my sister and parents had already checked in earlier). My family and my brother's family were both sent to different hotels in different directions.

I appreciate receiving the first night's stay courtesy of the Colonnade, but this really misses the point: We all wanted to be in the same hotel. We live in different cities and don't get to spend much time together.

I want to understand your policy. How can this happen? And what can you do for my family, who spent four days negotiating meeting times and taking taxis back and forth to see each other?

I look forward to your response,

Kevin Ward

FIGURE 25.1. Email complaint from Kevin Ward

As an experienced manager, you know to gather more information before responding to the guest. You discuss the situation with Kathy, the front desk agent on duty, who confirms Mr. Ward's version of the incident and tells you that his son "got the whole thing on video."

When you go back to your office, you do what you know must be done: search the Internet for posts about the situation. On TripAdvisor, you see a negative review about the hotel (see Figure 25.2).

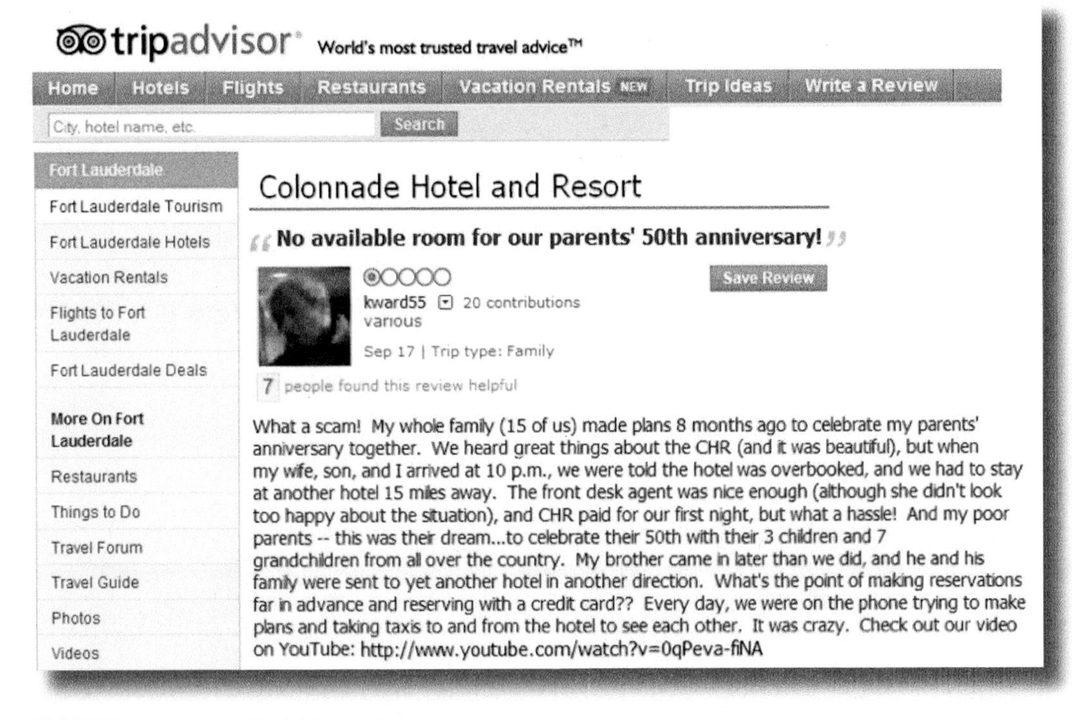

FIGURE 25.2. TripAdvisor review

On Twitter, you discover Mr. Ward's tweet (see Figure 25.3).

And on YouTube, you find the video that captured the front desk exchange (see Figure 25.4).

After taking in all of the negative social media reviews, you worry about the damage this incident will cause your hotel. You let out a big sigh and head out for another large coffee, knowing it is going to take some work on your part to fix this.

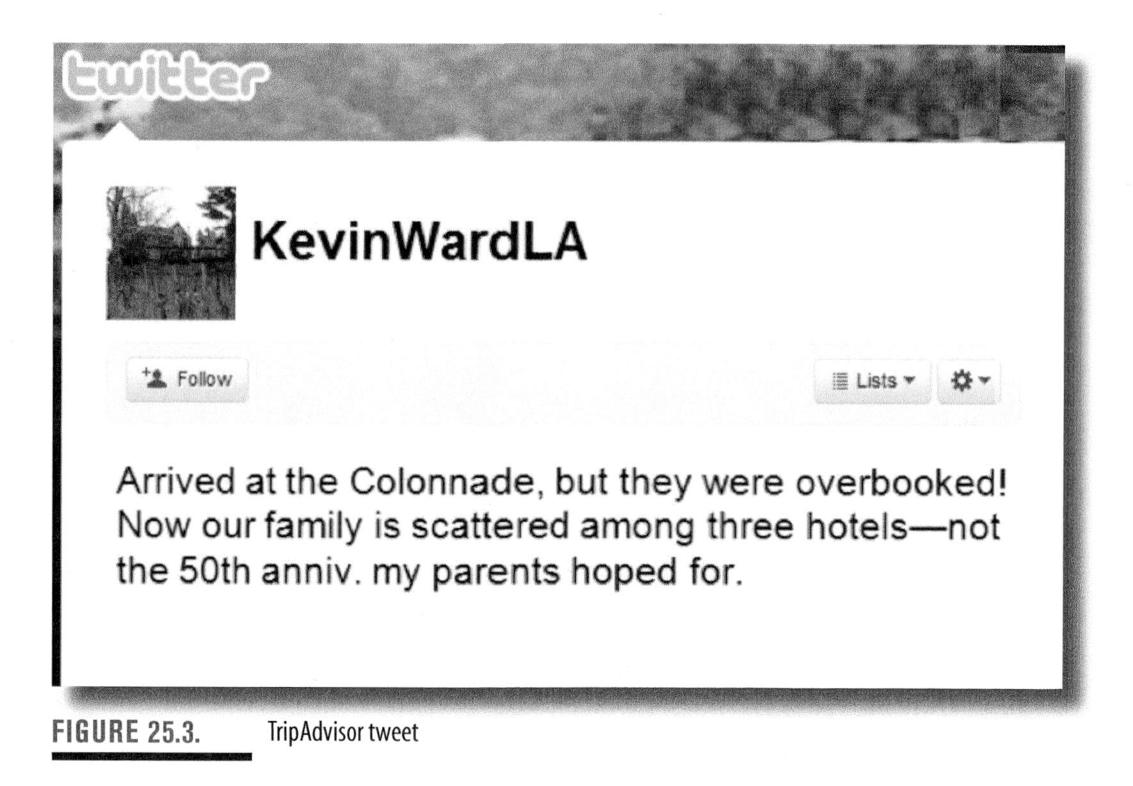

FIGURE 25.3. TripAdvisor tweet

FIGURE 25.4. **YouTube video**

To view the 1.53 minute video go to https://www.youtube.com/watch?v=0qPeva-fiNA

Background Information

For more context, read about the Colonnade, overbooking policies, and the hotel's procedure for "walking" guests.

About Colonnade Hotel and Resort

The Colonnade is a 250-room, 4-star hotel that caters to business and leisure travelers. The hotel has been open for ten years and recently went through a complete renovation. It is centrally located, close to both the major office centers and the well-known city parks and museums. The hotel has received fairly good reviews on TripAdvisor and other travel sites and has a high rate of repeat guests. In addition to the 250 rooms, the hotel has approximately 50,000 square feet of meeting space and three restaurants.

Overbooking Policies

Like most hotels, the Colonnade overbooks to protect itself against no-shows. The number of rooms the hotel overbooks depends on how many no-shows are expected each night. Management tries to predict this number with 100% accuracy, but they are not always successful, which is what happened when Mr. Ward arrived with his family.

The hotel asks guests to guarantee their reservation with a credit card. If guests with a credit card guarantee do not show up, their credit card is charged for the room. Customers are informed of this policy when they make their reservation, and it is reinforced in an email confirmation.

For more information about overbooking at hotels, read a *New York Times* article about the Waldorf-Astoria in New York City, "The Room Is Booked, Until the Hotel Says It Isn't" by Joe Sharkey
http://www.nytimes.com/2010/12/14/business/14road.html

Colonnade Hotel and Resort's Walking Procedures

The following information is communicated to employees in the Colonnade handbook:

> *As you know, even though we take credit card guarantees, we still have some no-shows. Because of this, we will sometimes overbook the hotel. This is usually not a problem, but there are times when you may need to relocate guests to other hotels. When this happens, be sure to follow the Colonnade Hotel and Resort Walking Procedures:*

- *Negotiate walk rates with referral hotels before walking guests.*
- *Obtain the selling status of comparable area hotels.*
- *Avoid walking Colonnade Prestige Club level members whenever possible.*
- *Inform the guests that the hotel cannot honor their reservation because of unexpected stayovers, but that we will pay for their room at a comparable hotel.*
- *Offer to forward all messages or calls to the guests at the other hotel.*
- *Provide relocated guests with a "walk letter" from the shift leader so that charges can be directly billed to the Colonnade.*
- *Enter the walked guests' name and the referral hotel's phone number in the property management system under the VIP guest section. This ensures that we provide the guests with exceptional service and do not walk them on a future stay.*
- *If the guests return the following day, provide an upgraded room, if possible. Also, a personal note of apology from the general manager should be placed in the guests' room.*
- *Send a VIP gift from the front office manager or general manager upon the guests' return.*
- *If the guests do not return to the hotel, send a personal note to their home address.*

Exercises

1. Write a Situation Analysis Memo analyzing the rhetorical context of this case.
2. Write an email response to Mr. Ward. In your message, you will want to acknowledge and apologize for the incident and invite his family to return to the hotel. If done well, your email will encourage Mr. Ward to post an update on TripAdvisor that presents Colonnade Hotel and Resort in a more positive light. You might consider how much detail to provide about the overbooking policy. Regardless of your rationale, overbooking may be difficult for even the most agreeable customer to accept.
3. Write a Project Assessment Memo describing your approach to the case.

Case developed by Amy Newman and Sherri Kimes, Cornell University, School of Hotel Administration

FOODBORNE ILLNESS ON FESTIVAL CASE

CHAPTER 26

The Foodborne Illness on Festival case involves managing the effects of tainted food and passenger illness on a cruise line. You will have to balance the need to inform passengers—and government agencies—with the interests of protecting the image of the cruise line and avoiding negative public relations. How much information should you reveal? How much information are you obligated to reveal (by law or professional ethics)? Weigh your options and choose your words carefully!

As the new food and beverage manager on Festival Cruise Lines ship *Enamorada*, you organize and oversee dining accommodations for the 2,667 guests on each voyage. You report to the cruise director, Carrie Anne Causwell, along with other department managers.

During the current 9 day voyage to several Caribbean islands, more than four dozen passengers have suddenly come down with some sort of illness.

"We can't afford the negative publicity and potential lawsuits of another massive food-borne epidemic on this line," Causwell tells you. "We've got to act fast and contain this before it gets out of hand. Find out what in our food stores is tainted; Dr. Lovelett should have analyzed the samples by now. We need to pull everything that might contribute to other passengers contracting the same symptoms. Then, redo the remaining menus to work around the stocks you pull. When you've got the potential damage under control, write up a report[1] to our CDC representative[2] on how this happened and why it won't happen again, and let all the kitchen staff know what we're dealing with. And make sure they aren't talking to the passengers about it."

You nod in agreement, and begin thinking about the task at hand. First, you need to figure out what foodstuffs the illness sprouted from. Then you'll need to work around that for the remaining meals. You head to Dr. Lovelett's small office and lab.

"Hey doc, any idea what we're up against yet?" you ask. "Looks like the Norwalk Virus.[3] We must've picked up some bad shrimp. The passengers who've been sick so far have all eaten shrimp. It'll all have to be pulled; not even cooking it will minimize the risks to the passengers" he replies. Two meals a day on board feature shrimp and other shellfish prominently. You ask if this virus might also be in the scallops, calamari, clams, oysters, and other seafood. "Hard to say," Lovelett replies. "It's possible, but more than likely it's just the shrimp."

Guiltily, you remember the last minute scramble to fill food and beverage inventory. A combination of last minute bookings and underordering to cut food costs prompted you to okay a new vendor without checking them out thoroughly, although you knew it was a relatively new company and had no major contracts yet with other lines. This type of outbreak does happen, though, and it might not have been the product, but the kitchen prep and storage at fault. You've heard of Norwalk, but you're not entirely sure how it spreads.

Whether or not this is the kitchen's fault, how much seafood can you really pull before the menu gets bland and repetitive? On the other hand, if this outbreak affects more passengers, you could really get into trouble for not nipping it in the bud. A small incident would definitely do less public relations damage than an entire cruise full of sick passengers. Is it better to be safe than sorry? You've got to get to the bottom of the outbreak, do some quick, efficient decision making to contain it, and reassure over 2,000 passengers that they're perfectly safe, and then explain and justify your response to the CDC.

Notes

1. The CDC Vessel Sanitation guidelines require cruise ships to report illness any time more than 2% of its passengers or crew complain of gastrointestinal symptoms (e.g., vomiting, diarrhea).
2. The Center for Disease Control (CDC) runs a vessel sanitation program (VSP), which was instigated in the 1970s. This is a cooperative program between the CDC and the cruise ship industry to help minimize the risk of gastrointestinal diseases. It has had great success in minimizing outbreaks over the last 3 decades. Twice a year,

VSP staff inspect over 140 participating cruise ships while they're in a U.S. port. Ships are scored on food, water, spas and pools, employee hygiene, and general ship cleanliness; only scores of 86 and above (out of 100) are considered passing.

3. Norwalk virus infection is an intestinal illness that often occurs in outbreaks. Norwalk and Norwalk-like viruses are increasingly being recognized as leading causes of food-borne disease in the United States. The viruses are passed in the stool of infected persons. People get infected by swallowing stool-contaminated food or water. Outbreaks in the United States are often linked to raw oysters. Infected people usually recover in 2 to 3 days without serious or long-term health effects.

Online Resources: Foodborne Illness

The following sources will help you consider the different audiences and language you'll need to use, as well as policies on how much should be disclosed to the government and passengers in cases of foodborne illness (FBI). Start with sites such as the following, but be sure to do your own background research as well:

CDC Vessel Sanitation Program (VSP)
http://www.cdc.gov/nceh/vsp/

Water Quality and Health: Sea Sick Article
http://www.waterandhealth.org/newsletter/new/winter_2004/sea_sick.html

Exercises

1. Write a memo to the kitchen staff reporting on the incident. Your memo must inform the kitchen staff and direct them on how to respond.

2. Write a CDC incident report. Your memo must inform the CDC of the incident and your company's remedial actions.

3. Write a letter to the passengers informing them of the situation.

4. Write a Situation Analysis Memo (SAM) addressed to your instructor in preparation for writing any of the documents you are asked to write for this case.

5. Write a Project Assessment Memo (PAM) addressed to your instructor explaining your approach to writing any of the documents you are asked to write for this case.

Original case contributed by Heather Lusty

LEGO BRICKS PROJECT

CHAPTER 27

Project Objectives

- Design and construct clear instructions for completing a technical process.
- Plan and carry out a multistage, collaborative writing project.
- Apply project management, research, and team writing techniques to the production of an effective technical document.
- Learn how to frame and tailor graphic and textual aspects of a document to the needs of a target audience.
- Conduct usability testing with intended users to determine the effectiveness of your document design and your step-by-step procedure.
- Analyze the extent to which a document achieves its purpose and revise the document for ease of use and improved guidance.

For this major collaborative project, you will work in teams of three to four students to develop, test, and revise instructions for a specific audience. You will assume the role of a "Building Instruction Designer" for Lego Brick Building Systems and act as if your work will actually be published directly to the user (children age 8–14). As a Building Instruction Designer (BID'er), you are committed to ensuring the consumer experiences an optimal building and playing experience by developing the various steps in creating a LEGO brick model.

Once you have constructed an acceptable model, you will move though a typical development process for creating a set of instructions to accompany your model. For this project, then, you will do the following:

- Create an instructions set that guides a user through the procedure of reproducing the LEGO brick model you have designed.
- Conduct usability testing with suitable participants to ensure the instructions enable the intended audience to construct the model.
- Analyze the participant responses received from the usability testing to revise your instruction document so it achieve its purpose effectively.
- Write a usability report for your design manager (instructor) addressing how the feedback you gathered during usability testing influenced your final instruction set.

Graduates in nearly all technical fields—computer science, engineering, and information technology—will be required to write instructions, often collaboratively with others, at some point in their career. However, for one of the most prevalent, easy, and seemingly intuitive forms of technical documentation, instructions are often the most poorly written. This project provides you with the opportunity to learn more about the proper design of these documents as well as hone the necessary rhetorical skills associated with it.

Deliverables

Document	Writer(s):	Audience	Length
1. Instructions	Group	User	12–15 pages (following good page design and layout) not including supplementary materials.
2. Usability Report	Group	Instructor	3–4 pages (following memo format); 12 pt., single spaced. Must attach copies of user test protocol and data collection sheets.
3. PAM	Individual	Instructor	1–2 pages (following memo format) including evaluation of self and team members. Failure to complete equals a letter grade reduction for report.

Designing a Model

Before you create your model, you must ensure that the development challenges for the model match the target audience group and that your final work meets the project frames for quality, budget, and time—that is, "Only the best is good enough." Even though the parents will ultimately be responsible for purchase, the model must appeal to their children age 8 to 14 who build and play with the model. In this regard, your model should be constructed with the following constraints:

- **The model must use no fewer than 50 blocks and no more than 75 blocks.** Larger models will be overly expensive and will ultimately deter parents from purchasing it, whereas smaller models will be overly simple and not appeal to the target audience.
- **The model must use at most three different colors of blocks and at least three different block shapes.** The model is a children's toy, so it should be both identifiable and appealing. The model should have consistency and uniformity in order to limit costs and to demonstrate the effort put into the design.
- **Your design manager (instructor) has veto power on the model if it seems either overly simple or overly complex.** Most models mimic objects in the real world such as trains, cars, or planes. Abstract and random creations will not meet the requirements for this project. For instance, a basic tower or a pile of random bricks is a poor choice for this age group.

Planning

Planning this project, like planning any technical document in a business setting, is a complex process and will involve:

- Establishing a plan to manage the project and accomplish the work
- Developing a suitable model for the target audience
- Learning the conventional elements of technical instructions
- Researching, analyzing, and compiling user testing information clearly and persuasively in a professional report
- Identifying potential issues with the instructions set and making appropriate adjustments

To begin, you will have to plan and set up appropriate activities with your group members, including usability testing. Structuring what and when particular activities take place is imperative if your group wishes to stay on schedule. You should distribute the workload evenly, and each member of the group should participate in all phases of document development. Even though all group members should contribute to all phases of the document development process, it does not mean each student needs to contribute to every section. Dividing the work based on skill set (i.e., graphic design, model construction, or Word layout) is an acceptable practice.

Instructions Set

You will need to collaboratively create the documentation that will guide the user through the process of building your model. This multipage document will display both written and visual directions and follow the principles of effective document design. Because the users for the instructions are children age 8 to 14, the document must be easy to follow. Visual consistency and readability are paramount. Your group will need to organize information in an easily accessible manner that coordinates the written and visual directions in a mutually supportive and usable form.

Format

The instructions set should generally contain the following sections:

- **Opening introduction:** The introduction should define the procedure about to be discussed and indicate any background or skills the audience needs to complete the model as well as the time it takes to build it. This section should also include a list of the number and type of materials (i.e., LEGO bricks) needed and include all special notices—notes, warnings, cautions, or danger notices—to alert the user to the possibility of damaging the building materials, messing up the procedure, or harming themselves.
- **Instruction body:** This section actually contains the instructions both visually and textually presented to the user. In constructing this section consider both the structure and format of those steps. The point of view for the graphics, as well as their structure of the written instructions, should always be consistent but not repetitive. The entire model building process should be divided into a manageable number of tasks with specific and fixed step-by-step guidelines for each task. These tasks should be appropriately labeled and clearly formatted to indicate differences between primary steps and nested or sub-steps. Be aware that supplementary discussion and explanation of information might be necessary in addition to telling the user how to construct the model. This is particularly important with steps that include more precise placement of bricks and careful construction.
- **Supplementary materials:** Any alternative steps, model changes, or additional explanatory information that is not directly pertinent to the actual construction of the model should be placed in this last section. Often LEGO brick instruction manuals include alternative models or construction choices for the included pieces. In addition, customer service contact information is provided in this later section to help with difficulties or deal with missing pieces. If the LEGO brick model is part of a set, the other models that belong to the set will be presented as well.

Instruction Set Checklist

Refer to the following checklists as you prepare your draft of the instructions set:

Do the instructions include the necessary core elements?

1. A General Introduction
2. The Step-by-Step Instructions
3. Conclusion

Do the instructions begin with an introduction that helps orient the reader?

1. Specify audience, purpose, and scope?
2. Provide an overview that tells readers what you will cover?
3. Include necessary background information?
4. List and describe any equipment that will be needed?
5. Make it possible to answer the following questions (assuming each question is necessary, some may not be):
   - **Who** should carry out the task?
   - **Why** should the reader carry out the task?
   - **When** should the reader carry out the task?
   - **What** safety concerns should the reader understand?
   - **Which** items will the reader need?

Do the instructions follow the "chunking" rule (7 +/- 2)?

1. Divides entire procedure into a manageable number of tasks?
2. Provides specific step-by-step instructions (with 5–9 steps) for each task?

Do the step-by-step instructions employ instruction conventions?

1. Each step sequence begins with a lead-in (aka "stem sentence")?
2. Each step defines a **single operation** the user can complete before moving on?
3. Each step uses **imperative mood** ("you" + command)?
4. Clearly distinguishes between **steps (numbered) and feedback/result** statements?
5. Uses graphics to enhance clarity/aid reader understanding?
6. Make appropriate use of special notices (notes, cautions, warnings)?

Does page design help reader navigate the page through the use of...

1. Titles?
2. First, second, and third level headings (as necessary)?
3. Numbered steps?
4. Bulleted lists?
5. Visuals/captions?
6. Page number (if necessary)?

Usability Report

Each group is responsible for submitting a collaboratively written report that presents the results of the usability tests as well as indicates what changes, if any, have been made to improve the model and the usability of the instructions set. The purpose of the document is to inform the Design Manager, your instructor, of the various changes your group has made to the instruction set and to justify those changes based on the feedback received from your users. You should write this document in memo format, not exceed four pages, and include one visual.

Format

Include the following sections in your report. As with all effectively organized reports, you are not limited to these sections and you may want to subdivide the sections to clearly indicate, for instance, global and sectional changes to the instruction set.

- **Overview:** Briefly summarize the purpose and content of this document. This report is focused on informing the reader of the revisions you have made to your instruction set.
- **Background:** This section reiterates to the reader, your instructor, the purpose of the project. You are providing a concise description of the rhetorical situation. Summarize how your team selected the final model and the layout for the instruction set based on your understanding of the users' needs and the needs of the project. Note any significant alterations in the scope or focus of the project including major changes with any individual members.
- **Findings:** This section describes the usability testing and summarizes the actual results of your user tests. Your group must compile the data gathered in a clear and readable fashion and then communicate it in a clear and concise way. In addition, you will need to include at least one visual. Consider the best way to present your findings graphically (e.g., charts, tables, images from instructions) and choose an effective way to highlight the problems discovered with the initial instruction set.
- **Revisions:** This section describes the changes made to the instructions based on the results found in the prior section and/or based on direct user feedback. You must clearly explain what specific adjustments were made and how those improvements directly correlate to the usability test results. While you can mention minor and otherwise unsupported changes in this section (e.g., font changes, graphic adjustments), the primary purpose of this section is to articulate how user interaction improved the final set of instructions.
- **Conclusion:** This final section reiterates the importance of the changes made to the final documents and provides a persuasive appeal about the usefulness of those improvements. However, you want to sound persuasive and honest. If applicable, discuss any limitations of the current study and any issues that remain unresolved or need further analysis or research.

Project Assessment Memo

As with other projects, each member of the group will need to individually submit a PAM that analyzes your group's thinking, your group's writing practices, and how the group worked as a team. You should write the document in memo format and not exceed two pages. The PAM contains the following sections:

- **Overview:** This section answers the question: *What is this memo about?* Include a statement of purpose and a summary of the main points of this document.
- **Context of project:** This section answers the question: *What was your understanding of the project?* Explain how you, as an individual, perceived both the task at hand and the requirements for completing it. You can also mention whether or not your group had any disagreements about the rhetorical situation or scope of the project.
- **Production:** This section answers the question: *What strategies or techniques did you use to produce/make/write your final documents?* Provide your instructor with the details about the steps your group went through to produce your documents. Recount the group's initial decisions about the project including how your group planned and finally constructed the documents.
- **Project objectives:** This section answers the question: *How did you accomplish/meet the project objectives?* You should provide an evaluative statement about the work your group completed and how well it meets the goals laid out for the project at the beginning of this chapter.
- **Course objectives:** This section answers the question: *How did my work on this project help me meet any or all of the course outcomes?* In particular, you should describe what you learned through the project and how that knowledge is in line with the course objectives.
- **Summary:** Conclude by evaluating the assignment holistically and how successful you were in tailoring the instructions to the users and, perhaps, discuss anything you learned from the project that goes beyond the objectives of the project. Always provide an open avenue for communication in the closing.

Online Resources

Resources for Model Ideas and Instructions:

- *Brick Instructions.com* by Jon May. http://lego.brickinstructions.com
- *The Brothers Brick* blog by Andrew Becraft. http://www.brothers-brick.com
- "Building Instructions" by Lego.com. http://service.lego.com/en-us/buildinginstructions
- "Create & Share Galleries" by Lego.com. http://www.lego.com/en-us/galleries
- *Brickset.com* by Huw Millington. http://brickset.com

Programs for Digital Model Design:

- "Lego Digital Designer" by Lego.com. http://ldd.lego.com/en-us/
- *SR 3D Builder* by Sergio Reano. http://sr3dbuilder.altervista.org
- *LeoCAD* by Leonardo Zide. http://www.leocad.org/trac/wiki

CHAPTER 28

JOB SEARCH PROJECT

Project Objectives

- Learn to tailor a resume and cover letter to a specific audience
- Apply page design principles to increase the readability and overall persuasiveness of the resume
- Write a persuasive cover letter that compliments the resume and enhances the overall application
- Integrate focused company research into the production of effective resumes and cover letters

In this project, each individual student completes the following tasks:

- Find an actual job advertisement for a position that you're reasonably qualified for
- Research the job and company thoroughly, using print and Internet sources
- Collect a minimum of four recent sources of information about the company, including one outside source (i.e., the company's website shouldn't be your only source of information about a company)
- Complete a **Job Analysis** planning exercise that articulates your researched understanding of the requirements of the position and how you best fit with those requirements
- Write a **resume** and **cover letter** following principles of effective job document writing. (NOTE: No Microsoft resume templates will be accepted for this project. If you want to base your resume on one of these templates, you need to justify the design choices in your PAM.)

- Write a **Project Assessment Memo (PAM)** that describes and justifies to your instructor your completed job search project

This is not a project about how to write a generic resume. While the generic resume has a place (e.g., at job fairs or for online resume banks), sending out 100 copies of the same resume is one of the least effective ways of getting a job, because these resumes can't compete with a resume that was carefully prepared to address the qualifications of a particular employer. To tailor the resume and cover letter, you need to research the company beforehand, rank important qualifications, determine key words sought by the employer, and then present only the information about yourself that demonstrates you meet the desired qualifications and persuades the employer that you are among the most qualified applicants. The **Job Analysis** exercise teaches the analytic process that should accompany the preparation of effective job documents.

Many students of business writing think they know how to write resumes. Most students already have a resume and may have even been required to write one for another class. But rarely are students taught the rhetoric of the resume, which is largely visual. The Resumes chapter that accompanies this project contains five basic page design principles in addition to covering the basic conventions of resumes. You'll learn to manipulate the page as a visual unit. Several kinds of visual design techniques, such as *the quadrant test* and *the vertical lines test* will be introduced, as well as page design vocabulary such as *white space*, *chunking*, *headings*, and *emphasis*. Visual design choices are rhetorical; they contribute to the persuasiveness of any document and are particularly important for resume and cover letter writing.

You may have never written a cover letter before, but it is customary to submit a cover letter along with the resume when applying to professional positions. The Cover Letters and Personal Statements chapter that accompanies this project covers the basics of writing cover letters, including how the cover letter should compliment the resume by including specific examples that elaborate key qualifications merely listed in the resume. Similar to an interview situation where an employer asks for an example or two of the applicant's relevant skills, the cover letter should go into more detail with specific examples. Since the resume is written in outline form, it can't go into as much detail as the cover letter can.

This project emphasizes the process of research and analysis that accompanies the preparation of effective employment documents. In this project, you will tailor your employment search to a real organization, for a specific job. Most experts believe that the most successful job seekers are those who target a handful of specific companies at a time, adjusting their documents to suit the needs and requirements of each particular audience. More typical strategies, like answering classified ads or mass-mailing generic resumes, are less effective.

In "How to Apply for Fewer Jobs (But Land More Interviews)" Kristen Walker advises to follow the "9 out of 10" rule: If the job doesn't rank a 9 out of 10 in terms of "both your level of excitement and competency, then it's not worth applying to." This way, you produce fewer, but better quality applications. See https://www.themuse.com/advice/how-to-apply-for-fewer-jobs-but-land-more-interviews.

Find an Actual Job Advertisement

Don't know what you want to be when you grow up? Here's three strategies for finding a job advertisement for this project:

1. Identify job titles for jobs you're interested in by searching on job-hunting websites and looking at career planning information.
2. Consider what kinds of work you'd want to devote your life to get paid for. Too many students rule out careers in entertainment, music, and sports because they think the chances of breaking into those fields are too remote. Well, now's your chance to dream, and find that ideal internship or entry-level job.
3. Browse the U.S. Department of Labor's *Occupational Outlook Handbook*, which details various industries, giving job descriptions and prerequisites, working conditions and industry prospects, and includes information about salaries and promotion opportunities at: http://www.bls.gov/oco/.

You must first identify an actual advertised position related to your professional goals. There are three basic paths to consider when choosing the type of project you want to pursue:

- **Entry level:** If you're a senior and graduating shortly, you can use this project to prepare for a post-college job related to your degree.
- **Internship/scholarship:** sophomores and juniors, you won't meet the minimum qualification for an entry-level job, a college degree. Since employers prefer to hire college grads with internship experience, sophomores and juniors should use this project to prepare applications for internships and scholarships.
- **Graduate school or promotion:** If you are planning on going on to graduate school or already have a job, you have a couple options in this project. You can use this project to prepare a graduate school application, writing a resume and personal statement. The Researching Graduate Schools box contains resources for researching graduate schools and writing personal statements. The graduate school option is for seniors only, since younger students don't have enough college experience to write a convincing personal statement. If you have a job already, you should look for a better job, either within your company or at another company.

Researching Graduate Schools

- US News & World Report's (usnews.com) "education" section includes graduate school rankings and other helpful information at http://grad-schools.usnews.rankingsandreviews.com/best-graduate-schools.
- Peterson's is another reputable graduate school ranking and research service at http://www.petersons.com/graduate-schools.aspx.

Personal Statements for Graduate Schools

Personal Statements for graduate applications are similar to cover letters, but there is more emphasis put on the writer's qualifications. The format is also slightly different, as personal statements are written more in academic paper format and not usually in business letter format.

- "Writing the Personal Statement" by Purdue OWL at http://owl.english.purdue.edu/owl/resource/642/01/.
- "Personal Statements and Application Letters" by Indiana University at http://www.indiana.edu/~wts/pamphlets/personal_statements.shtml.

You must locate an actual job advertisement, using either online sources, such as your university's career center website, or more traditional sources, such as notices from your academic department or current employer (see Figure 28.1). Don't select a job where you clearly do not meet the minimum qualifications or would have to fabricate credentials. While you are not required to actually submit your application to the employer, that obviously could be one useful outcome of this project. The more authentic the position is for you, the more effective your job documents will be.

FIGURE 28.1. **UNLV's CareerLink.** A job-hunting resource devoted to UNLV students, CareerLink includes a database of jobs and other tools, such as a resume posting and on-campus employer recruiting sign at https://unlv-csm.symplicity.com/students/index.php

Where to Search for Jobs

- **"Top 10 Best Job Websites" by Allison Doyle.** Includes Indeed.com as the #1 site, CareerBuilder.com, and LinkedIn. See the others at http://jobsearch.about.com/od/topjobsdb/tp/top-10-job-sites-2015.htm.

If there is a company you would really like to work for, yet it is not advertising any openings, you can still target the company, submitting your resume and cover letter as an unsolicited application. If you choose to write an unsolicited application, you still need to find a job ad for a similar position at another company, which will help you analyze the requirements for your unadvertised position.

If you are absolutely content with your job status, you still need to learn how to research and write effective job documents: Research suggests you'll change jobs between 5 to 15 times in your lifetime. Choose a realistic job that you *could* apply for, treating this project more as an academic exercise. Also, please *do not* choose a job you have held before, and *do not* write to an employer you've already met.

If you are unclear of the best path for you, talk about options with your classmates and instructor.

Your goal in this assignment is to design documents that will persuade an employer to grant you an interview for a professional position. But remember this: job documents usually never get you the job; they get you the interview that may or may not get you the second interview (and so on) that may eventually get you the job.

Research the Company

In today's tight job market, you need every advantage possible. One method for doing this is thoroughly researching the company or companies that you're most interested in. Anything you can learn about a company will assist you in designing a resume and cover letter that make you appear to be the most qualified person for the job. In fact, research shows that targeting 10 to 15 companies at a time through formal research and follow-up is a more effective strategy than mass-mailing your resume to 100 companies or posting your resume on an Internet job bank.

To develop your application for a specific company, you will need to research the company to familiarize yourself with the most current information about the job, the working environment, pay scale, benefits, and so on. Look for the following to help you write your resume and cover letter:

- Job pay, benefits, promotion opportunities
- Description of company services or products
- Mission statement
- History
- Career opportunities
- Annual reports
- Press releases
- LinkedIn, Twitter, Facebook, and any other social media accounts
- Yelp customer reviews
- SWOT analysis (Strengths, Weaknesses, Opportunities, Threats)

Company Research Resources

- "Best Ways and Places To Research Your Target Employers" by ExecutiveCareerBrand.com at http://executivecareerbrand.com/best-ways-and-places-to-research-your-target-employers/.
- "Top 8 Sites for Researching Your Next Employer" by Rich Hein at http://www.cio.com/article/2387201/careers-staffing/top-8-sites-for-researching-your-next-employer.html
- "Careers" Libarary Databes by UNLV Libraries http://guides.library.unlv.edu/careers

Outside Research

You should not rely on the company as your only source of information, since it is a biased source. Few companies will promote negative information about themselves. That is why you should also look to outside sources of information such as recent Internet, newspaper, or magazine articles about:

- What the company has done lately
- What's going on in the industry
- What the prospects are for the company
- Why this is/isn't a good company for you

Industry or Career Research

You won't always find lots of information about a given company, especially if the company is small or new. In cases like this, you'll have to be more creative, looking for information about:

- Similar or competitor companies
- Industry trends, or issues that all companies like your company are dealing with
- Career trends, or what general issues face people in similar positions as the one you're targeting

Once you've collected a minimum of four sources of information, including at least one source from outside the company, use them to determine what the company wants:

- What are the minimum qualifications needed for that specific job?
- What additional qualifications are desired for the position?
- What additional employee attributes are desired by the company?

A primary feature of the job search process is to critically analyze ads and other company information. Try to look for key words or concepts in the text as clues about what the company is looking for and values. In this way you attempt to identify specific adjectives or skills from the company literature to answer the following questions:

- What do the key words and concepts imply about the company?
- What do the key words and concepts imply about the company's goals?
- What do the key words and concepts imply about employee/company relationships? About employee/employee relationships?
- What do the key words and concepts imply about the company's expectations for employees?

Once you've developed a better knowledge of the desired qualifications for that particular position based on your analysis of the job ad and company information, **match your background to what the company wants**. Connect your skills and experience to the qualifications sought by the company. Ask yourself:

- What qualifications is the company looking for? (What do they want?)
- How do I match the minimum qualifications for this position? (Do I have it?)
- Do I have any additional, desired qualifications? (What are my strengths? What sets me apart?)

Completing the **Job Analysis** for this project will help you synthesize your research and articulate a deeper, more informed understanding of the rhetorical context of your targeted position. Does this mean you'll have to write job analyses for every job you apply for? No, but you should go through the same research process. One advantage of writing your analysis down is that you'll have notes you can consult as you prepare your job documents and, hopefully, later for an interview.

Job Analysis Planning

The Job Analysis provides an opportunity to articulate the results for your job and company research. An effective way to synthesize, or make sense of, researched information is to summarize it in your own words. By thinking on paper, you will increase your chance of remembering it. You will also have a record to refer to later, such as when you prepare for your job interview.

After you've completed your company research and gathered a minimum of four sources of information (including one outside source), begin planning and writing your Job Analysis. Include the following sections:

- Job description
- Company research
- Company description
- Important qualifications
- Job analysis
- Submission format

Job Description

State the position you have chosen and summarize the responsibilities of the advertised position. Answer the question: What is the position title, at what organization is it located, and what are the duties, responsibilities, or tasks will you be expected to perform?

Use your own words. Be sure to translate any technical language or professional jargon. For example, if a Test Engineer job ad states that "You will be responsible for application testing, and certification of new SAN (Storage Area Network) configurations, while working closely with both sales and development teams," don't just cut and paste that statement, but explain—to yourself and your instructor—just what's involved in "application testing," "certifying new SANs," and so on.

If your job ad doesn't provide a detailed job description, consult other sources such as similar job advertisements or career planning information to construct a description of the job's main responsibilities.

The position I am applying for is an entry-level accountant for Parsons. If I were to obtain this position I would be reporting to the controller. There are many responsibilities that come with the position of accountant, including the following:

- Auditing: audit accounts, sub-job and other financial information.
- Generating reports: creating periodic financial reports, statistical reports, and invoice recapitulations
- Maintaining records: oversee journals, general ledgers, and subsidiary accounting records.
- Collecting data: collect data to perform tax reports.

Sample company description

Company Research

Answer the question: What four sources of recent information about the company, including one outside source, did you find?

Once you find a position, you must research the company through online sources such as the company's website, social media platforms, and library periodical indexes to learn about general background information and recent news. You can also use primary, or first-hand, sources such as interviews with current employees and reviews by customers on sites like Yelp. **You must find a minimum of four sources of information about the company, including at least one unbiased outside source.** Include the URL (complete website address) for online sources and a brief description of what you learned about the company from the source. Subpages of a company website can count as separate sources, but the job ad itself does not count as a source.

1. http://money.cnn.com/magazines/fortune/best-companies/2012/snap-shots/11.html

Zappos.com is ranked #11 in CNN: Money's Top 100 Best Companies To Work For in 2012.

2. http://www.bjtonline.com/business-jet-news/zappos-tony-hsieh

Interview with Tony Hsieh, CEO of Zappos.com and his plans for the future.

3. http://www.lvrj.com/business/panel-predicts-crush-of-people-vertical-construction-for-downtown-las-vegas-172750561.html

Zappos' plan to reconstruct downtown Las Vegas coming soon in 2013

4. http://downtownproject.com/

"Led by Zappos.com CEO Tony Hsieh, we are passionate people transforming Downtown Las Vegas into the most community-focused large city in the world."

Sample company research

Company Description

Answer the question: What company would you be working for if you get this job?

Use your own words to summarize what you learned about the company from your company research. Do not copy and paste text directly from your sources. This information will come from both the job ad and the company research you've collected.

Your summary should answer as many of the following questions as possible:

- What is the company?
- What are its products or services?
- What is the company's mission?
- What are the company's core values?
- How big is the company?
- Where is it located/headquartered?
- When was it established? How old is it?
- How many employees does it have?
- Who is its leader or leaders?
- What has the company done lately?

Use the company description to note anything else about the company you've learned from your research that you think relevant to planning your job application documents.

MGM is one of the leading and most respected companies in the world when it comes to gaming, hospitality, and entertainment. MGM owns a total of 21 properties located in Nevada, Mississippi, Michigan, New Jersey, Illinois, and Macau. Eleven of these properties, including the City Center project, are located on the Las Vegas Strip.

ARIA Resort and Casino is one of five new buildings at MGM's new project City Center in Las Vegas. The vision the project developers have is to let visitors experience the excitement, people, action, and thrills a city has to offer. The development is 76-acres located right in the center of Las Vegas. ARIA will have approximately 4,000 rooms and be 61 stories with a 165,000 square foot casino. City Center plans to also have 225,000 square feet of convention and meeting space that can accommodate up to 5,000 people. Other specs on the meeting and convention space of the ARIA Resort and Casino include:

- 51,000 sq ft ballroom with a 25 foot ceiling that can be divided into 10 meeting rooms
- Two 38,000 sq ft ballrooms that can be divided into 8 meeting rooms
- 20,000 sq ft ballroom that can divide into 3 meeting rooms
- 38 breakout rooms that range from 800 to 2000 sq ft

MGM Mirage and Dubai World worked together to bring the $8.5 billion City Center project to Las Vegas. During tough economic times the project was about to go bankrupt. Even though this project was saved, the company itself is facing bankruptcy due to a $13.5 billion long-term debt. Despite the setbacks that City Center has faced MGM has City Center slated to open in late 2009.

Sample company description

Qualifications

Answer the question: What are the qualifications sought by the company?

In the qualifications section, summarize what you think the company wants, or is looking for. Prioritize what skills, abilities, knowledge, and values are most desired by that company for that particular position. A well written job ad includes "minimum" and "additional" qualifications. What are they and which do you think are most essential to doing the job as described? What other qualities, not listed in the job ad, are valued by the company?

Here you can copy and paste directly from the job advertisement and possibly add more information based on your research or review of similar job advertisements. Arrange the qualifications in order of importance.

The position for Lead Convention Porter requires the applicant to have the following qualifications:

- Two years previous experience as convention porter
- At least two years previous catering and banquet experience
- Working knowledge of banquet set-ups
- The ability to read and understand MEOs and BEOs
- Excellent guest service skills
- Ability to communicate in English
- Ability to complete work in a timely and efficient manner
- Ability to work under stressful conditions

Sample qualifications list

Job Analysis

Answer the question: How do your skills or experience match the specific qualifications listed for this job?

As concretely and as detailed as possible match your skills and experience to those desired by the company. If the job ad asks for "3-5 years customer service experience," consider how you meet this criteria. Think carefully and critically about how what you've done or learned in the past that might apply. Be creative and inclusive about fitting your skills to those sought in the job ad. Think about how you meet the minimum qualifications and what additional desirable qualifications you have.

Make a persuasive case that you're qualified for the position by answering point by point the list you made in the previous "important qualifications" section. What do you think you should emphasize in your job documents? Spend a good amount of time explicitly connecting your skills to those listed in the job ad. The key is to base your answer on the criteria set forth in the job ad itself. The more examples and detail you brainstorm, the more content you generate for your job documents, particularly the cover letter.

Company wants: Two (2) to Three (3) or more years of experience in a similar role preferred

I have seven years of customer service experience/administrative assistant and five years as a front desk representative.

Company wants: Previous experience in a high-end luxury Gaming or Hospitality environment is desired

I have worked for the SLS Las Vegas for almost a year now, answering guest questions, providing information and assessing any request they had even if it the request was not of my department. I made sure the guest was satisfied with the service. I answered phone calls with a friendly and professional manner to make the guest feel comfortable and important.

Company wants: Must have strong interpersonal and organizational skills and must have an outgoing personality and possess strong attention to detail skills

I get along and communicate with my associates very well. Never had a confrontation and I encourage them to keep going and find success. I am able to work well independently as well as a team. I enjoy the company of my coworkers and I maintain an organized workplace. I like to pay strong attention to details. I won't stop doing an assignment until I finish it because I don't like to leave things for later.

Company wants: An intermediate to proficient understanding of Computer systems such as: Opera, GoConcierge, HotSOS, Microsoft Word, Excel & Outlook is preferred

I use Go Concierge and LimoBiz on a daily basis to schedule transportation reservations for VIP guest like casino players, invited guest and DJs. I am proficient in Opera, which helps me in finding a room by guest first name or last name as well as see comments and instructions. I use HotSOS to input bellman orders like deliver luggage to room, check-in, check-outs, flower delivery, envelope delivery, bell assistance, or to input any other request from the guest to a different department. I use Microsoft office to create signs or any document needed in Excel and Power Point. I check and send daily emails on Outlook.

Company wants: Multi-lingual is desired

I am proficient in written, spoken, and reading Spanish. I have worked with hundreds of Spanish speaking people through my past jobs. I have an extended Spanish vocabulary. There has been a few times where I have helped guest requesting their luggage. A guest was very upset because there were charges in her account that she was not happy about. Even though it was not my department's concerned, I called Front Desk to explain the situation and asked if there was something they can do about the charges. The lady was frustrated because she could not understand what the FD agent was saying to her and I was the only Spanish speaking person she had talked to. I explained the reason for the charges to her and she was satisfied.

Company wants: A valid Food Handler's & Alcohol Awareness Card are required and must be obtained before entering this position
I already have a TAM card which has helped me know when a person is intoxicated and what to do in these cases.

Company wants: Able to work a flexible schedule, including weekends and holidays
I am very flexible in terms of workings weekends and holidays because many of them are not relevant to me and I am willing to help.

Sample job analysis

Submission Format

Does the position advertisement specify whether the application should be mailed, faxed, emailed, or submitted as part of an online form? What contact information is given? Is any particular file format requested (e.g., MS Word, PDF)? Should the resume be formatted for an Applicant Tracking System? If you are given choices, which submission format will you choose, and why? Determine the actual submission requirements and adjust your plans for writing the cover letter and resume accordingly.

Answer the question: How should you submit your resume and cover letter?

Checklist for the Job Analysis

- Does the writer include the following sections: *job description, company research, company description, important qualifications, job analysis, and submission format?*
- Does the writer identify and describe the job in detail and in his or her own words?
- Does the writer list and briefly explain four recent sources of information about the company including at least one non-biased outside source?

- Does the writer demonstrate specific and detailed knowledge of the company, including the company's structure and history, its products or services, and what it has done lately?
- Does the writer clearly list the qualifications for the job and arrange them in order of importance?
- Does the writer describe and provided detailed and convincing examples demonstrating how he or she meets those qualifications?
- Does the writer consider the specific instructions and/or requirements for submitting the application documents?

Resume and Cover Letter

After you have determined how your background matches the qualifications in the position you seek, and thought about which of your qualifications you wish to highlight, you will then write your resume and cover letter.

You should write your resume first. You may have an existing resume to work from, but you should quickly decide if it is in a format (chronological, functional, or combination) that suits this new position. You may need to start over with a new format. Also remember that a resume is an **outline of relevant qualifications**. That is, you can not simply update an existing resume by adding new details to it without removing irrelevant or old details, particularly if you are targeting a new field. If you have a resume for your service jobs, you will likely need a whole new resume if you are applying for a professional internship or entry-level position. Non-traditional students starting a new career will also likely need to start a new resume from scratch.

As you plan your resume, follow the guidelines from the Resumes chapter. You know you will have your contact information at the top, but what details will you include in the body and how will you arrange those details. What resume format you choose will help you answer these questions. Do not worry about length or design at first. Err on the side of too much detail, but know you will have to cut down to get to one page eventually (for most resumes). Once you think you have added all the relevant details (and removed any irrelevant details) focus on visual design. Visual hierarchies can help you create easy-to-read levels of information. Use white space and chunking to balance the information vertically. Go for a simple, classic, readable look. When you think you are done, seek feedback from peers, friends and family who have experience reading resumes, and set up an appointment for a resume critique at your campus career services center.

Since the cover letter highlights and elaborates key details that are listed in the resume, it should be written after you have drafted your resume. Remember that resume does not merely repeat what is in the resume. It elaborates the key qualifications with some specific

Checklist for the Resume

- Does your name stand out? Do you give only one phone number and email contact information? Did you include a link to social media?
- Is the resume opening focused on clearly summarizing how your skills match the reader/employer's needs (e.g., headline, profile, professional summary)?
- Is the resume as a whole effectively designed, i.e., does it make a pleasing visual first impression, is it easily scanned, does it use white space effectively, does it pass the quadrants and vertical columns tests?
- Do the sections support the claims in the resume opening? Are sections arranged in order of importance, with most important content at the top of the resume?
- Do the descriptive words support the claims for skills or knowledge and focus on keywords important to the audience?
- Is the style in outline form, using bullet lists and sentence fragments instead of complete sentences and paragraphs? Is punctuation minimized, are action verbs used to describe skills and experienced, are the verb lists written in parallel, or consistent verb tense, form?
- Is the length of the resume appropriate, usually one page for internships and entry-level jobs?

examples that are either briefly listed on the resume or not listed on the resume. A good cover letter is like a substitute interview. Imagine the employer reading your letter and wondering how you meet his or her company's qualifications. The body of your cover letter is an answer to this question: "I believe I am qualified for X, Y, and Z reasons." But your cover letter does not just claim to meet the qualifications, it demonstrates or backs up the claims with concrete evidence, in the form of detailed examples, which are usually arranged in order of relevance. Your most convincing qualification should be list first (after the standard opening paragraph).

Just as with the resume, do not worry about length at first, but know you will eventually have to cut your letter to about three-quarters of a page. Saying all you want to say about yourself in less than one page is an excellent exercise in writing concisely. After you have the main details and organization down, then work on eliminating unnecessary words. Consider the cover letter's checklist and consult your peers and people who have experience reading cover letters. Your campus career center can give you feedback on your cover letter, as well.

Checklist for the Cover Letter

- Does the letter as a whole have all the necessary parts (return address, mailing address, date, signature and typed name)? Address a specific person? Attend to the audience's concerns and interests? Look professional?
- Does the introduction state the specific job? Connect with the company? State the applicant's qualifications?
- Does the body have focused paragraphs? Support any claims made about the writer or the job? Avoid merely listing experience? Persuasively describe qualifications? Make the qualifications clear?
- Does the conclusion end courteously? Offer information for further communication?
- Has the writer shown and used knowledge of the company? Carefully proofread the letter? Used appropriate language? Exaggerated any claims? Used her/his most relevant, persuasive experiences or qualifications? Presented him/herself professionally? Used an effective design/page layout?

Project Assessment Memo

Your instructor may ask you to write a Project Assessment Memo (PAM) for your Job Search Project. Your PAM should provide specific information on your approach to preparing your resume and cover letter. Include the following sections:

Overview: Answer the question: What is your purpose for writing the PAM?

Context of Project: Briefly review for your instructor what position you chose and why. You could mention whether or not you are serious about your choice.

Documents: What composing decisions did you make to write your cover letter and resume? You can subdivide this section into:

Resume: Discuss organization, content, design, and style choices. This section could consider questions such as: How does your objective statement focus on job and company needs? How did you make the document easy to scan? How did you target a specific audience? How is your design effective and persuasive? How do your sections support your claims? How did you hierarchies information? Why did you do it that way? Describe your font and highlighting strategies

Cover Letter: Discuss organization and content. This section could consider questions such as: What was the goal for your introductory paragraph? How did you connect with the company? What specific examples did you use to highlight your experience? Are these the most relevant and persuasive? Why? How did you focus the body paragraphs? How did you support your claims? What persuasive tools did you use in describing your qualifications? What avenues did you open for future communication? What tone/style did you use? How did you show and use your knowledge of the company?

Production: How did you plan and write your documents? *Mention specifically what sources of information about your company you collected and what specific methods you used* (e.g., which library indexes and Internet search engines) to find your information. Discuss how this research influenced your planning and writing.

Summary: How well do you feel you met the project objectives? Evaluate how successful you were tailoring your job documents to a specific situation and perhaps what you learned from the project.

STAFF DEVELOPMENT

CHAPTER 29

Project Objectives

- Learn to use writing as a tool to prepare effective oral presentations
- Apply principles of visual design to presentation visual aids, including slides and printed handouts
- Learn to maintain professional development/communication skills

Staff development, otherwise known as training or in-servicing, is a common practice at most businesses. Many companies devote a significant amount of their staff development resources to building their employees' communication skills, from training in the latest industry-related computer software to raising consciousness of workplace gender and race issues. Translation: effective communication skills, broadly defined, are valued in the workplace.

For this project, you will be expected to work in a team to prepare and deliver a 10-12 minute presentation with accompanying visual aids (a slide deck and a handout). Your group's presentation will be a subject that interests you and that will benefit the class as a whole. You're encouraged to choose a subject that you're already versed in, from previous experience or casual interest, but you will also be expected to augment your existing understanding with some formal research from current published sources and/or expert opinions. However, being well versed in a topic area is not a requirement for joining a particular group. As a non-expert, you can offer valuable insight to the rest of your group as you prepare the workshop.

Your class will discuss possible topics. Presentations can (1) introduce some rhetorical principles/compositional practices related to business communication, (2) focus on instruction of a very specific feature of a particular piece of computer software, or (3) focus on describing communication practices in a specific field. Possible topics include:

- Instruction in computer software (e.g., intermediate to advanced features of Word; web authoring/using a content management system; making charts in Excel or Visio; using a slide authoring application such as PowerPoint, Google Slides, or Prezi; creating handouts in MS Publisher)
- Business communication trends (e.g., use of non-sexist language; use of social media in a job search; doing research on the Internet)
- International communication (e.g., cross-cultural communication)
- Influence of technology (e.g., rise of the email genre or the future of mobile applications.)
- Business communication in the small/medium/large company (e.g., exploring the differences)
- Remember that we're developing our "staff," so your presentation must include a practical element. In other words, your presentation must ultimately be deemed useful and interesting by your primary audience, your fellow classmates.

Even though business writing classes aren't speech classes, most of the basic rhetorical principles emphasized in this course apply to presentations in business contexts. In fact, you can't be successful without them. These skills include using writing and research to plan and organize your presentation; using effective design techniques for visual aids that support your presentation (e.g., handouts, overheads, slide decks) and using rhetorical principles for delivering presentations. These skills can be used, for example, to maintain your professional ethos at even the smallest, seemingly most informal of business meetings. Visual aids direct your audience's attention during such meetings and increase their retention of information afterwards.

Team Presentations Principles

The following chapters contain useful guidelines for planning and producing presentations and team-written documents:

- "Presentations"
- "Collaboration"

Deliverables

To complete the Staff Development project, each student will participate in groups to produce the following:

Component	Length and Medium	Elements
Staff Development Proposal	Length: 800 to 1,000 words, not including attachments Submit print copy to instructor on due date	Team-written proposal must include: a **summary** of what you want to present an **audience analysis** a **rationale** for how the class will benefit a list of **outcomes** a plan for **visual aids** a list of background **sources**. Attach a detailed **script** including a presentation outline, plans for visual aids, designated speaker roles Attach drafts of **slide deck** and **handout**
Staff Development Presentation	12 minutes maximum (If you type a script, about two double-spaced pages	An effective presentation will reflect research, include clear intro/overview/body/ closing equals a five-minute talk) structure, be easy to follow (include clear transitions), and give practical strategies/tips that the audience (fellow classmates) finds useful and interesting
Visual Aids	Slide deck and 1-page handout. You can use other visual aids as appropriate (e.g., projecting actual software application using overhead LCD projector)	Your visual aids will vary according to your presentation and will be more or less textual, based on role you see for your visual aid. Your visuals should, however, be appropriate for your purpose, professionally produced, easy to read, and leave the reader with key information, including sources
Project Assessment Memo	500 words or less individually written proposal	Evaluate each team member's contributions to the project, including your own
Audience evaluation	Completion of presentation evaluation for each presentation	Audience members are participants too. Failure Failure to complete evaluations will hurt your final grade

Planning

Your initial thinking about the rhetorical context for the Staff Development Project might start with the following:

- What will the audience, your classmates, find useful and interesting about your topic?
- What might your audience, your classmates, already know about your topic?
- How can they best understand your topic in a workshop environment?

Initial Writing Decisions

- Why are you developing these materials?
- What do you hope to accomplish?
- What is the purpose for this workshop?
- What are the benefits for using and understanding this material?
- What is the purpose of the handout?

Content Decisions

- What information must be present?
- What information would the audience find most useful?
- What major audience questions do you need to answer?

Design Decisions

- Which types of visuals would be most meaningful for your workshop participants?
- Which types of visuals would reinforce your expertise?
- How might you design the materials to maintain the participants' attention?
- How can you design a handout that the audience can use in the future?

Questions About the Rhetorical Context

- Are you an expert? How much do you know? What else do you need to know?
- How can you best show your expertise?
- How will you present yourself as both legitimate and valid?
- Who is the primary audience? What do you know about them?
- How much do they know about your topic?
- How can you reach them most effectively?
- What form should the materials take? Will they work together? Or will they be separate?
- How will you organize the materials?
- How can you assure that your materials will be understood?
- How can you assure that your handout will be useful?

Proposal

The staff development proposal is a collaboratively written document designed to help your team plan its presentation. The proposal also serves to inform your instructor of your plans and persuade him or her that your presentation has been carefully tailored to the audience (i.e., classmates) and will meet the requirements of the project.

In a 2- to 3-page memo addressed to your instructor, provide a detailed summary of your team's topic, a rationale for why it choose the topic and particular emphases of the presentation, an analysis of the audience's knowledge and attitude about the topic, a list of outcomes, a description of your visual aids, and a list of sources. Your team should discuss the design of its script, PowerPoint slides, and handout. You must also attach drafts of these documents to the proposal for the instructor's review.

This document is a proposal in the sense that your primary reader, your instructor, wants to know what you plan on presenting and how exactly you plan to do it. Thus, the proposal is not simply about a rough idea of what your team is thinking of presenting, but rather it should include a thorough and careful discussion of your presentation. This will enable the instructor to assess the effectiveness of your plans and provide feedback on anything that needs reconsideration or revision before your team presents.

Format

Your proposal should include the following sections:

- Topic Summary
- Audience Analysis
- Rationale
- Outcomes
- Visual Aids
- Sources

Topic Summary

Answer the question: *What will your team present and how will it present it?* Discuss the overall purpose of your presentation and the points you will cover to accomplish your goals. When planning the content of your presentation, remember that it should

- fit within the 10- to 12-minute time limit
- expand the audience's existing knowledge and skills related to the topic
- give practical advice that the audience finds useful and interesting
- be informed by outside research
- fit within the standard introduction/overview/body/closing structure
- be easy to follow, including clear transitions

The topic summary should review the basic outline of your presentation. It should also reference your script, including how your team divided speaking responsibilities.

Remember that a well written script provides the following information in an easy to read format: detailed outline, verbatim speaker dialogue, directorial cues, plans for visual aids, and a timeline.

Audience Analysis

Answer the question: *What is your audience's existing knowledge and attitude toward your topic*? Be as concrete and detailed when assessing your audience's familiarity and receptiveness to your topic. First develop a profile of the demographic makeup of your audience. Then, given this general profile, what can you assume they already know about your topic? What can you assume they should know or would like to know about your topic? Also, consider how receptive the audience will be toward your topic. If you think they may be hostile or disinterested, what might make them more interested?

Rationale

Answer the question: *Why will the class find this topic relevant and how did your team tailor the information to suit the audience?* For instance, justify why you choose to include certain information and exclude other information. Refer to your audience analysis when deciding what topics to cover in your presentation. Outside research may help you establish the importance of your topic for students.

Outcomes

Answer the question: *What should audience members know and/or know how to do as a result of participating in your staff development presentation?* This should be a list that articulates the objectives of your presentation. Such a list may or may not resemble the basic outline of your presentation. For instance, a presentation on how to use PowerPoint might include the following list of outcomes:

> *This presentation aims to introduce the class to the basics of Prezi, including the following:*
>
> - *how to use the interface*
> - *how to create a slide deck*
> - *how to use animation effects*
> - *how to avoid Prezi mistakes in business*

Visual Aids

Answer the question: *What visual aids does your team plan on using for its presentation?* Discuss in the choices your team made regarding visual aids, including what the particular design schemes are for the slideshow and handout, and why the team choose these designs. Discuss elements of design such as content, organization, format, style, and visual design, including use of graphics. Remember that the team is not limited to PowerPoint and a handout as visual aids. For instance, if the team is doing a presentation on how to use PowerPoint, the software application itself should be used as a visual aid. In this case, the team may elect to use PowerPoint visuals to introduce and overview the topic, but then ask students to launch PowerPoint at their workstation and follow along with a hands-on tutorial presented by the team.

Sources

Online guide to Chicago Style:

- University of Wisconsin Writing Center guide to Chicago/Turabian Documentation Style at http://writing.wisc.edu/Handbook/DocChicago.html.

Answer the question: *What information did your team consult to inform its own understanding of the topic and decide what to present to the class?* You should provide **at least four (4) sources** your team referenced while planning the presentation. For the purposes of this assignment, use a *Chicago Style* format for documenting your sources accurately and consistently.

Proposal Evaluation Checklist

You can use the following checklist to evaluate the effectiveness of your proposal prior to submission.

Proposal

- Is the proposal in memo format, 2 to 3 pages in length?
- Does it include all required sections: overview, summary, rationale, audience analysis, outcomes, visual aids, sources?
- Does the summary describe topic and main points in sufficient detail?
- Does the rationale describe why this topic is useful to students in sufficient detail?
- Does the audience analysis describe who your audience is, what their current knowledge is about your topic, what their attitude is toward your topic, and what information you should emphasize based on your audience analysis?
- Does your outcomes section list in detail what skills and knowledge the audience will gain as a result of your presentation
- Do you have a list of 4 or more sources for your project (formatted consistently using a specific style for documenting sources)?

Script

- Do you have one master script that combines all parts?
- Does your script include an outline of main points, directorial cues for "who does what, when," verbatim dialogue for each speaker, timeline estimates, and text of visuals?
- Does your script plan for a 10-12 minute presentation? (No more or less)

Slide Deck

- Do you have a title slide, overview slide, body slides, and a conclusion slide? (NOTE: software demonstrations omit body slides)
- Are your slides visually appealing and formatted following design principles of consistency, balance, visual hierarchy, and simplicity?
- Do your slides use text appropriately (size, parallel lists w/fragments and minimal punctuation, readability, contrast w/background)
- Do you have appropriate visuals/graphics where appropriate in your slides that enhance the text (charts, graphs, pictures, photos, etc)?

Handout

- Do you have a short handout that gives audience information they can take away from your presentation and reference later?
- Does your handout include a title, list of presenters, date, and context (e.g., business writing class)?

- Does your handout include details not appropriate for slides (e.g., big chunks of text, definitions of technical terms, complex charts or graphs, lists of sources for future reference)
- Is the handout visually appealing and formatted following the design principles of consistency, balance, visual hierarchy, and simplicity?

Purposes

- Does the proposal (incl. attachments) inform the instructor of your team's presentation plans and persuade the instructor that your group has a well thought out plan for its presentation?
- Does the proposal include evidence that the presenters will give a sufficient amount of information about the topic?
- Does the proposal include evidence that presenters are aware of the primary audience's (i.e., classmates') needs and interests?
- Does the proposal include evidence of thorough and appropriate research?

Product

- Does the proposal include the following sections: topic summary, audience analysis, rationale, outcomes, visual aids, and sources?
- Does the proposal reference within the document and attach to the document drafts of the following: script, slide deck, and handout?
- Does the script indicate a clear introduction, overview, body, and conclusion structure?
- Does the script contain an effective introduction that introduces topic, team members, and includes a hook to establish significance and relevance of topic for audience?
- Does the script contain an overview, following the introduction, that forecasts the main points of the presentation to the audience?
- Does the script use transitions effectively (e.g., "Now, Jane will discuss…)?
- Does the script include a summary before concluding?
- Is there evidence of plans to use adequate and appropriate visuals (slideshow and handout)? Are the visuals easy to read (appropriate fonts, use of specific headings, uncluttered design, balance of text/images/white space)?
- Does the handout include information, including resources/references, for take-away use?

Process

- Does the proposal, script, slide deck, and handout exhibit evidence of careful planning, drafting, revising, and editing?

Online Resources

Presentation Scripts

- Sample Script: "Universal Access: Electronic Information in Libraries," University of Washington. http://www.washington.edu/doit/UA/PRESENT/scintxt.html

Presentation Techniques

- "Delivery Habits to Avoid" URL: http://www.effectivemeetings.com/presenting/delivery/avoid.asp
- "Top 12 Most Annoying PowerPoint Presentation Mistakes" http://thevisualcommunicationguy.com/2013/09/24/top-12-most-annoying-powerpoint-presentation-mistakes/

Design Tips

- "10 Tips on How to Make Slides that Communicate your Idea, from TED's In-House Expert" http://blog.ted.com/10-tips-for-better-slide-decks/
- *"Tips on Designing Brochures" by Robin Williams* http://www.informit.com/articles/article.aspx?p=20718 &rl=1

MS PowerPoint and Publisher

- Microsoft's PowerPoint page https://products.office.com/en-US/powerpoint
- Microsoft's Using PowerPoint Links http://www.microsoft.com:80/office/powerpoint/using/default.asp
- Microsoft's Publisher page https://products.office.com/en-US/publisher

PROCEDURES PROJECT

CHAPTER 30

Project Objectives

- Learn to design, produce, and communicate clear instructions for doing a technical process
- Learn to use writing as a tool for producing effective presentations, including writing scripts and designing visual aids (PowerPoint and handouts)
- Learn new computer technologies that will be useful for this class as well as the workplace

In groups of 3 or 4 you will run a brief (~ 25 minutes) hands-on software tutorial workshop for the class. In addition, your group must provide written instructions for the steps you cover in class.

Most graduates in technology, engineering, technical graphics, and professional writing will be required to write instructions, or collaborate with others writing instructions. Writing effective, accurate documentation has both economic and legal ramifications:

1. The better the instructions, the fewer calls/emails to customer service staff.
2. Documentation that results in harm to users (either unclear or lacking proper warnings) is liable (every year, many companies are sued over faulty instructions).

The class will negotiate five or six software programs that most of the class is unfamiliar with but believes would be useful to learn. Your group will sign up to present a tutorial workshop for the one program that you are most interested in learning in-depth. You are not required to have mastered the software you will be presenting. This is your opportunity to learn more about software you're unfamiliar with.

Project Options

Your goal is to make the tutorials as interactive as possible. Rather than lecturing, have your classmates actually perform the tasks at their computers when possible. Before choosing your topic for this project, make sure the software is supported by your school's computer lab. Since you need to plan a hands-on demonstration, you'll need to make sure that the application is available in your computer classroom. The following are some possible applications and functions that fellow students would find interesting and useful:

- **Intermediate to advanced Word functions.** Word has over 2,000 features. Some useful but little known functions include using tables, using textboxes, inserting graphics, setting page attributes, setting up form-fields (including mail merge), etc.
- **Presentation applications such as PowerPoint, Google Slides, or Prezi.** Knowing how to choose and edit a basic slide show or how to use more advanced features such as inserting images and animation effects should be very useful.
- **Web authoring or content management platforms.** Your workshop could illustrate some basic codes, how to make simple pages, or how to use a popular web authoring tool or content management system (e.g., Wordpress). Make sure you choose a computer application that is available in your campus computer classroom, or web browser. This workshop works well when the entire class creates its own simple page and publishes it using their student computing user account. (Most school's have instructions for students hosting web pages via their student computing accounts.)
- **Photoshop or Paintshop Pro.** Paintshop Pro is a less expensive alternative to the popular image editing software Photoshop. Workshop could demonstrate how to create and/or download buttons, backgrounds, and graphics for documents or web pages. Capturing and manipulating *screenshots* is a key function related to image editing software, one of great use in preparing the written instructions for the Procedures Project.
- **MS Excel.** Show students how to perform basic to advanced calculations or how to create charts and graphs and insert them into Word documents. This last feature will come in handy for the final Technical Report Project.
- **MS Publisher.** Show students how to create professional quality brochures, newsletter, reports, etc. using Publisher, Microsoft's desktop publishing software.

Deliverables

The following are documents your team will produce for this project:

- **Brief Proposal** in memo format outlining what program or function your team wants to demonstrate and why other teams should find it valuable. Your proposal also needs to assess who your audience is, including its needs, knowledge, and concerns about your topic. You will also attach drafts of your tutorial script, instructional handout, and your PowerPoint slides.
- **Tutorial Script** outlining exactly what you will be presenting to the class during your tutorial. The script should include not only who plans to say what and when, but also what will be shown on overhead/projector screen, checkpoints for evaluating whether or not your audience is keeping up, and other notes. The script should be attached to your proposal.
- **Instructional Brochure or Reference Card** documenting your tutorial. Your team will design a handout for reference during your tutorial and so the audience can remember and apply what your team teaches them afterwards. The handout must be carefully designed to fit a maximum amount of useful information into a relatively short amount of space/pages. You can decide between brochure or reference card layouts. Length of handout will be up to your team's discretion. Just remember, the longer the handout, the higher the risk of overwhelming your reader with information. But you need to provide enough information to have usable instructions, obviously. A draft of your handout should be attached to your proposal.
- **Slide Deck.** While your primary visual aid during the presentation will likely be the actual software you are demonstrating and asking fellow students to follow along with, you will also want professional quality slides to help facilitate your tutorial, usually for the opening and closing of your tutorial. Drafts of any slides you plan on using should be attached to your proposal.
- **NOTE**: Each group must also submit final copies of their script, instruction handout, and PowerPoint slides the day of their presentation.

Steps for Completing the Project

Follow these steps to complete the Procedures Project:

- **Step 1: Do Background Reading:** Read the chapter on Presentations in this textbook. Since one of the main goals of this assignment is to learn how to compose effective instructions for print and oral delivery, you should also consult the online guides for writing instructions offered by Jonathan and Lisa Price, experts on technical writing and writing for the World Wide Web. These guides will familiarize you with basic concepts of writing instructions, such as determining how to divide a task into discrete steps, how to organize the steps, and how to write effective instructions.

Online Resources for Writing Instructions

- "How to Organize Step-By-Step Procedures" by Jonathan and Lisa Price http://www.webwritingthatworks.com/CPATTERN procedures.htm
  - "The title is a Menu Item" by Jonathan and Lisa Pric http://www.webwritingthatworks.com/DPatternPROC01.htm
  - "Intros are Optional" by Jonathan and Lisa Price http://www.webwritingthatworks.com/DPatternPROC02.htm
- "Put Instructions into Discrete Steps" by Jonathan and Lisa Price http://www.webwritingthatworks.com/DPatternPROC0 3.htm
  - NOTE: The following are all linked from the bottom of "Put Instructions into Discrete Steps"
    - "Format Steps So They Stand Out" http://www.webwritingthatworks.com/EPatternPROC01.htm
    - "Write Short Energetic Steps" http://www.webwritingthatworks.com/EPatternPROC02.htm
    - "How Many Steps?" http://www.webwritingthatworks.com/EPatternPROC03.htm
    - "Separate Explanations From Steps" http://www.webwritingthatworks.com/EPatternPROC04.htm

- **Step 2: Form teams and select a topic:** Two groups can present on a similar topic, but the presentations should not be identical, e.g., one group can present on PowerPoint basics and another team can present on intermediate or advanced topics. Which skills are basic and which are advanced depends on the group's estimation of the background knowledge shared by the majority of classmates.
- **Step 3: Do some background research:** Each team must cite a minimum of 4 sources in their proposal. If you're topic is PowerPoint, e.g., try using search engine terms such as *PowerPoint tutorial* or *PowerPoint basics* or *PowerPoint advanced* to find example instructions. You can use any samples you find as models to write your own tutorial, e.g., determining what are considered basic skills by the majority of tutorials you find.
- **Step 4: Analyze your rhetorical situation:** In a prewriting exercise and class discussion, consider the following questions:
  - Who is your audience?
  - What do they already know about your topic?
  - What should they know about your topic/what will they find interesting and new?
  - What topics should you include, given your audience and allotted time?
  - How should you arrange your topics, given your audience and allotted time?
  - Who on the team will present which topics?

  - What will the other team members do while one is presenting? (e.g., work keyboard, roam room and act as helper if people in audience gets stuck)
  - How will the print documentation/instructions be integrated into the presentation (i.e., passed out at the beginning and used to follow along with, or passed out at end and used as take-away reference)?
  - What should the print documentation/instructions contain?
  - How should the visual layout of the print documentation/instructions be designed?
  - How will the print documentation/instructions get produced?
- **Step 5: Draft your Proposal:** Your proposal for this project should inform your instructor of how you plan on presenting your tutorial. By the time you submit your proposal, your deliverables for this project—the proposal, script, slideshow, and handout—should be nearly, if not completely, finished and ready to go.
  Use the guidelines from the **Staff Development Project Proposal** to format and organize your proposal.
- **Step 6: Practice Your Presentation:** You can practice as a team nearly anywhere if someone has a laptop (provided it has the necessary software), or your group can meet in a campus computer lab and quietly rehearse your script.
- **Step 7: Make Any Necessary Adjustments:** Make sure your proposal fits within the allotted time limit. If when you practice as a time your presentation is too short, you will need to consider what other topics should be added. If your presentation is too long, you will have to decide what to cut. Revise your proposal accordingly.
- **Step 8: Submit Your Proposal and Practice Some More:** By submitting a proposal that represents as close to your final deliverables as possible, your instructor can review your plans and comment on whether or not any changes need to be made before the actual presentation day.

CLIENT PROJECT

CHAPTER 31

Project Objectives

- Write proposals, progress reports, and formal reports
- Identify and solve organizational problems through writing and research
- Plan and carry out a multi-stage, collaborative research and writing project (known as "document cycling" in the workplace)
- Research the specific needs of the audience(s) to which you will be writing and tailor your writing process and products to these needs
- Analyze and present a large body of information generated by your research
- Design and draft effective documents and presentations, including final and intermediate oral and written reports, memos, letters, email, and various types of visual representations of data
- Manage a complex document production process which includes merging files, re-purposing existing documents, using style sheets and templates, and conducting multiple review and revisions

This project puts you in the role of a research analyst working to provide actual organizations—clients—with useful information. Student teams work with a client organization to identify a need or problem that can then be addressed through research. The final product of the project is a recommendation report that synthesizes relevant information (e.g., expert opinions, published research, industry trends) and offers feasible solutions. The project introduces you to the process necessary for producing group-written professional quality reports, including how to organize a team, manage a project, conduct primary and secondary research, and write preliminary documents that contribute to a final report.

What organization you work with and what problem you research depends in large part on the organizations and organizational needs identified by you and your client. Acceptable organizations include university organizations and local profit/nonprofit companies. Some student projects have included:

- Exploring ways to increase awareness, gain volunteers, and increase donations for a local animal shelter (Client contact: shelter manager, recruited by student)
- Recommending fundraising options to raise sufficient funds and to motivate member participation in UNLV's business fraternity, Alpha Kappa Psi (Client contact: club president, friend of student)
- Developing recommendations for a literacy program at an area elementary school (Client contact: student's mother)
- Comparing payroll processes and payroll software systems for a medium-sized electrical engineering firm (Client contact: student's mother)
- Developing a more structured and organized volunteer program for a nonprofit service organization (Client contact: student's boss)
- Evaluating web service options for a local apartment management firm (Client contact: student's boss)
- Researching web marketing strategies for an area mortgage company (Client contact: student's father)

For more help choosing a client, see the first part of the Planning section titled **Identifying a suitable client**.

The Client Project is a service learning activity. To learn how to adapt your writing skills to new contexts, the class will act like a consulting company. You will work together with your classmates and instructor to provide clients—actual local organizations that have been selected by you—with information that will potentially help them provide better products and better services. The project creates an opportunity for real world experience, but at the same time provides the support and structure of the classroom. Unlike real world work situations, where the goal typically is to "get the job done," the main goal of this project is to help you learn to apply your writing skills to new situations. (Learning in the workplace is often only a secondary goal. Sink or swim, as they say.)

Your identity as a writer in this project is rather complex, but very authentic. You will primarily be representing yourself as a student doing a project for your business writing class. But since you will be interacting with actual members of the local community, you also must be conscious of your role as representative of your college or university, your instructor, and your fellow classmates. You must also consider the values of your client organization, its clients or customers, as well as the local community.

Deliverables

This project will extend over several weeks, focused on a research task performed for actual clients identified by you. It can include a variety of documents depending on the time allotted and instructor's discretion:

- **Client Proposal:** Each individual student is responsible for identifying an appropriate client and then writing a research proposal to the instructor. The class, with help from your instructor, will then choose the most appropriate projects for teams of 3 or 4 students to work on (e.g., from 24 or so proposals, about 6 are actually chosen as client projects).
- **Letter to Client:** After projects are selected, each individual student will write an acceptance or rejection letter informing his or her client about the status of the proposal.
- **Project Plan Report:** Each team will write a detailed plan for completing the research and producing the final report. The team may be asked to share the plan in an informal presentation to the rest of the class.
- **Progress Report:** Each student will individually prepare, about mid-way through the project, a report that explains her/his individual and team progress within the project.
- **Final Client Project Report:** Each team will write a report that presents the client, the primary audience of the report, with the results of the its research and recommendations for addressing the client's problem.
- **Oral Presentation to the Client:** Each team may be asked to deliver a professional quality oral report based on the team's final written report.
- **Project Assessment Memo:** Each individual will write an individual email PAM that evaluates how each team member contributed to the project.

Planning

Planning the client project, like planning any informative report in business settings, is a complex process that involves:

- Identifying a suitable client (individual)
- Framing the problem or need as a research project (individual, group)
- Proposing a client research project (group)
- Establishing a plan to manage the project (group)
- Determining how the final report will be produced (group)

Identifying a Suitable Client

Your first responsibility in the client project is to identify an actual organization that meets the criteria for a suitable client. Once you've found a client organization and obtained the contact's consent, you need to work with that client to define a suitable need or problem that calls for researched information.

Client Selection Guide:

1. List at least three organizations (businesses, nonprofits, clubs) that meets one or more of the following criteria:
   - That you work for, belong to, or participate in (like a company, volunteer group, student club)
   - That you used to work for or belong to
   - That your family member or close friend belongs to
   - That you would like to belong to or work for (If you're not involved outside of classes and your survival job, you need to get involved. Now might be a good time to make contact with a group that could help advance your career goals)
2. Which of these organizations do you have most **access** to?
   - Is there someone you are close-enough to that you can approach about this project and that, if selected, would be available for interviews? Local organizations are generally more accessible, but with email and phones, it is possible to have high access with out-of-town clients.
3. Which of these organizations has a high degree of **willingness** to participate?
   - Who would be most interested and eager to volunteer their time to the project? Who would feel their organization could really use the help (as in, helpful information)?
4. Which organization will **benefit** most from the project?
   - Which organization has minimal resources and couldn't ordinarily afford to pay outside consultants for information? Nonprofits, small businesses, and student organizations are better choices than larger corporations.
5. Which organization can yield a project of most **interest and significance** to students?
   - Which organization would you and your fellow students find interesting/enjoyable/rewarding to work with or which organization could most contribute to you and your fellow students' educational goals (e.g., choosing a online record store because students are into music, choosing a finance company because many business majors plan on going into finance, or choosing any kind of company that has needs related to major areas, like a Youth Sports Little League that needs a fundraising strategy)?

In your proposal, you need to persuasively establish that your client is a good choice using the criteria identified above.

Framing a Problem or Need as a Research Project

Once you've identified a client and received informal consent from the contact, you should begin a fact-finding dialogue with them about how this project can help improve the mission of that organization.

1. List the services, products, mission, activities, or characteristics of this organization
2. Interview the client to determine which aspects of the organization could be improved
   - Which aspects have problems or do not meet the expectations of the client, the client's customers, industry expectations, etc?
   - Which aspects of the organization could be improved, rethought, or expanded? Research can be pro-active too, or help an organization to strategically improve its services, define new goals, or adapt to industry and technological trends. There's no such thing as a perfect organization.
   - If you have trouble identifying a problem about the status quo, or the current state of the organization, begin by exploring what should be, or how to improve seemingly problem-free aspects of the organization.
     - *How* do other organizations do "X"? (e.g., use the web)
     - *Should* the organization do "X"? (e.g., have an e-business strategy)
     - *What* would improve "X"? (e.g., our current e-business strategy)
3. Define the problem as a research question: Once your client identified a problem- or need-based topic, you need to articulate a researchable problem. A researchable problem is one that is framed as an open-ended **guiding research question**. The *how*, *should*, and *what* questions above are examples. Defining the problem as a research question helps you focus your research activity on seeking information that will best answer the question. This straightforward, systematic way of conducting research has helped scientists for centuries. Scientists use the scientific method to articulate a hypothesis in the form of a question that is then verified through experimentation.

Guide to Formulating a Primary Research Question

Write out three versions of a question focused on exploring the problem or need you identified. Try using different journalistic question words—who, what, where, when, why, and how.

How	Prompts procedural questions	"How can Acme improve its mail sorting system?"
What	Prompts comparisons and lists	"What electronic mail sorting system is best for Acme?"
Why	Prompts cause/effects questions	"Why is the Acme mail system not working?"
When	Prompts timing questions	"When should Acme upgrade its mail sorting system?"

(continued)

Where	Prompts questions of location	"Where should the mail sorting duties be located in Acme?"
Who	Prompts lists of players, participants	"Who is affected by Acme's inefficient mail system?"

- Be sure the from of the question includes reference to the client organization
- Be sure the question is the *primary-question* (secondary questions are those questions that need to be answered before the primary, or overarching question can be answered)
- Be sure each question is open-ended, or not easily answered by yes or no
- Be sure the question is not biased, or implying a preferred answer (e.g., Why should Acme Insurance choose PowerPoint over Google Slides? *This question is biased; it assumes PowerPoint is the better choice!*)

4. Choose the form that you feel will yield the best information. Once you've chosen a question, re-evaluate it using the criteria of access, willingness, benefit, interest, and significance. Be sure the problem is **manageable** in the time allotted for the project, that it is: (1) not too simple or too complex for a group of students, and (2) can be researched adequately in timeframe of the project.

You will need to argue the feasibility of the problem as you've defined it in your proposal using the above criteria.

Proposing a Client Research Project

Once you have identified a client and a problem, your instructor may ask you to write a formal client proposal that informs your instructor and classmates about your client, the client's need or problem, and the feasibility of the project. The instructor may choose only a small number of clients from the lot of individual proposals or students may be pre-assigned into groups and tasked with writing a proposal for each team's project. Either way, your goal is to persuade your readers that you have selected an appropriate client and articulated a very doable research project based on the parameters of the client project.

The client proposal should be written in memo format and contain the following sections:

- **Proposal abstract:** A concise 75- to 100-word summary of the proposal. Include the main points of the proposal—client description, client need or problem, and project feasibility. The abstract is written after the main sections of the proposal have been completed.

- **Client description:** Provide information on the following: Who is the client? What is your relationship to the client? Where is the client located, what products/services does the client provide, how large is the client's scope? Include anything else you deem relevant about the client.
- **Client need:** Describe the problem or need the client currently has that you propose investigating. If the problem or need concerns some procedure or practice, then describe that procedure. Articulate the problem as a specific research question or set of questions. Consider what types of research you imagine need to be done to address the questions you have outlined and some possible solutions. Include a guiding research question.
- **Project feasibility:** Evaluate the chances of success for your project. Does the proposed client meet the criteria outlined on page 340: access, willingness, benefit, interest, and significance? Have you obtained the necessary consent? (Reference the attached client consent form.) Is this a researchable project? Will there be sufficient access to information? Is the project complex enough to support the work of three or four students, at the same time focused enough so that it can be completed in a semester? What are the benefits of the project for the students and the client organization?
- **Conclusion:** Reiterate your recommendation and appeal to the goals of the client project. Answer the question, Why should we do this particular project?
- **Enclosure:** Attach the signed client consent form to your memo.

Before you submit your proposal, you must contact the targeted client, discuss the project, and get permission. Your proposal *will be considered incomplete without a completed and attached client consent form.* There are several approaches to getting the client's consent depending on your relationship to the client. For example, it may take an introductory meeting where you leave the consent form document for their review and signature, or you may email the form to the client after an initial phone conversation. You might then arrange a meeting to pick up the consent form and discuss possible client needs.

Remember that, above all, this project is an academic project for a college course. Put into those terms, you can generally persuade a prospective client contact to see the value of the project for all parties: for you and for them—they will potentially benefit by having a team of students focused on offering researched recommendations for improving communication practices for their organization, cost free (with perhaps the exception of some time commitment to be available for consultation).

The consent procedure that accompanies the proposal is designed to (1) underscore the fact that you're to find an actual client; (2) increase the client's interest and participation; and (3) initiate a dialogue between you and the client about the client's organization and possible needs.

Your proposal will be written for multiple audiences: your instructor, the client, your fellow classmates—who want to know background information about the client before deciding upon a work team.

Establishing a Plan to Manage the Project

Once you have been assigned to a client project team, which may or may not be the project you proposed, your group needs to work together to develop a plan to finish the project.

Read the chapter on Collaboration, particularly the part titled Document Production for principles on managing the production of a team-written document

Your team will need to work together to write the **Project Plan** report. This involves developing:

- a research design plan
- a project plan

Research Design Plan

Why research?

- To find out about the client's needs/problems (client research)
- To find out client's ideas for solving problems or meeting needs (what will be acceptable to client) (client research)
- To find out what is feasible/realistic for client (e.g. what can the client afford) (client research)
- To find out about options that may help the client (external research)
- To discover industry-related practices which address similar needs/problems (external research)
- To discover industry-related trends which might influence needs/problem (external research)

Research is essential for addressing your client's problem or need. Research may have acquired a negative association for you in school from lots of research term papers that didn't seem worthwhile or enjoyable to you. But the fact of the matter is that anytime you need information, you are conducting research. Have you ever done any comparison shopping before buying a cell phone plan, car, or electronic device? You were researching. Businesses conduct research all the time to collect information about their target market, competitors, new products and services, human resources, etc. Businesses rely on information so much that people make careers as business analysts who specialize in researching data and making sense of it.

Developing a Research Plan

1. Download the research design template to plan your research
2. Identify secondary questions that need to be answered
3. Develop a strategy for answering the questions:
   - Primary sources (interviews, observation, surveys)
   - Secondary sources (printed/published material)
4. Estimate how long each research task will take to complete
5. Assign research tasks to individuals
6. Create a Gantt chart depicting tasks and responsibilities

A research design plan is your team's plans for collecting information that will help address the need or problem identified by the client. At this state of the project, you team will brainstorm what information it needs to collect to answer the main question. This information can also be articulated as questions. For example, if your team is researching the question, "What is the best accounting software for a small company?" one obvious question that needs to be answered is: "What are the available accounting software programs for small businesses? You might need to answer the question: "What is my client company's criteria for choosing the software?" You might also want to know what other similar companies use and if there are other alternatives, such as hiring an accounting company. These are all **secondary questions** that need to be addressed to answer the main research question.

You also need to think about what are the best sources of information for answering your secondary questions. Should you look interview the client, look up information in trade journals or newspapers, interview experts, investigate competitors? As Figure 31.1 shows, there are two main types of information: (1) information that comes internally from the client and (2) information that comes externally of the client, from first-hand interviews with experts or customers, to second-hand information published in magazines, newspapers, and journals.

After you team brainstorms its secondary research questions and determines the best sources of information for each question, it will next have to estimate how long it will take to answer each question. It may take several weeks to conduct a survey of customers, or it may take a few hours to interview the client. Don't forget the time it takes to prepare surveys or interview questions when estimating the duration of a research activity.

Finally, your team will need to decide who on your team will complete what research. The project planning phase will help your team when it comes to determine who will be responsible for particular parts of your group project.

What sources of information will yield best results?	
Client /Internal	**External**
Interview. Efficient way to get individual viewpoints. But reliance on isolated viewpoints can be limiting. **Survey.** Allows researcher to collect large number of viewpoints BUT time consuming to develop, distribute, and collate survey. **Observation.** Allows researcher to gain first-hand knowledge and provides means of verifying personal views ("what really happens" vs. what people say or think is happening") BUT can be limiting and time consuming. **Records.** Allows researcher to gain historical perspective on problem and provides "hard" evidence for certain kinds of problems BUT records don't always provide evidence of problems.	**Other companies (interview, direct observation).** To get information about what other companies are doing, to help the client. **Faculty/Professionals (interview).** To get expert opinions and suggestions for sources; strategies for doing research. **Product Vendors (interview, brochures, product literature).** To find out about available options and to get product information. **Internet (online research/browsing).** To track down on-line information of relevance to your subject; corporate websites may provide some information (or the name of a contact person for an interview).

NOTE: Never rely on only one source of information. Use at least a couple of different sources and a couple of different methods to cross-check and verify your understanding. Do as comprehensive a job of research as time and resources allow.

FIGURE 31.1. Internal vs. external research

Project Plan

A project is any multi-step process that involves the coordination of people, organizations, resources, and time. Project management and project planning grew out of the defense industry in the 1950s and 1960s as a method for planning and implementing complex projects that involve scores of people and often multiple companies.

Six Steps to Effective Project Management

1. Establish the project
2. Build the project team
3. Organize the work
4. Assign tasks (and estimate resources)
5. Set the project schedule
6. Complete the plan

The common method for visualizing a project plan is called a Gantt chart, a horizontal bar graph that depicts a project's discrete tasks and the estimated time it will to take to complete each task. Major deadlines, called *milestones*, are depicted with symbols such as diamond or triangle. Because of the importance and prevalence of project management for getting information-based work done, there are lots of software applications available for creating Gantt charts and other project visuals, e.g., Microsoft Project.

> Gantt charts visualize project plans, including tasks, responsibilities, and time (milestones/ deadlines)
>
> - "Gantt Charts: Planning and Scheduling Complex Projects" http://www.mindtools.com/pages/article/newPPM_03.htm

Once your group has determined its specific research tasks, it should then identify the discrete deliverables that are due to complete this project as a classroom assignment. If your instructor is following this project as written, each team will have to produce the following documents:

- Project plan
- Client letter
- Progress report
- Draft of recommendation report
- Draft of presentation
- Final report
- Final presentation

When the research tasks are combined with the project deliverables, the project plan is complete (see Figure 31.2 below).

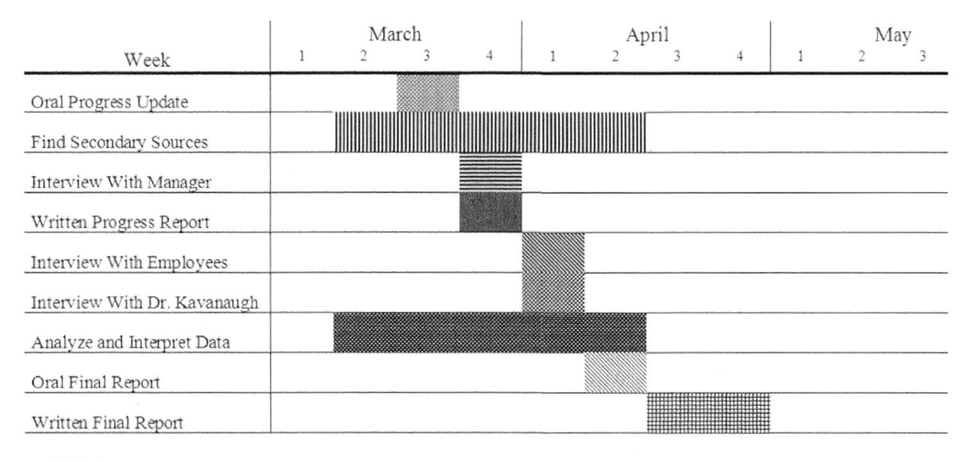

FIGURE 31.2. Client Project Plan Depicted in a Basic Gantt Chart

The **Project Plan** report will be a memo written to your instructor summarizing him or her of your group's plan for completing your Client Project.

Determining How the Final Report Will Be Written

How will your team coordinate the production of the final report? One of the biggest challenges is bringing the work of multiple authors into one, common format. To do this, your team needs to decide ahead of time issues about production responsibilities and deadlines, content plans, and style guidelines (e.g., how to format text, consistent spellings, etc).

Tips for Collaboratively Producing Reports

- Discuss general purpose and form of report
- Make an outline and develop headings and subheadings, include plans for visual aids (called a "storyboard," see below)
- Design a template/develop a "style sheet"
- Import "boilerplate" text from preliminary reports (proposal, plan, progress reports)
- Assign sections to various individuals (after group has collaboratively determined outline, including recommendations
- Write distinct sections individually
- Exchange sections and do revisions
- Merge separate files into one master file (and make backups!)

Storyboarding, like scripting presentations, is a practice borrowed from artistic designers. Like comic book, cartoon, television, or movie storyboards, report storyboards lay out in advance a detailed picture of a plan for the report, including the report's title, organization, section headings and subheadings, content, and graphics. It also indicates individual responsibilities.

Use the storyboard template to plan your final report

Follow the guidelines for writing business reports discussed in the Reports chapter of this textbook. Business reports are closer to published books than academic papers. That is, business reports often have front matter" elements such as a cover and table of contents. Business reports also pay more attention to visual design, as it is understood the visual "look" of the report contributes to the organization's image and the persuasiveness of the report.

You report must include at least two visuals in your report, including one table and one figure. This is an arbitrary number to push you to practice using visuals as a way of presenting information clearly and quickly. You can include more than the minimum.

Visuals are an important component to reports because they:

- Promote understanding/clarify complex ideas
- Add emphasis to key points
- Attract the eye
- Save space
- Arouse interest
- What deserves a visual?
  - points that need emphasis
  - points you want to draw reader's attention to
  - points that need clarification (e.g., numbers, stats, processes, abstract ideas)

Remember that each type of visual is good for showing certain types of information, so think carefully about choosing the most appropriate type of visual for the information you wish to display.

What visual should you use?

- **Bar charts** show comparison and contrast
- **Line graphs** show trends
- **Pie charts** show percentages/parts of a whole
- **Photographs/drawings/screenshots** give exact or facsimile representation
- **Organizational charts** show structure or hierarchy
- **Flow charts** show steps in a process
- **Diagrams** show concepts and processes
- **Tables** array information in rows and columns; good for showing lots of data, abstract juxtapositions, comparison/contrast (e.g., comparing several computers according to various specifications like cost, processing speed, memory, RAM, etc.)

Also remember that *visuals don't speak for themselves!* They must be explained in the text. Do not just write, "Here is a chart showing our third quarter profits." Instead write, "As Figure 1 shows, our third quarter profits rose by 12% overall, but our main services were down by 2%, 4%, and 1% respectively. This data indicates that our business is evolving and we should reconsider how we allocate resources to our various services."

Refer to the Reports chapter for more information on choosing the best visual and how to integrate the visual into your final report.

There is also more information on the report format for the Client Project under the Final Report section of this project.

Project Plan

Each team will be responsible for co-writing a project plan report memo and presenting it to the class as a brief oral progress report.

The project plan is a preliminary report that:

- Defines the team's project
- Articulates a plan for completing the project, including research
- Persuades the reader, your instructor, the project is well thought out and feasible.

The project plan answers the questions: How will you proceed to provide the client with relevant and useful information? Have you thought the client's need or problem thoroughly enough to proceed with the support of your instructor and the client? The project plan also helps the project participants themselves, for whom such a report acts as a guide, almost a contract, to specify who will do what and when. If done thoroughly enough it will also serve as a draft of your methodology section of the final report. For more information, refer back to the Establishing a Plan to Manage the Project part of the Planning section.

Your team's project plan should be written in memo format and include the following:

- **Client Description:** Repurpose your client description from the project proposal.
- **Client Problem:** Repurpose from the project proposal, unless the team has decided to refocus the problem in some way.
- **Methodology:** Describe the team's research methodology, which includes the main research questions the team will pursue, what methods (both primary and secondary) the team will employ, and a justification of the methods you've selected, i.e., why are you using these methods? Why are they appropriate given your client's problem?
- **Project Schedule:** Describe of the team's plan for completing the project, including a discussion of who is going to do what, by when. Also include major milestones like progress report and final report draft/final deadlines. You must include a **Gantt chart** visualizing the team's schedule (but remember, the visual alone is not enough; you must explain the visual in your prose discussion).
- **Closing:** Brief statement about status of project. If there are any questions, obstacles, or problems the team is facing, you can mention them (optional).

Oral Presentation of Project Plan

The oral report's purposes are to inform the instructor and other teams of your group's research design and project plan. After a brief review of the client and problem, focus on describing in some detail your methodology, or the research methods your group will use accompanied by a rationale or justification for those methods (i.e., detail both *what* methods you're using and *why* you're using those methods).

The oral report should be approximately 5-8 minutes and be supported by at least 2 visuals. One visual must be a **Gantt chart** depicting the team's production schedule or a **Research Design Chart** depicting the team's research design. The other visual can be whatever the team deems most appropriate to support their discussion, such as textual outlines of the oral progress update, diagrams of the client organization, or the team's analysis of the client's need/problem (possibly using the research design charts). Teams are free to choose the medium for visually supporting their reports (e.g., overhead transparencies, Word files, PowerPoint, etc.).

Client Letter

After your class has selected clients and formed project teams, your instructor may ask you to write a letter to your client informing him/her of the status of your proposal and thanking him/her for participating in your education.

For some, this will mean writing an acceptance letter, telling the client your proposal was among those selected and informing them of the next steps in the project (i.e., contacting the client to set up a meeting with the group).

For most others, the client letter will be a rejection letter, telling the client that their project was not among those selected by the class. The rejection letter is more challenging to write, because it must deliver "bad news" in a way that maintains good will.

Write a short (less than one page) acceptance/rejection letter in proper business letter format:

- Inform client of status of proposal
- In the case of rejection letters, try to politely indicate why letter wasn't selected (e.g., "...only 4 of 21 proposals were accepted")
- In case of acceptance letter, remind them next step is meeting with group to discuss client's ideas for meeting needs and develop suitable research plan
- Thank client for his or her time and interest
- Remember to maintain a professional tone (include a buffer, an indirect lead in, in the opening of rejection letters)

Sample letters?

There's no template for this assignment. You'll have to use your best judgment to craft your client letter. You can, however, think about adapting the basic formats and styles of acceptance and rejection letters to your purpose and audience.

- "Declining Job Offer" letter
  http://www.quintcareers.com/sample_declining_letter.html

Progress Report

Each team member may also be asked to submit an individual or group progress report individual progress reports inform the instructor of an individual's contribution to the overall team project. An individual report places one's own efforts in the context of the whole group's progress. A group progress focuses on the group's progress overall, but should discuss how individuals have been doing completing the individual research tasks. The report should be approximately 1,000 words, so maintaining clarity and concision will be important.

Organizing the Progress Report

Progress reports are common in business. The genre is designed to justify work done, evaluate that work, and also to forecast work remaining. As a genre, the progress report is conventionally divided into two major sections, work completed and work remaining. Context is provided by overview, background, and conclusion sections.

Include the following sections in your progress report. You are not limited to these sections; you may want to subdivide the sections, for instance, to report on different work completed or remaining:

- **Overview:** Briefly summarize of purpose and content of document. This report is focused on informing the reader of the status of the Client Project.
- **Background/Context:** Reiterate for the reader the purpose of research project, including client, client need/problem. Outline initial project plan, including project schedule, designated tasks and sources of information. Answers questions: What is the project? Is the project proceeding as planned? Have there been any significant changes in the scope or organization of the project?
- **Work Completed (or Results to Date):** What data do you have to report at this point? What preliminary (tentative) conclusions can you offer? You should describe actual results; not just merely say, "I've made lots of progress." This reporting should resemble the "Methods" section of a final Recommendation Report, justifying

your selection of sources and choice of research methods. Yet, like the findings and recommendation sections, it will also report on the significance of the data. It is the opportunity to articulate initial insights gained from the research.
For example, concerning interviews, answer the following questions: What were your plans for conducting the personal interviews? Why were you doing them/what do you hope to accomplish with them? Who (name and title) in the organization have you decided to interview and why? What questions did you plan to ask? (For surveys, you may even attach a copy of your questions.) How long did it take? Are these one-time only interviews or do you plan a follow up as well? What were the results? What kind of notes did you take? What information was valuable? Why was it significant? How does this help answer your research questions or solve client's problem?

- **Work Remaining:** This section should describe explicitly tasks which remain to be completed by you and the group before the completion of the project. Considerations resemble the work completed section (i.e., justifying tasks, sources and methods) though there is no data as yet to report.
- **Conclusion:** Reiterates status of project and assesses the chances for its success. As a writer, you want to be persuasive as well as honest. Your tone should be positive, but if there are any concerns with the project, you should not avoid mentioning them. Your decision-making readers need to be given an honest report so that they can offer assistance, redirect resources, etc.

Final Report

The following information describes the features of your Final Recommendation Report and Oral Presentation for the Client Project. Business reports are further discussed in the Reports chapter of this textbook.

On the day the Final Recommendation Report is due, you'll need to prepare two (2) copies of the report, one for your instructor and one for the client. Your final written report must include all of the elements discussed below. Please pay close attention to the following throughout your planning and writing process.

- Goals for the Client Recommendation Report
- Report Format
- Review Checklist for Final Report
- The Oral Presentation

Goals for the Client Recommendation Report

Your overall aim is to help the client improve their business communication practices so that the organization can function more effectively and fulfill its goals. The client will use this report to help them decide what changes in their organization are necessary to make.

Analyze the Rhetorical Situation of the Client Report

Purposes

- Why are you writing this report? What are the main purposes and what are secondary purposes?
- What are your goals for this report? Which sections should be clearly related to the goals of the report?
- What different types of evidence are available to you?
- How can these be used effectively?

Writer(s)

- Who are the writers of the report? Are they experts?
- What is their relationship to the readers?

Reader(s)

- List all possible readers (both primary and secondary) and describe their relationship(s) to the material.

Text

- What form/design should the report take?
- How can this form/design help you achieve your goals?
- How many different ways can you arrange the report?
- What material will you choose to keep?
- What material will you choose to delete? Why?

Report Format

Each recommendation report should be between approximately 5–7 pages per person and **include at least two visuals**, one table and one figure (see the Using Visuals in Your Report section of the Reports chapter). These page guidelines are just that—guidelines. Remember that professional quality business reports use appropriate page design, so page lengths may vary according to your particular project, how many visuals you use, how much white space you have in the margins and between sections, etc.

Ordinarily, students prepare two copies, one for teacher and one for client. Formal reports are usually bound. You should make an effort to have the report nicely bound by your campus copy center or some other commercial copy store, for appearance does make an impact in formal reports. (Your instructor may or may not require this.)

Your report must include the following elements in some form:

- Front Matter of Report
  - Letter of Transmittal
  - Cover/Title Page
  - Executive Summary
  - Table of Contents
- Major Sections of Report
  - Introduction
  - Background
  - Recommendations
  - Methods & Findings
  - Conclusion
- Back Matter of Report
  - Bibliography
  - Appendices

Front Matter of Report

Letter of transmittal

The letter of transmittal is a cover letter that delivers completed projects to external audiences. Generally, the letter should follow standard letter format and be limited to one page. These letters can either be bound with the report or delivered with the report as a separate document. The letter should contain the following elements:

- **Opening Paragraph:** statement of purpose, the date the work was commissioned, statement that the report has been delivered, and the title of the report.
- **2nd Paragraph:** briefly summarize your recommendations and the pertinent findings from your research, including one or two significant details.
- **3rd Paragraph:** briefly describe your research methods, as well as acknowledging any key contributors beyond the consulting group and important sources.
- **4th Paragraph:** close courteously, express a willingness to work further with the client, offer to answer any questions, and provide an open avenue for communication.

Cover/Title Page

Generally, the cover includes names of the writers, title of report, name of recipient, and date of report. The cover page should have the logo at the top of the page. Separate this information with a border. The title of the research comes next, followed by the date. At the bottom of the page, you need to list the following information: Instructor's name and address; the names of each member of the research team; the name of the client, client company, and company address.

How should you choose your title? The title should be brief, but must also convey something of the subject of the report to the reader. It's helpful to think about describing the purpose or objective of the report, such as "A Usability Study of MySportStat.com's Content" or "Web Advertising Strategies: A Recommendation Report."

Executive Summary/Informative Abstract

An executive summary is a mini-version of the final report, designed for decision-making audiences who lack the time to read the entire report closely. This document should be written in lay terms, designed for managers who may not have the technical expertise of other readers. It should be no more than two (2) pages and is placed before the body of the report. Include the following parts:

- Clear description of the context that led to the research performed, including a discussion of the problem. Be sure to use specific terms and evidence to show the problem exists.
- Brief description of the methods used for research and the research findings would also be appropriate here.
- Detailed list of recommendations with some justification. A bulleted list of detailed recommendations is particularly effective.

Table of Contents

The table of contents provides readers with a way to find the information that they want quickly and efficiently. List all major headings and all subheadings.

Microsoft Word can generate a table of contents automatically using the INSERT> REFERENCE > INDEX AND TABLES menu. For this to work, all major headings and subheadings need to be formatted as headings. You can search Word's Help menu or find tutorials on the Internet for more instructions.

Major Sections of Report

Introduction

Like any opening, include a descriptive summary reviewing the gist of the report, focusing on the statement of the problem, pertinent background/history of project, and results/recommendations. Also include a summary of the report sections.

Background

Describe the context and purpose of the report in sufficient enough detail to justify report. Include:

- Client description
- Statement of client's need or problem
- Statement of project objectives

Be sure to convince the readers that there are possible problems by describing causes. Provide statistics, interview quotes, and any other evidence you can find to back up your points. Speculate about the problem/need: that is, if nothing is done, what will the future hold? Create a sense of urgency to convince the readers that they need to take action and implement the information that you have gathered. Remember that the primary audience is the client, so pay attention to the principles of maintaining goodwill and finding positive ways to express negative ideas when describing the client's problem (i.e., don't bash the client in the report, use more tactful language to discuss deficiencies in the client's organization).

Recommendations

Following the managerial organization pattern, the recommendations should come after the report introduction and background section (see Reports chapter). The recommendation section should:

- Give an answer to your guiding research question based on what the client will find feasible.
- Describe how the information you have gathered can best be implemented for the benefit of the organization.
- Convince your audience that your information is valid and important for the organization.

Make arguments following the general criteria for evaluating options listed below, and be sure to support your claims with evidence from your research

Effectiveness	• Will your recommended plan be effective? • Will it work? • Will the recommendations solve the problem?
Resource Feasibility	• Can client implement the recommendations? • Do the recommendations require technology or resources (e.g., personnel) that are unavailable? • What will have to be acquired to make the option work? • Will the recommendations require suspension of services or production? • Will there be any training involved? Hiring new personnel?
Desirability	• Will the organization want to implement the proposed options? • Will anyone have personal objections to the option? Do the recommendations harm or threaten anyone (either staff, clients, or customers)? • Is the option legal? Ethical? • Does the option have desirable effects? Potential undesirable effects • What are they?
Affordability	• What does the option cost to implement? To maintain? Is the cost reasonable? • Is the cost justifiable given the level of need or severity of the problem?
Preferability	• Is the option better than others? Why? • Develop and explain the criteria you use to make comparisons (e.g., choosing a computer system based on criteria such as: cost, processing speed, software package, ease-of-use, technical support)

Source: Leslie Olsen and Thomas Huckin. *Technical Writing and Professional Communication* (2nd ed., McGraw-Hill, 1991).

Methods and Findings

The methods and findings section, sometimes referred to as the methodology section, follows the recommendations section. It should include the following sections in your methodology section:

- **Findings** section that discusses the results of your research. List and provide details about what you have found out, including what your surveys, personal interviews, user testing and any secondary source material revealed (this can be a subsection of Methodology or a major section of its own).
- **Methods** section that describes and justifies your research techniques

Critical to any primary research report is the data that you collect. You present it in various tables, charts, and graphs (see the section on creating, formatting, and incorporating graphics into your reports). Call attention to relevant results (don't leave it up to reader). Less important findings can go in appendixes if they are so big that they interrupt the flow of your discussion.

Conclusion

Include a persuasive closing, convincing your audience to accept your study and recommendations. Suggest benefits of accepting report's claims.

Back Matter of Report

Bibliography

Cite your outside information/sources using the Chicago Manual of Style footnotes and bibliography format and follow it consistently by imitating examples for citing various sources.

Appendices

An appendix is a place to put information that just will not fit in the main body of the report but still needs to be in the report. Big tables of data, large maps, forms used in an organization, or verbatim transcripts from interviews—these are good candidates for an appendix. Each one is given a letter (Appendix A, B, C, and so on). You can include photocopies of related information. You can also include survey or interview results, but you should never include "raw" data; always present your results so that your reader does not have to sift through large amounts of information. Data should always be processed, condensed, and tabulated for the reader's benefit. Reader are free to disagree with the report writer's interpretation of the data, but the report writer and readers should never disagree on what data is being presented.

Remember that most charts and graphs should be placed as close as possible to the relevant discussion in the body of the report. You may not need to have appendices in your report if you only have visuals but not any supporting documentation to include.

Review Checklist for Final Report

Format

- Is the audience responsible for seeing that the findings are used and the recommendations implemented?
- Does the report use effective, informative headings? Are they significant of the content of the section?
- Is it easy to distinguish between main sections and sub-sections?
- Is white space used effectively? Does the page design look choppy or busy?
- Are the paragraphs and sections focused and persuasive?
- Has the writer used a professional tone?
- Is the report paginated?
- Does your executive summary review key points from each section of the report, informing readers of recommendations and conclusion?

Organization

- Are the sections arranged most logically and effectively for understanding the information and its relationship to the purpose and thesis of the report?
- Does the report have an introduction, which includes the purpose and forecast of the document and conclusion?
- Does the background section summarize the problem/situation?
- In what ways does each section support or relate to the main point?
- Are the research methods clearly explained?
- Does the conclusion restate the main argument of the report? Does it indicate what the audience should do with the findings? Does it provide contact information from the writers?
- Does the conclusion create a sense of urgency?
- If there are appendices, is a list of their titles provided immediately following the conclusion?

Findings

- Are the findings presented as conclusions about specific issues?
- Are results used to support, contradict, or qualify other findings?
- Are tables or figures used to clarify complex data?

Recommendations

- Are the recommendations explained in persuasive detail?
- Do the recommendations meet the needs of the organization?
- Are the recommendations feasible in terms of the organization's constraints and goals?
- Are the recommendations supported with findings?
- Does the discussion provide a brief suggestion of how the recommendations could/should be implemented?

Tables or Figures

- Have you included at least 2 visual/graphic aids?
- Are your graphics consistently formatted (e.g., have titles and labels) and appropriated placed?
- Do the graphs visually support the discussion of them? Are their titles indicative of their content?
- Is each visual the most effective in representing the data?
- Are other graphs needed to make a discussion more clear?
- Is the interpretation of the graphs in the discussion well developed and clear so that graphs would be unnecessary?
- What results do the graphs reveal that the writers did not discuss?

Sources

- Are you citing sources using Chicago Style? This includes footnotes for in-text citations (making references whenever you're quoting, paraphrasing, or summarizing) and the bibliography at the end of the report.
  URL: http://writing.wisc.edu/Handbook/DocChicago.html

Project Assessment Memo

In a formal business memo no longer than 500 words, evaluate the success of your project, the final versions of documents, and how your team performed as a group.

This PAM is primarily a group performance evaluation. Your main purpose is to inform your instructor of how your group performed throughout the project and whether or not there were any problems that affected the outcome of the project. Remember, each individual receives a separate grade for this project, and your instructor will use the final project evaluation, your PAMs, and his or her familiarity with your group to assess final grades for each team member. Team members do not automatically receive the same grade. Poor participation could hurt your grade.

Be sure to assess the performance of each team member, including yourself, according to the following criteria:

- Contributed his or her fair share to all phases of the project?
- Participated actively in meetings?
- Was dependable, prompt, and courteous as a group member?

For each team member, including yourself, provide an overall rating for his or her contributions to the group. You could use a scale of 1 to 5 (be sure to identify which is best) or assign letter grades.

If you feel one or more persons did not contribute equally, be sure to provide concrete reasons why they didn't contribute. Likewise, if you feel some members of your group are dissatisfied with your performance, you might want to provide an explanation from your point of view.

Don't forget to comment on the quality of your final project.

USABILITY PROJECT

CHAPTER 32

Project Objectives

- Practice writing proposals, progress reports, and technical research reports
- Apply project management, research, and team writing techniques to the production of effective reports
- Plan and carry out a multi-stage, collaborative research and writing project
- Research, analyze and present large amounts of information clearly and persuasively in professional quality reports
- Understand concepts of usability and user testing in effective document design
- Identify manageable problems or needs within organizations and articulate them as researchable questions

For this major collaborative research project, you will work in teams of 3 to 4 to evaluate an existing website that belongs to an actual person or organization. You will assume the role of a usability consultant and act as if an actual client, selected by your team, wishes to improve his or her organization's website to meet the needs of a specific target audience. As a usability consultant, you are committed to the idea of user-centered design—that is, designing documents by focusing on potential users from the very beginning, and checking at each step of the way with these users to ensure that they will like and be comfortable with the final product.

Your job will be to help a client, whom you may or may not actually interact with, by conducting primary research about the website and providing feedback and design advice for a more user-centered website. For this project, then, you will:

- Identify an actual client organization that has an interactive website
- Describe in detail the client organization and target demographic for the website. The more information that you can begin with, the better your research design and analysis will be
- Design a research plan to elicit relevant information for your client about the usability of its website
- Perform tests on the existing site with intended users to determine its effectiveness in specific areas
- Analyze the extent to which your client's site achieves its purposes as well as those of the users
- Write a usability recommendation report addressed to the client organization that (1) provides the client with information about your assessment of the users' needs and the usability of the site, and (2) recommends ways the client might revise, upgrade, or improve its web-based communication strategy

Each group articulates the results of its evaluation in an analytical report addressed to the client. This project introduces students to the *report writing cycle*, or the process of documents and deadlines by which collaboratively written business reports are produced. Students write a variety of documents, including a project plan and progress report, culminating in the final report. Students also submit individual team performance evaluations that assess the contributions of each team member.

What Does Usability Have to do with Learning How to Write?

This project familiarizes you with collaborative report writing techniques by asking you to act as part of a web design consulting team. Each team will research targeted groups of users and, based on this information, make recommendations about how to upgrade web-based documents to better perform for these target groups. Usability consulting is the shorthand for this service. Refer to the User Testing chapter for an overview of user testing.

The essential idea of usability, that the effectiveness of important documents can be improved with input from actual intended users, is older than web-based documents and can be applied to the production of any document, print or online.

While most of you won't go into careers as web designers, knowing how to identify good, usable web design will help you produce better documents. Like all effective business writing, a good website takes into account rhetorical issues of audience and purpose. Because the Internet allows for the presentation of information in multi-media, issues of visual design are critical.

By acting as a usability consultant, you will also have to apply all you've learned in business writing to produce your final recommendation report, including how to plan carefully, research thoroughly, think critically, and communicate effectively. Because you'll be working in a team—which mirrors how most reports are written in the workplace—you'll also have to utilize teamwork, project management, and group writing skills.

Deliverables

Document	Writer(s)	Audience	Format
1. *Design Plan Report*	Group	Instructor	3–4 pages; 11 pt font, single spaced; Must attach copies of user test protocol (data collection sheets)
2. *Progress Report*	Group	Instructor	850 words. Informs instructor of findings from user tests and status of project
3. *Final Technical Report*	Group	Client	12–15 pages (following good page design and layout); 12 pt. single spaced, plus appendices and references. Must include at least 3 visuals (incl. 1 table, 1 chart, and 1 graphic, usually a screenshot). Teams must write a report draft approved by instructor beforehand.
4. *PAM*	Individual	Instructor	500 words. Focuses on evaluating self and team members. Failure to complete equals letter grade reduction to final report grade

Choosing a Client Website

To prepare for usability project team selection, brainstorm a list of three possible clients. Keep in mind the following criteria for choosing a good client:

- **Can be a national, local, or university organization.** Examples of past student projects that involved national organizations include Ebay.com, MySportStat.com, or Fandango.com. Local organizations include area businesses or nonprofits, like a relative's accounting or law firm, or an animal shelter or youth foundation. University organizations are good choices, like your college and department websites. Most student services, clubs, and organizations have websites too.
- **Must be an interactive website.** The website must involve interactive components that require the user to actually do something on the site, like move from page to page to find information. Most website have specific functions too, like ticket sales, book listings, and airline reservations. A basic website with just one long, scrolling page of information, which requires little user interactivity, is a bad choice.
- **The more authentic and more possibility for client interaction, the better.** You should consider choosing a client that you have some existing relationship with, so you can talk to them about the website and your project. This usually makes for a better, more meaningful project. Your initial analysis is also made easier because you're getting this information firsthand from the client rather than having to estimate these elements yourself. Maybe your mom is trying to sell her cookies online (an actual student project) or your friend has a website promoting the local music scene (another actual project).
- **Should target class demographic.** You want to choose a client that lets you use classmates as the test subjects who provide user feedback about your website. The project is set up so that you will have time in class to work on planning your project, including conducting the actual website evaluation/usability tests. Obviously, if you want to test intended users in class, then they should already be represented in the typical business writing class. It might be hard to do in-class testing on a medical website whose target audience is senior citizens, but it will be easy to test the university's Registrar website. (You are not prohibited from testing out of class. Students often elect to conduct more tests than in-class time allows.)

Think twice about choosing a high end, professionally produced website. A good user test can reveal strengths and weaknesses to any website. But since you're not required to be a usability expert for this project, you might want to avoid popular websites that have probably already been extensively user tested. The more popular the website, the more you'll have to know about web design and usability to offer substantive recommendations. Think about websites you've visited recently and think if you're aware of any usability problems from your own experience using these sites.

Design Plan Report

In a memo 4 pages or less, not including attachments, your team must inform your instructor of the following:

- Your choice of website for this project
- Your initial analysis of the website
- Your plan for evaluating the website
- Your group's schedule for completing the project

One of your primary purposes is to persuade your instructor that you have a well thought out plan for completing the research phase of your Usability Project. You need to argue your plan for evaluating the website. Your Design Plan should clearly communicate your goals, present an initial analysis of the website, and articulate your plan for conducting the user tests.

The Design Plan is an important first step in the report writing cycle. By writing this planning document—a form of a proposal—your team is articulating its objectives, which will serve as a blueprint for completing the project. It's much easier to successfully complete a project after you've articulated what your goals are, what needs to get done, who's going to do it, by when. You can use your plan to monitor the team's progress and stay aware of deadlines. By putting the plan in writing, the Design Plan also includes information that can be re-used later in the final report. This practice of using text from one document for other documents is called boilerplatting and is commonplace in business writing.

Format

Your Design Plan must include the following information:

- **Opening:** As with all business correspondence, open your memo with a brief statement of purpose and summary of main points. Do not use a heading for the opening section.
- **Background:** Who is your client, i.e., what is the website? Discuss the site's primary and secondary functions/purposes and its target audience/user groups. For example, a travel site's primary purpose may be to book travel reservations. Its secondary purposes might include providing destination information and instructions on booking reservations online. The people that would use this site include Internet savvy adults that prefer to make their own travel plans. They likely comparison shop at other travel sites. Most have probably booked travel online, etc. Consider if your website has multiple audience. For example, user groups for university sites include prospective students, current students (who have different needs than prospective students), visitors, athletic boosters, alumni, faculty, etc.
- **Preliminary Evaluation:** What is your initial evaluation of the website? What are the strengths and weaknesses of the site's navigation, content, visual design, and functionality? Give reasons to support your evaluation. To help you derive some basis for analysis, include comparisons of your website to similar/competitor sites (e.g., "Compared to competitor sites X and Y, we believe the site…."). Comparing the client site to competitor sites is an excellent way to discover the site's strengths and weaknesses.

NOTE: You might want to consider adding a more formal comparison to other websites as part of your method. In this approach, you would develop a more detailed comparison to one or two sites in your final report's findings section, which will allow you to base some of your recommendations on not only user test results, but also your own analysis of competitor sites.

- **Testing Goals:** What areas of the website will you evaluate? This should follow from your preliminary evaluation. For example, if an analysis of a business website reveals that it is (1) difficult to navigate, (2) hard to find information about the company's services, and (3) hard to read, you might articulate the following project goals:

We will evaluate the following areas:

1. *Navigation--Can the user browse the website without complication?*
2. *Content--Can users find specific information regarding the company?*
3. *Readability--Is the text size and background color making the information hard to read?*

As the example above shows, state the project goals as research questions. This will make it easier in the Method section to create specific tasks that address the areas you aim to study.

- **Method:** Describe your team's plans user testing. You must attach a draft of your user test protocol/data collection sheets so your instructor can review them. In this section, be very detailed in your description of test procedures: How many users will you test? (NOTE: testing 5-10 users is sufficient for usability analysis.) What tasks will you ask them to perform? What data will you collect (before, during, and after the tests)? Discuss your procedure for conducting the tests. Will you administer a written protocol? Will it be interview-based? See the Usability Testing chapter for instructions on planning your test methods.

  If your team has elected to include a comparative analysis as part of the method, be sure to also discuss in this section (1) which sites you will be comparing and why you chose these sites, and (2) the criteria you will be applying in your analysis.

  Briefly describe why the methods you chose are appropriate. Why are you asking certain questions in the pre-test portion of the user test? One reason, for example, is that you want to verify that the people you test match the typical user profile of the website. Or another example, you plan on collecting information about people's online shopping habits because you want to gauge their familiarity with using e-commerce websites. Why are the tasks you plan to administer appropriate? For example, the direct tasks of your test will allow you to determine how difficult it is for users to access information, complete a purchase, register for a login account, etc. Why are you only testing 5 to 10 people? (Jackob Nielsen writes that testing just five users can identify 80% of major usability problems.)
- **Schedule:** Describe your plan for managing the project. Include description of discrete tasks, individual/team responsibilities, time estimates, and major milestones/ deadlines. Be sure to note key project milestones: When will you administer your

tests, collate the information, analyze your data, draft your reports, and submit the final version? Illustrate your schedule discussion with a Gantt chart. Remember, the Gantt chart alone is not sufficient—you must reference it and explain it in a prose discussion of your schedule.

- **Conclusion:** Conclude with an argument about the doability your project. Make sure your design plan is reasonable for the scope and timeframe of the project.
- **Attachment**
  Attach a copy of your team's test protocol, i.e., any questionnaires you plan to distribute to users during the tests and any other data collection sheets you plan to use.

Design Plan Review Checklist

- Is the Design Plan written in memo format, with proper heading, opening, body, closing, and end notation?
- Does the Design Plan follow the organization and heading-subheading structure outlined above?
- Has the website's purposes and target audience been described in detail?
- Does the design plan offer a preliminary analysis of the websites strengths and weaknesses? Is the preliminary analysis derived in part from comparison to other similar/competitor websites?
- Are the group's testing goals clear? What does the group want to learn from the testing about the user, the website, the way the website is used/read, etc.?
- What methods will be used to meet the testing goals?
  - How will test subjects be selected?
  - When and where will the testing and/or interviews be conducted?
  - What specific tasks will the users be given? Is there a protocol for the test(s)?
  - What materials will be needed? (e.g., recording devices, product, documents, etc.)
  - Who will do what for the research and the testing?
- Is there a discussion of the group's schedule and does it reference a Gantt chart illustrating major tasks, milestones, and timelines?
- Is your team's user test protocol/data collection sheet attached?

Progress Report

Each team is responsible for submitting a group-written progress report that presents the results of the usability tests. It should be written in memo format, not exceed four pages, and include at least two visuals (table, graph, chart, or picture/screenshot).

Organizing the Progress Report

Remember that a progress report informs readers of the status of a project. It is designed to justify work done, evaluate that work, and also to forecast work remaining.

As a genre, the progress report focuses on work completed and work remaining—with context provided by overview, background, and conclusion sections. As with all effectively organized reports, you are not limited to these sections. For example, you could use the types of research completed as subsection headings.

- **Overview:** Open your report with a brief summary of purpose and content of document.
- **Background:** The next section reiterates for the reader the purpose of the research project. Summarize what website your team selected and your project objectives. (*Hint*: Boilerplate this information from your original Design Plan Report.) Also, if there have been any significant changes in the scope or organization of the project, you should note them.
- **Work Completed:** This section describes actual results to date. For this project, you should be summarizing the results of your user tests. Compile your data and present it in a clear and readable fashion. Just writing "we've made lots of progress" without a detailed discussion to support your claim would not convince a reader that your group is progressing.

  For the usability project, this section should resemble the findings section of the final report by summarizing the results of your user tests. You must include at least two visuals in this section. Consider the best way to present your data graphically now, so you can re-use any visual aids in your final report. Will a table work? Should you use a pie or bar chart? How will you summarize and present the results of any open-ended questions you asked? Is there a way to present this information visually (e.g., categorizing and tallying types of answers in a table). Or will you just summarize results and quote exemplary answers?

  Another useful visual aid when it comes to analyzing websites is the screen shot, literally an image of the computer screen that shows the relevant part of the website captured using the Print Screen key. As Seth Gordon recommends, "highlight problematic areas through a combination of narrative and screen captures. The screen shots are especially effective for communicating the context of the problem. It's often easy to see the problem but extremely difficult to document it." Gordon also suggests mentioning significant positive and negative feedback and any interesting or unexpected findings when discussing results of the user tests. Finally, according to Gordon, "a good evaluation captures respondent feedback, and the best way to communicate that is through respondent quotes"[1]
- **Work Remaining:** This section should describe tasks that remain to be completed by the group before the completion of the project. For instance, what type of recommendations does the group foresee making? How have the test results impacted your plans for the report? What work have you done towards drafting the report so far? And what work, specifically, remains to be done?

- **Conclusion:** Your final section should briefly reiterate the status of your project and assess the chances for its success. You want to be persuasive as well as honest. Your tone should be positive, but if there are any concerns with the project, you should not avoid mentioning them. Your reader, your teacher or manager, needs to be given an honest report so that s/he can offer assistance, re-direct resources, etc.

Usability Report

Refer to the Reports chapter for more information on formatting business reports. The usability report is its own sub-category, or genre, of business report and a simple Internet search should locate several examples. You can use these examples as guides, but keep in mind that not every usability report presents information in the clearest and most effective way.

The main sections of the report should be as follows:

- Overview
- Recommendations
- User Test Results
- Conclusion

Don't forget to also include the following conventional elements of reports:

- Front matter
  - Letter of transmittal
  - Cover
  - Executive Summary
  - Table of Contents
  - List of Figures
- Back matter
  - Bibliography (if applicable)
  - Appendices (if applicable)

Usability report recommendations are often written in a problem/solution format, whereby each problem area identified by the user testing is noted (with concrete references to user test data as evidence) and then suggestions or recommendations for addressing the problem area are given.

Solutions to most usability problems are usually fairly obvious, once an actual user identifies something as a problem. Solution ideas might come from user feedback/suggestions or a comparison to other websites, particularly competitor sites. Since you are not expected to be a usability expert for this project, your report should include recommendations for improving weak areas but will not necessarily be evaluated on the technical effectiveness of the solution.

As the Progress Report section in this project and the chapter on Reports indicates, screenshots are almost essential when conveying the results and recommendations of your user tests to a reader. A screenshot, or screen capture, is an image of whatever is displayed on a computer monitor at the time of the screen capture. Callouts, or textboxes with arrows drawn to parts of the screenshot, help clarify what is being illustrated in the screenshot. Screenshots are useful for pointing to flaws in the client website or pointing out other websites that have better design elements. The Report chapter has more information on how to take screenshots and insert them into your document.

Usability Report Evaluation Checklist

Refer to the following checklists as your prepare your final report:

Formatting

- Are all report parts included: cover, transmittal letter, table of contents, list of figures, executive abstract, body, references (optional), appendices (optional)?
- Does the report use frequent, effective headings and subheadings, and is it easy to distinguish between main sections and sub-sections?
- Are headings (L1, L2, L3, L4) formatted consistently from section to section and from front matter to body?
- Is the report paginated following guidelines for page numbering, e.g., lower-case Roman numerals in front matter, Arabic numerals in body?
- Does the same header and footer consistently appear on all pages of report from front matter to body, except for letter of transmittal?
- Do you have a List of Figures page separate from the table of contents?
- Is white space used effectively, or does the page design look choppy, blank, or busy? Make sure format (e.g., headings, header/footers) is same for front matter as report proper

Recommendations

- Are recommendations subdivided with "talking" subheadings (i.e., starting heading with an action verb, e.g. "Recommendation #1: Minimize busy graphics…")
- Are recommendations developed in sufficient detail including (1) support from references to user test results, comparison to other websites, and/or expert opinion, and (2) suggestions for improving the problems, including possible screenshot illustrations?
- Do the recommendations in Executive Summary correspond to the recommendations in the main body of report?

Background and Methods

- Does background and method information correspond to content from the design plan, i.e., did you boilerplate your design plan content (minus the Schedule section) into the report?
- Is the "Method" section written in past tense (e.g., we asked users to…)

Graphics

- Is the minimum three graphics included (one chart, one table, and one screen-shot)?
- Are all graphics labeled with a caption, including a consecutive number and clear title that conveys the main point of the visual?
- Are all graphics referenced in the text (e.g., "As Figure 28.1 shows,…")?
- Are all graphics placed as close as possible after the initial reference in the text? (Don't place graphic before; if it can't fit on the page that it is first mentioned, but it at the start of the next page and let reader know: "see page #.")
- Is the graphic simple, uncluttered, and easy to read, i.e., is visual unclear, either because of design of visual, inappropriate visual choice, or trying to convey too much information in one visual?

Style and Editing

- Does the report follow a more formal style than business letters and memos, e.g., does it avoid references to the reader as "you" (except in transmittal letter, where you should use it)? Are informal, slangy word choices avoided, such as "things" and "report is broken down by"?
- Are pronouns used appropriately and consistently? Don't use *you* as a third-person reference, write *users* or *one* instead. Check for pronoun agreement, e.g., "we asked one user what they thought" is incorrect because the subject *one* is singular and the pronoun *they* is plural.
- Are words spelled consistently, e.g., how did your team decide to spell W-E-B-S-I-T-E?
- Have you eliminated *telegraphic writing*, where words such as articles (a, an, the) are missing or the tense is wrong?
- Are there no stacked headings, i.e., does the report flow from one section to another when read aloud, or do more transitions from section to section need to be added so the reader doesn't get confused (e.g., "In this section, we discuss the methods we used…")?

Project Assessment Memo

See the instructions for writing the **PAM for the Client Project**. As with that project, the PAM for the Usability Project should be an evaluation of your group's members. In a formal business memo no longer than 500 words, evaluate the success of your project, the final versions of documents, and how your team performed as a group. Remember to speak specifically about how each member contribute to the project, including yourself. If here is any reason you believe someone should not receive the same grade as other students, you should mention it and provide reasons for your assertion. Likewise, if you believe other members in your group might single you out for some reason, you should indicate why you believe other team members might rate your contribution low.

End Note

[1] Seth Gordon, "User Testing: Reporting Your Findings," Builder.com, last modified October 6, 2003, http://www.techrepublic.com/article/user-testing-reporting-your-findings/

INTERNATIONAL PROJECT

CHAPTER 33

Project Objectives

- Learn the business report genre
- Practice the report writing cycle
- Understand and apply principles of project management
- Research and integrate information into business reports
- Evaluate data and information based on the needs of specific audiences
- Develop an understanding of cross-cultural communication

The International Project puts you and your group in a situation as researchers who will offer recommendations to the Bellcom corporation so that they may create materials to better orient their employees to the customs and features of their new international surroundings. To fulfill the requirements for this project, you will need to recommend the kinds of materials for a complete package that prepares employees for the move.

Our class will take on the persona of a consulting group, a company which provides a wide variety of research services. Our most recent client, the Bellcom Corporation, is expanding its services internationally, and has had trouble recently with employees traveling, living, and conducting business abroad for the first time (see chapter 22, A Business Faux Pas Case). At this time, Bellcom is opening new offices in the following cities:

- Amsterdam, Netherlands
- Athens, Greece

- Brussels, Belgium
- Dublin, Ireland
- Mexico City, Mexico
- Nairobi, Kenya
- New Delhi, India
- Rio de Janeiro, Brazil

While these sites are ideally suited for their business purposes, Bellcom wants to prepare the employees who will be working in these cities for the kinds of cultural differences they will face. Our research company has been hired to make recommendations about the kinds of materials that Bellcom should have developed to aid their employees preparing to move to their new surroundings. Your primary contact at Bellcom for this research is Sharika Jones, the vice president of human resources.

Your major collaborative project in this class will be to collect information so that you can make recommendations for developing a travel guide. While Bellcom wants to be comprehensive in developing their travel guides, they also want to make sure that they are not offering unnecessary or redundant information. Your goal is to create a recommendation report that describes and argues for the most important information that employees will need upon moving to a particular city. By performing this assignment, you will develop skills that are valued in the business world and, therefore, provide you with certifiable experience.

Deliverables

This project will include the following documents:

- **Research Design Plan:** 3–5 page memo informing instructor of your plan for effectively completing the project
- **Progress Report:** 3–5 pages e-memo informing your instructor of your initial findings, analyses, and work to be completed
- **Final Report:** 8–12 pages (good page design and layout) report addressed to Sharika Jones of Bellcom (include appendices, references, and at least three visuals)
- **Project Assessment Memo:** 500-word email evaluating self, team, and overall project. Failure to complete equals letter grade reduction to final report grade

Readings

For the International Project, you should read the following selections in this textbook:

- Usability Project Design Plan
- Usability Project Progress Report
- Client Project Final Report

While the above readings are not direct correlations to the kind of writing that we are doing for this project, they will give you a rough overview of the kinds of documents that you need to create.

Planning

Each group will be responsible for the following in order to complete this project:

- Determine your goals as a group
- Decide how your group will accomplish the work
- Create a research agenda
- Perform research and evaluate information
- Develop most appropriate recommendations for situation
- Construct, evaluate, and revise your report
- Organize your report so that information can be found quickly and easily

Project Planning

To begin evaluating the material requested by our client, you will have to plan and set up appropriate research activities. You should distribute the workload evenly, and each member of the group should participate in all phases of document development. To gather good-quality information, you need to carefully avoid setting up activities that give irrelevant or even misleading information; therefore, as a group you should develop criteria that will assist you in analyzing and evaluating information so that you can gather the best information possible. To help avoid research problems, you should consider the following:

- who the client is and what kind of information they need
- the strengths and weaknesses of each team member
- main topics for research
- appropriate criteria for evaluating information
- research sources and strategies

- potential obstacles or problems in gathering information
- a Gantt chart or other organizational document to insure that the group stays on task

Research

You'll need to determine the most important information that Bellcom employees (most of whom have families) will need to know.

- What kinds of information can best help Americans adjust to the differences in culture? How do you define cultural information? Social/Political/ Economic? More specifically?
- What are the most important pieces of information?
- What are the best sources of information?
- What are common stereotypes about the country? What values do these stereotypes suggest about the country? How can these be counteracted?

Some possible topics include:

- preparing to move
- general information about the city
- weather
- single-life/family life
- housing
- money
- transportation
- food
- health care
- schools/education
- entertainment
- culture and cultural differences
- language

Document Production

Purpose(s)

- What are the purposes for writing?
- What are your/our goals?
- How much is enough information to make a credible argument?
- Does the information anticipate and answer the audience's questions? Why or why not?

Readers

- Who are the primary readers?
- Who are possible secondary readers?
- Describe your potential readers in as much detail as possible and explore the ways these readers will effect the documents that you are creating?

Writer

- Is the writer's identity important? Why or why not?

Text

- What form/design should the report take?
- What style/tone would be the most appropriate?
- How can this form/design help you achieve your goals?
- How many different ways can you arrange the materials?
- How can the information be organized most effectively?
- What kinds of graphics/graphs/pictures can you provide? How are these items helpful?

General Evaluation Criteria for the International Project

- When evaluating the documents that you create for this project, here are some general criteria for you to consider:
- Is there sufficient information?
- Will the reader feel they can make informed decisions in helping their employees prepare to move to this country?
- Are the cultural elements defined?
- Is the style/tone consistent?
- Are the documents easy to read?
- Is the information arranged effectively?
- Does the material provide sources for more information?
- Is the material arranged hierarchically?

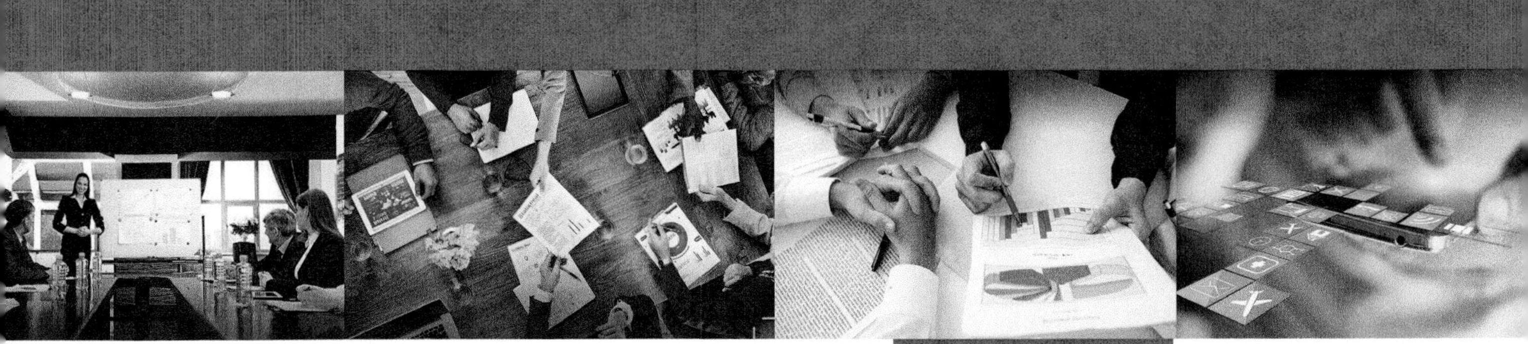

INDEX

NOTE: Page references in *italics* refer to figures.

D

T